Water Resources

Challenges in Water Management Series

Editor:
Justin Taberham
Independent Consultant and Environmental Advisor, London, UK

Titles in the series:

Urban Water Security
Robert C. Brears
2016
ISBN:978-1-119-13172-4

Water Resources: A New Water Architecture
Alexander Lane, Michael Norton and Sandra Ryan
2017
ISBN: 978-1-118-79390-9

Water Resources

A New Water Architecture

By

Alexander Lane
Swindon, UK

Michael Norton
Long Newnton, UK

Sandra Ryan
Bristol, UK

WILEY Blackwell

This edition first published 2017

Registered Offices
John Wiley & Sons, Inc., 111 River Street, Hoboken, NJ 07030, USA
John Wiley & Sons Ltd, The Atrium, Southern Gate, Chichester, West Sussex, PO19 8SQ, UK

Editorial Office
The Atrium, Southern Gate, Chichester, West Sussex, PO19 8SQ, UK

For details of our global editorial offices, customer services, and more information about Wiley products visit us at www.wiley.com.

Wiley also publishes its books in a variety of electronic formats and by print-on-demand. Some content that appears in standard print versions of this book may not be available in other formats.

Library of Congress Cataloging-in-Publication Data
Names: Lane, Alexander, 1986- | Norton, Michael (Michael Robin), 1952- | Ryan, Sandra -1979
Title: Water Resources : A New Water Architecture / by Alexander Lane, Michael Norton, Sandra Ryan.
Description: Chichester, West Sussex : John Wiley & Sons, Inc., 2017. | Includes bibliographical references and index.
Identifiers: LCCN 2017007294 (print) | LCCN 2017008973 (ebook) | ISBN 9781118793909 (cloth) | ISBN 9781118793954 (Adobe PDF) | ISBN 9781118794074 (ePub)
Subjects: LCSH: Water-supply--Management. | Water demand management.
Classification: LCC TD215 .N67 2018 (print) | LCC TD215 (ebook) | DDC 333.91--dc23
LC record available at https://lccn.loc.gov/2017007294

Cover Design and Graphics: Courtesy of Lisa Mellis

Set in 10/12pt WarnockPro by SPi Global, Chennai, India

10 9 8 7 6 5 4 3 2 1

Contents

Series Editor Foreword – Challenges in Water Management

The World Bank in 2014 noted:

'Water is one of the most basic human needs. With impacts on agriculture, education, energy, health, gender equity, and livelihood, water management underlies the most basic development challenges. Water is under unprecedented pressures as growing populations and economies demand more of it. Practically every development challenge of the 21st century – food security, managing rapid urbanization, energy security, environmental protection, adapting to climate change – requires urgent attention to water resources management.

Yet already, groundwater is being depleted faster than it is being replenished and worsening water quality degrades the environment and adds to costs. The pressures on water resources are expected to worsen because of climate change. There is ample evidence that climate change will increase hydrologic variability, resulting in extreme weather events such as droughts floods, and major storms. It will continue to have a profound impact on economies, health, lives, and livelihoods. The poorest people will suffer most.'

It is clear there are numerous challenges in water management in the 21st Century. In the 20th Century, most elements of water management had their own distinct set of organisations, skill sets, preferred approaches and professionals. The overlying issue of industrial pollution of water resources was managed from a 'point source' perspective.

However, it has become accepted that water management has to be seen from a holistic viewpoint and managed in an integrated manner. Our current key challenges include:

- The impact of climate change on water management, its many facets and challenges – extreme weather, developing resilience, storm-water management, future development and risks to infrastructure
- Implementing river basin/watershed/catchment management in a way that is effective and deliverable
- Water management and food and energy security
- The policy, legislation and regulatory framework that is required to rise to these challenges
- Social aspects of water management – equitable use and allocation of water resources, the potential for 'water wars', stakeholder engagement, valuing water and the ecosystems that depend upon it

This series highlights cutting-edge material in the global water management sector from a practitioner as well as an academic viewpoint. The issues covered in this series are of critical interest to advanced level undergraduates and Masters Students as well as industry, investors and the media.

Justin Taberham, CEnv
Series Editor
www.justintaberham.com

Foreword

I grew up in the valley of Wastwater and the River Irt in Cumbria in England. I was fascinated by water. Where did it come from? How was it that the Irt kept flowing even during dry weather? How did water influence the shape, width and depth of streams, rivers and lakes? Why was water such a haven for flora and fauna?

That fascination endured through my childhood and into my university education, where hydrology and geology were far and away my best subjects within a crammed civil engineering curriculum. I embarked upon a career in civil engineering, starting as a graduate trainee with a small English Water Board designing water supply infrastructure. My career subsequently took me far and wide to work on over 100 projects in more than 20 countries. The fascination with water has never faded, and I have contributed to solving problems to a wide range of water challenges, from water resources through water supply and wastewater, to managing floods and providing irrigation systems.

In 2008 my employer Halcrow Group Ltd entrusted me with a review of 'water scarcity' on the basis that it was perceived as a global mega trend that was going to impact most infrastructure and environment sectors in the early twenty-first century. I undertook that review with gusto and, along the way, learned more of the vital role that water plays in the biosphere; I was able to calibrate that learning against a background of having worked in a wide range of institutional settings.

It was during this time that the seeds of New Water Architecture (NWA) were sown in my mind, an idea that stemmed from Integrated Water Resources Management and was infused with elements of the emerging concepts of the water-food-energy system, virtual water and urban sustainability. My enthusiasm to develop new ways of thinking about water has been, and continues to be, shared by many colleagues, but two young people have stood out: Alex Lane at Halcrow as co-incubator of NWA; and Sandra Ryan at Amec Foster Wheeler as a kindred water spirit in an Oil & Gas-orientated company. When in 2012 I put to Sandra and Alex the idea of writing a book, they jumped at the opportunity. I am deeply grateful that they did.

Michael Norton
Long Newnton

Preface

We, the authors of this book, have been profoundly affected by our experiences in both our careers and our personal lives that are increasingly highlighting the very real problems of water scarcity, stress and mismanagement. Most worryingly, these problems appear largely disregarded or unheard of by those that hold the necessary power to take action.

We live and work in an age where we see water problems growing and accelerating; the pressure is tangible, and yet there remains inertia within organisations that have responsibilities and duties to govern and protect our water environment and its resources. Some of that inertia may be a consequence of the size and complexity of the challenges that face us. We suspect there is an element of 'heads buried in the sand', of hoping that water problems will somehow work themselves out while we continue to manage according to the *status quo*.

Recent years have seen a proliferation of advocacy groups and enlightened businesses that see things differently and argue that change is not just a good idea, but essential. Old, traditional concepts such as perceived 'rights' to unlimited water are being questioned and challenged by these groups and businesses, but governments seem to be lagging behind. New ways of thinking about water, such as water footprinting and virtual water, are gaining ground in terms of application and acceptance and are helping us to understand the complexity of our water problems. Slowly but surely, we believe a paradigm shift is occurring. Water management is no longer a responsibility ring-fenced by public authorities and taken for granted by water users. Businesses and their investors are experiencing the impacts of water stress and, gradually understanding the threats (and potential opportunities) these introduce, have taken the lead, beginning to embrace principles of water stewardship and increasingly forcing governments to follow suit.

Our careers have enabled us to work on many water projects, some at a very local scale. Such projects have given us a grassroots knowledge of hydrology and water management systems and an important understanding of the perceptions of, and demands for, water from businesses and industries. We have observed and are helping to direct the gradual change in the attitudes to water held by many corporations. We have also worked with national governments and with regional transboundary institutions and partnerships to analyse water issues at strategic international scales, and to assess the role of water in the development of national and international economic policy. We can see that the opportunity landscape for water professionals is beginning to look very different to that which has passed.

While we see the recent positive progress and the valuable lessons that are being learned, it is the numerous remaining barriers and obstacles that triggered us to write this book. The overlooked importance of water in so many of our daily activities – in the electricity and fuel we burn, the food we eat and the products we consume – is the dominant theme running through this book. Populations are changing, climates are evolving and our natural water systems are responding in complex and interlinked ways which we don't yet fully understand. 'More of the same' as a management approach is therefore not going to resolve our problems. While some of our current water resource management techniques and approaches are effective and, with continued improvement, can provide beneficial outcomes or even become cornerstones of a new water management paradigm, others have run their course and are now not fit to address the challenges we face. Having the courage to question the validity of well-established approaches and taking the necessary steps to enable change is difficult but necessary.

This book does not claim or intend to give all the answers. It presents the authors' collective views on how a new water management paradigm, a New Water Architecture, could look and feel. Its purpose is to stimulate the much-needed critical thinking on attitudes and approaches to water management that will drive real progress.

We believe that mankind must face up to the consequences that will be experienced if water management is not improved, if availability and access to safe and secure water resources does not improve, and if the detrimental impacts that arise from poor water management continue to become more widespread and intense. History presents us with several examples of civilisations whose collapses can be traced to water. People have always followed water and its associated riches and, while we may increasingly try to make water follow us, it is only resilient water management systems that will be able to sustain the global distribution of people in secure environments.

We urge all water professionals to measure their success in terms of their contribution to improving the water environment, in securing appropriate and fair allocation of water resources, and in increasing the perceived value of water in the eyes of all those with a stake in sustainable water management. By this, we are not referring to only political and corporate leaders, but to everyone.

Acknowledgements

We have received much direct and indirect support during the journey from initial idea to publication of this book.

First and foremost we wish to thank Professor John Bridgman of Birmingham University for suggesting to Wiley that we might be interested in writing such a book. Wiley asked a number of people to review the book proposal, and we wish to thank them. One of those reviewers has sadly since passed away, Professor Raúl Galindo of the Universidad Técnica Federico Santa María in Valparaiso Chile, and so we pass our thanks to his family.

We wish to thank Justin Taberham for his enthusiastic review of our final manuscript and his many suggestions of how we might disseminate our key messages. Thanks also to Laura Polito for her insight and advice on how to utilise the many media available to us to publicise and promote this book.

For production of the book we thank Lisa Mellis for her work with many of our original illustrations, copy-editor Elaine Rowan and all those too many to mention in Wiley who helped us on our journey.

Our families have played their part, not least putting up with times when we worked late or over the weekend as we tried to balance our day jobs with book writing. These include Shirley Norton, Cheryl and Peter Lane, and Caroline and Michael Ryan.

When we embarked upon the journey, our employers gave us permission to write this book: Amec Foster Wheeler for Sandra and Michael, and CH2M for Alexander. Our thanks go to them.

Finally, we wish to thank those friends, colleagues and others who have knowingly or unknowingly given inspiration to one of more of us, before and during the journey, including: Tony Allan, Tony Allum, Rodolfo Aradas, Frank Chasemore, Roger Falconer, Chris Fawcett, Mark Fletcher, Arjen Hoekstra, David Johnstone, Harlan Kelly, John Lawson, Bill McCall, Alex Mung, Ken Newnham, Isabella Polenghi-Gross, Susie Roy, John Readman, Rebekah Rice, Stacey Sabol, Barry Walton, Dominic Waughray, Marvin Williams and David Yaw.

List of Abbreviations

The following acronyms and abbreviations are used in this book.

Acronyms/Abbreviation	Term
AgMIP	Agricultural Model Inter-comparison and Improvement Project
ASR	Aquifer storage and recovery
AWS	Alliance for Water Stewardship
CAP	European Union Common Agricultural Policy
CAPEX	Capital expenditure
CBM	Coal-bed methane
CCS	Carbon capture and storage
CDP	Carbon Disclosure Project
CHP	Combined heat and power
CSO	Combined sewer overflow
CSP	Concentrated solar power
DEFRA	England and Wales Department for Environment, Food and Rural Affairs
EPA	US Environmental Protection Agency
EU	European Union
FAO	UN Food and Agriculture Organization
GCM	Global climate model
GDP	Gross domestic product
GM	Genetically modified
GWP	Global Water Partnership
ICDPR	International Commission for the Protection of the Danube River
IEA	International Energy Agency
IMF	International Monetary Fund
IPCC	Intergovernmental Panel on Climate Change
IWRM	Integrated Water Resource Management
MAR	Managed aquifer recharge
MDG	Millennium Development Goal
MENA	Middle East and North Africa
NETL	United States National Energy Technology Laboratory

Acronyms/Abbreviation	Term
NGO	Non-governmental organisation
OECD	Organisation for Economic Co-operation and Development
OPEX	Operating expenditure
PPP	Public-private partnership
PV	Photovoltaic
PUB	Singapore Public Utilities Board
RCM	Regional climate model
RO	Reverse osmosis
RWR	Renewable water resource
SAR	Shallow aquifer recharge
SUS	Sustainable drainage systems
TEEB	The Economics of Ecosystems and Biodiversity
UN	United Nations
UV	Ultra-violet
WASH	Water, sanitation and hygiene
WBCSD	World Business Council for Sustainable Development
WCD	World Commission on Dams
WEC	World Energy Council
WEF	World Economic Forum
WFN	Water Footprint Network
WHO	World Health Organization
WRG	World Resources Group
WRI	World Resources Institute

Units and Conversions

The following table provides conversions between the units of measurement commonly used in this book.

Unit	Conversion
1 megalitre (ML)	1,000,000 litres (L)
1 gigalitre (GL)	1,000 ML
1 cubic kilometre (km^3) or 1 cubic gigametre (Gm^3)	1,000,000,000 cubic metres (m^3)
1 m^3	1,000 L
1 km^3 or 1 Gm^3	1,000,000 ML

Glossary

Terms describing water resources

A variety of terms are used to describe water resources of one form or another and these are often used interchangeably. They are typically quantified as an annual total in cubic kilometres per year (km^3/yr). For ease of comparison between case studies, this book typically adopts the renewable water resource (RWR) term.

- *Renewable water resource*: The long-term average annual inflow and runoff that feed surface water catchment areas and groundwater aquifers.
- *Non-renewable water resource*: Groundwater bodies that have a negligible rate of recharge over the human time-scale.
- *Internal renewable water resource*: Volume of renewable water resource generated from precipitation within a defined territory.
- *External renewable water resource*: Volume of renewable water resource generated outside of a defined territory.
- *Natural renewable water resource*: The total amount of a country's surface and groundwater water resources (internal and external), generated through the hydrological cycle.
- *Non-conventional water resource*: Volume of water obtained through technologies such as desalination.
- *Actual renewable water resource*: The sum of internal renewable water resources and external renewable water resources.
- *Exploitable water resource*: That portion of water resources considered to be available for use. This varies from location to location dependent on physical, socioeconomic and environmental factors.
- *Blue water*: Fresh surface and groundwater, in other words, the water in freshwater lakes, rivers and aquifers.
- *Green water*: The precipitation on land that does not run off or recharge the groundwater, but is stored in the soil or temporarily stays on top of the soil or vegetation.
- *Grey water*: Freshwater that is required to assimilate the load of pollutants based on natural background concentrations and existing ambient water quality standards.
- *Virtual water*: Water consumed or polluted in order to produce a product, measured over its full production chain.
- *Improved drinking water source*: One that, by nature of its construction or through active intervention, is protected from outside contamination, in particular from contamination with faecal matter.

- *Improved sanitation facility*: One that hygienically separates human excreta from human contact.

Terms describing pressures on water resources

- *Water stress*: The effects felt, by humans and/or the environment, when the quantities and qualities of water available in a given location are insufficient to meet the demands placed upon them.
- *Water scarcity*: Water scarcity occurs when the demand for water is greater than the available resource. Water scarcity can be defined in three ways:
  1) *Physical water scarcity*: a physical shortage of water of an acceptable quality with respect to aggregate demand;
  2) *Infrastructural water scarcity*: where even though the actual renewable water resource may be sufficient to meet demand, there is inadequate infrastructure to get the water to where it is needed; and
  3) *Institutional water scarcity*: where institutions, legislation and/or regulation fail to ensure that water is supplied in an affordable and equitable manner. This type of water scarcity is sometimes referred to as economic water scarcity.
- *Water footprint*: The water footprint is an indicator of freshwater use that considers both direct and indirect water use of a consumer or producer. The water footprint of an individual, community or business is defined as the total volume of freshwater used to produce the goods and services consumed by the individual or community or produced by the business. Water use is measured in terms of water volumes consumed (evaporated or incorporated into a product) and/or polluted per unit of time. A water footprint can be calculated for a particular product, for any well-defined group of consumers (e.g. an individual, family, village, city, province, state or nation) or producers (e.g. a public organisation, private enterprise or economic sector). The water footprint is a geographically explicit indicator, showing not only volumes of water use and pollution but also the locations.
- *Direct water footprint*: The direct water footprint of a consumer or producer (or a group of consumers or producers) refers to the freshwater consumption and pollution that is associated with the water use by the consumer or producer.
- *Indirect water footprint*: The indirect water footprint of a consumer or producer refers to the freshwater consumption and pollution incurred in the consumption or production of products. It is equal to the sum of the water footprints of all products consumed by the consumer or of all (non-water) inputs used by the producer.

Terms describing different approaches to water management

- *Integrated Water Resource Management*: A process which promotes the coordinated development and management of water, land and related resources, in order to maximise the resultant economic and social welfare in an equitable manner without compromising the sustainability of vital ecosystems and the environment.

Terms describing the different ways in which people use water

- *Withdrawal/abstraction/extraction*: The taking of water from a natural source such as a river.
- *Consumed water*: The amount of water taken up in the process of a use and not returned to the environment.
- *Returned water*: The amount of water withdrawn from a source that is surplus to the amount needed for the use, or that water which is rejected or wasted, and that is returned to a waterbody.
- *Managed aquifer recharge*: The recharging of water into an underground aquifer for storage and subsequent withdrawal.
- *Aquifer storage and recovery*: One of the most common forms of managed aquifer recharge: injection of water into the target aquifer via a well.

Other terms

- *Hydroclimate*: The physical and chemical characteristics that define a particular aquatic habitat.
- *Primary energy source*: An energy source that has not undergone any transformative process. Primary energy sources can be renewable, such as wind and solar, or non-renewable such as coal, oil and natural gas.
- *Final energy*: That energy available to a user once a primary energy source has undergone conversion. Final energy types include electricity, gasoline and diesel oil.
- *Green infrastructure*: The use of natural ecosystems to deliver outcomes more commonly delivered by built (or grey) infrastructure.

Part I

Setting the Scene

1

Water Resources in the Twenty-First Century

In Earth's 45th millionth century a global crisis of freshwater scarcity is looming, a crisis that is accelerating thanks to our unbridled development and our burgeoning demand for food and energy, and as a result of the effects of climate change. Just 0.1% of the total global water volume of 1.4 billion km^3 is accessible freshwater; we are already withdrawing one-quarter of our accessible renewable water resource (RWR) however, much of which is already needed to sustain our ecosystems and biodiversity, themselves vital for our survival.

In this book, we argue that the world faces water security challenges of a scale previously unseen and largely unsuspected by its population. Estimates suggest that we need four times the current global rate of investment in new water supplies if we are to successfully meet projected water demand in 2030 (2030 WRG 2009). To have any chance of meeting future water demands, we believe there is a compelling need for water professionals to emerge from their comfort zones and to engage with politicians, decision makers and those stakeholders with influencing power. While we can and should continue to develop cost-efficient water technologies, water professionals must grasp this moment to put themselves at the centre of the often-siloed disciplines of science, technology, politics, environment and economics. New models of integrated water management are required to address complex multi-stakeholder demand patterns and water-related responsibilities.

1.1 A Looming Crisis

On 31 October 2011, a baby girl born in Manila was chosen to symbolise the 7 billionth human being on the planet. Although the rate at which the global population is growing has almost halved since the 1970s, in the last 40 years the world's population has still doubled. Alongside this increase, strong economic growth has seen standards of living rise dramatically in the developed world. Forecasts of population growth suggest that by 2050 there may be 9.5 billion humans sharing the planet, most of them living in our ever-expanding cities. We have already reached a point where more than half of all people live in urban areas, and this proportion is expected to rise to two-thirds later this century. The influence of these demographic trends on water resources in discussed further in Section 2.3.3 and in detail in Chapter 4 'Live'.

Water Resources: A New Water Architecture, First Edition.
Alexander Lane, Michael Norton and Sandra Ryan.

Significant volumes of research have been carried out and continue to be conducted into potential scenarios of climate change and their projected impacts on RWR and water demand. The evidence is strong that the influences are real and that the impacts are already with us and set to intensify (Intergovernmental Panel on Climate Change (IPCC) 2013). Very broadly, predictions are for increased rainfall and runoff in higher latitudes and reduced rainfall and runoff in tropical and mid-to lower latitudes. The volumes of water stored in glaciers are expected to fall, thereby reducing annual melt-water flows and in turn affecting water supplies in dependent areas such as Peru and California. Higher temperatures will exacerbate water pollution problems in many rivers and lakes, and will increase evaporation from open waterbodies and soil. More intense rainfall events will result in more frequent stormwater flooding in urban areas as well as from rivers.

1.2 Human Interactions with Water in the Biosphere

It is estimated that the world's total RWR is between 33,500 km^3 and 47,000 km^3 per year (Millennium Ecosystem Assessment 2005). Vast amounts of this resource are, for all practical purposes, unavailable due to their remoteness relative to demand (for example in the Amazon Basin, Canada, Greenland and Russia). It has been estimated that only around 50% of the global RWR can be accessed (Millennium Ecosystem Assessment 2005).

Currently, we withdraw around 4,500 km^3 of our accessible RWR (2030 WRG 2009). In the last 40 years, global water withdrawals have almost tripled and this growth rate remains strong, increasing by over 60 km^3 each year. Despite these increases in withdrawals, demands for water are growing even faster and are expected to reach 6,000 km^3 a year by 2030 (2030 WRG 2009). Even with our increasing water supply rates, and allowing for more efficient use of water, meeting this demand is believed by many authors to be unlikely (2030 WRG 2009). It can be argued that even now we are reaching what some observers are calling 'peak water', the concept of the safe water withdrawal limit that must not be passed if we are also to leave enough water in our rivers to maintain their aquatic ecosystems and biodiversity, a vital and much underappreciated resource in their own right.

Now that more than 1 in 2 people live in urban environments, the need to address the pressures that urban lifestyles exert on water resources is paramount. Urban water managers already face challenges of aging water infrastructure, large energy demands, high maintenance and treatment costs, and increasingly stringent environmental regulations. Many are also facing population growth, and the impacts of climate change on water demand and on urban stormwater runoff.

Water management in cities and urban settings has experienced many developments in thinking in recent years. The International Eco-Cities Initiative identified as many as 178 significant so-called 'eco-city' initiatives at different stages of planning and implementation around the world (Joss *et al.* 2011), and most of these initiatives include a water management component. Examples include Curitiba (Brazil), Auroville (India), Dongtan (China), Masdar (UAE), Freiburg (Germany) and Stockholm (Sweden). The evolving aim is to move from urban systems which are heavy users of non-renewable resources and generators of waste to urban systems which reduce their water demand, use renewable resources and recycle their wastes into valuable products (see Figure 1.1).

Importantly, this aim applies as much to the resources of food, energy and other materials as it does to water; water is at the heart of urban sustainability, however. Already, most urban water utility managers are implementing measures which can be loosely classed as 'demand management': promoting the uptake of household appliances which use less water, advocating garden rainwater harvesting and considering the recycling of treated wastewater, for example. They also wish to minimise the costs and carbon footprint of their primary water supply systems, seeking water from sources which cost less to secure and at the same time offer resilience against the potential future impacts of climate change and weather extremes.

It is projected that future water withdrawals required to grow and process our food will reach 4,500 km^3 by 2030, compared to around 3,100 km^3 in 2010, unless significant efficiency gains are realised (World Economic Forum 2011). These withdrawals are around seven times higher than those for drinking water. At the current time, around 30% of the food eaten worldwide is grown under irrigation, accounting for 70% of all water withdrawals (World Economic Forum 2011). Irrigation underpins crop production, particularly commercial cropping, because it significantly increases crop yields over and above those which can be achieved by rainfall alone. While there are still vast tracts of cultivatable land on the planet with regular rainfall, the growing trend for crops to be grown under irrigation shows no sign of abatement. A special report in *The Economist* in February 2011 concluded that of all the constraints to 'feeding the 9 billion', that of finding sufficient water is the most intractable. The relationships between water and food are explored in detail in Chapter 5 'Eat'.

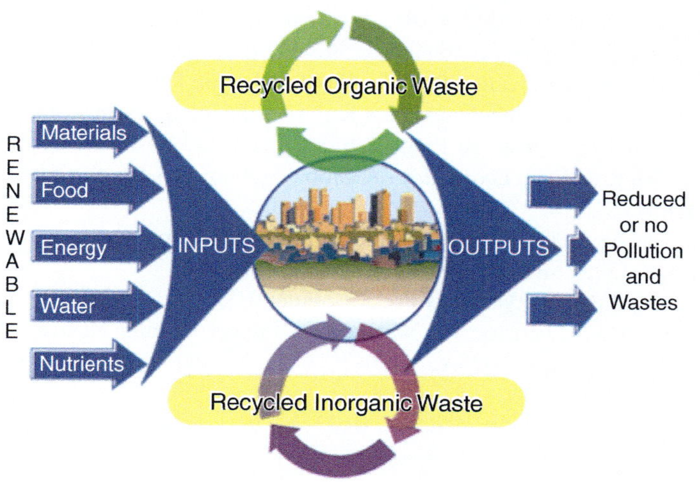

Figure 1.1 Inputs and outputs in an idealised urban resource system. *Source:* adapted from Rogers (1998).

1.3 An Inspiring Challenge

In his 2010 BBC Reith Lectures, UK Astronomer Royal Professor Sir Martin Rees said "This is a crucial century. The Earth has existed for 45 million centuries. But this is the first when one species, ours, can determine – for good or ill – the future of the entire biosphere".

This is a profound statement and one that has inspired the authors of this book. We believe that the future of the biosphere as a sustainable habitat for mankind will be framed by how effectively we manage our water: water in our rivers, lakes and aquifers; water in our soils; water which sustains our incredible biodiversity and ecosystems; and, most of all, the water that we humans use to live, eat and consume (Part II: chapters 4, 5 and 6, respectively).

In the subsequent chapters of Part III of this book we describe the fundamentals of water resources, the current state of water stress through our live, eat and consume activities, and how current policy, regulation and water management seek to address water scarcity and increasing water insecurity. In Part IV, our final collection of chapters, we propose a new way forward characterised by conceptual, physical and institutional integration of all aspects of the management of our planet's water, an approach which transcends current valiant yet largely unsuccessful attempts to implement Integrated Water Resource Management (IWRM). We term this new approach a New Water Architecture.

References

2030 Water Resources Group. 2009. Charting Our Water Future: Economic frameworks to inform decision making. 2030 WRG, London, UK.

Intergovernmental Panel on Climate Change (IPCC). 2013. Summary for policymakers. In Stocker, T.F., Qin, D., Plattner, G-K., Tignor, M., Allen, S.K., Boschung, J., Nauels, A., Xia, Y., Bex, V. and Midgley, P.M. (eds) Climate Change 2013: The Physical Science Basis. Contribution of Working Group I to the Fifth Assessment Report of the Intergovernmental Panel on Climate Change. Cambridge University Press, Cambridge, UK and New York, USA.

Joss, S., Tomozeiu, D. and Cowley, R. 2011. Eco-Cities - A Global Survey: Eco-City Profiles. University of Westminster, UK.

Millennium Ecosystem Assessment. 2005. Ecosystems and Human Well-being: Current State and Trends. World Resources Institute, Washington DC, USA.

Rogers, R. 1998. Cities for a Small Planet. Westview Press, Boulder, USA.

World Economic Forum. 2011. Water Security: The Water-Food-Energy-Climate Nexus. World Economic Forum, Washington DC, USA.

2

Fundamentals of Water Management

2.1 The Planetary Picture

This chapter explains the role of water in the biosphere, how water resource systems have evolved in different parts of the world, how water underpins all aspects of life and economic development, and why water has to be actively managed. The rationale behind the evolution of traditional approaches to water management is explored and the emergence of water-related social, environmental, and economic challenges are outlined to emphasise the urgency of the need for reform: reform both in the way in which water is valued and reform in how water allocations are prioritised and managed. Crucially, this chapter also explains why the past is no longer a suitable blueprint for how we must manage water going forward.

2.1.1 The Blue Planet

Viewed from space, the Earth appears as a blue planet with only a few relatively small green, brown and white patches. Over 70% of its surface is covered by the oceans and there are roughly 1.4 billion km^3 of water held in its seas, rivers and groundwater systems (Gleick 1993). It is difficult to appreciate scale when we start to talk about billions of cubic kilometres. The view of the Earth from space, or even that witnessed when taking a flight across an ocean, gives some perspective of the sheer scale of how much water there is on Earth. To try to visualise this, a volume of 1.4 billion km^3 would have a depth of some 800 km if spread over the area of the South American continent.

Once we consider the volume of freshwater on Earth, the numbers start to fall dramatically. Excluding the water which is frozen within ice caps and mountain glaciers, Earth has around 10.6 million km^3 of freshwater (USGS 2014), a volume that could fit within the shallow Mediterranean Sea basin just one and half times over. Currently, around 7 billion people need to share this relatively small amount of water, a number that increases by more than 228,000 people every 24 hours (Population Institute 2010). Sharing what is already a relatively scarce resource is made all the more difficult by the unequal distribution of water across the Earth's surface, and because so much of it is difficult to access, both physically and economically.

Figure 2.1 and Breakout Box 2.1 illustrate how small our shared freshwater resource is compared to the volume of water in the oceans, the tiny proportions stored as ice, groundwater, surface water and in organisms in the biosphere, and the volume in the atmosphere at any given time.

Water Resources: A New Water Architecture, First Edition.
Alexander Lane, Michael Norton and Sandra Ryan.

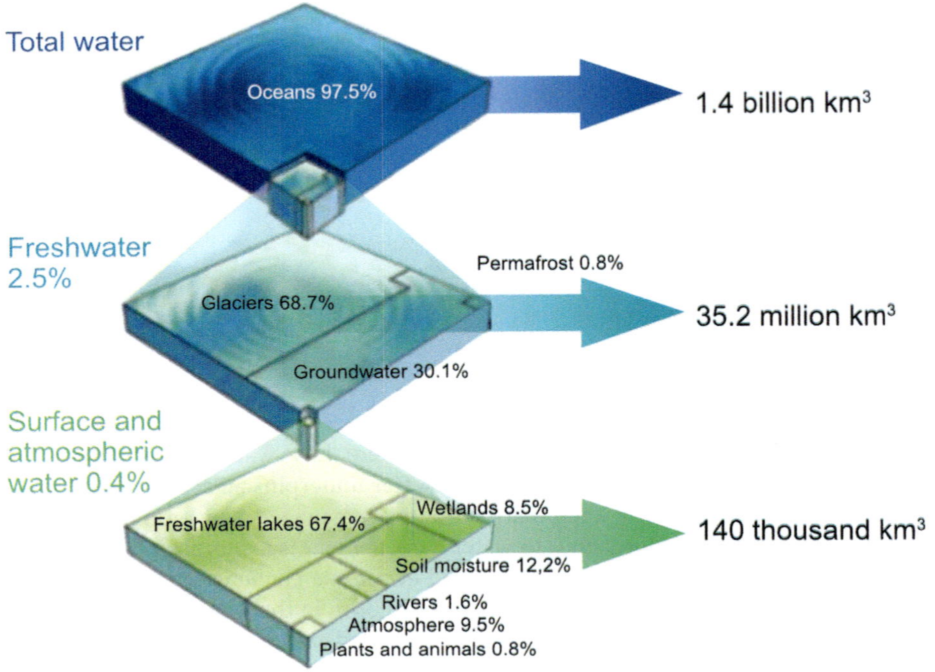

Figure 2.1 Freshwater as a proportion of the total global water volume.
Source: UNESCO The United Nations World Water Development Report 2, Section 2: Changing Natural Systems, Chapter 4, Part 1. Global Hydrology and Water Resources, p.121. Data from Shiklomanov and Rodda (2003). Freshwater has a global volume of 35.2 million cubic kilometres (km^3).

Breakout Box 2.1 Global water facts

Only a tiny fraction of the water on Earth is fresh (between 2% and 4%, approximately 35.2 million km^3) and with around 70% of that amount frozen in icecaps and glaciers, an even smaller proportion is present as liquid freshwater: around 10.6 million km^3.

Polar and glacial ice is sometimes referred to as freshwater that is 'locked up' or trapped and essentially unavailable. However, this definition ignores the important role that these ice resources have in influencing atmospheric conditions, ocean currents and meltwater which are all fundamental to sustaining the conditions through which the vital liquid freshwater resources are derived.

Approximately 30% of the Earth's freshwater is stored underground in aquifers. Significant proportions of this groundwater are inaccessible because the geology of the rock makes exploitation too difficult. Although this situation may change over time as technology evolves, it is this type of freshwater that may more appropriately be described as 'locked up' or trapped.

2.1.2 Water and the Biosphere

Life on Earth is sustained by the movement of water within what is essentially a closed system. All the water on Earth has been here for over 45 million centuries, changing

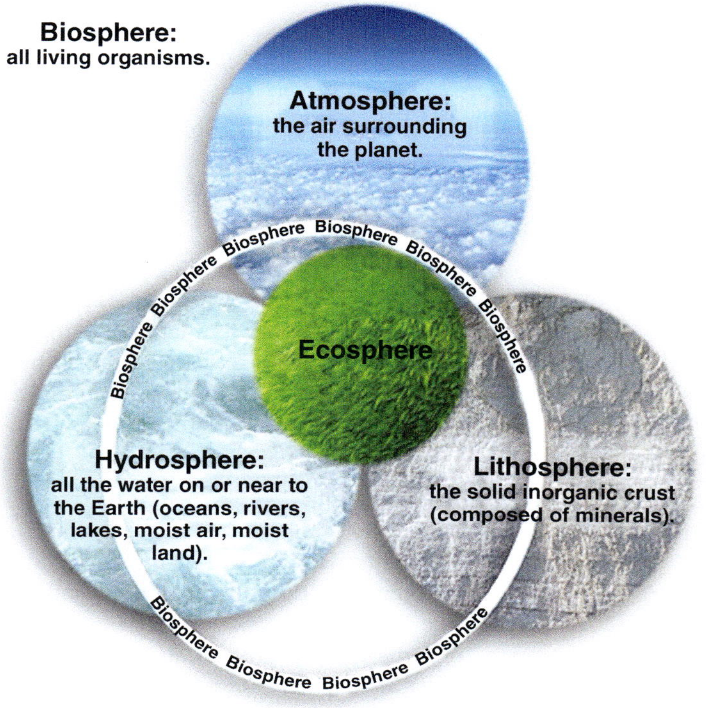

Figure 2.2 The spheres supporting life on Earth.

state from gas to liquid and solid and shifting between the atmosphere, the hydrosphere, the lithosphere, and within the biological structure of plants and animals. That part of Earth that we recognise as the 'living zone' is the biosphere and the finely balanced interactions between the different 'spheres' support life (see Figure 2.2).

Biosphere functionality is governed by the availability and quality of water. The biosphere also serves as an important part of the water cycle in its own right, regulating atmospheric chemistry, influencing the movement of water and purifying its quality.

Where there is water, life in some form is invariably found. Life can even thrive in hostile environments as long as minimum water requirements are met. The abundance and biodiversity of life is generally greatest at low altitude in low latitudes, and some of the planet's greatest biodiversity is found in the ancient rainforests where the climate has been warm and wet for a very long time (Bass *et al.* 2010).

Life can tolerate hyperalkaline, saline and even concentrated arsenic conditions. Flamingos have been found flourishing in the extremely hostile conditions of Laguna Diamante inside the active Cerro Galán volcano in Argentina, for example (Belluscio 2010). Despite the connotations of its name, the Dead Sea is in fact home to algae and bacteria that have adapted to the hypersaline and magnesium-rich conditions. Life (in the form of multi-cellular organisms) has even been found more than 2 km deep in the Earth's crust, at depths previously thought incompatible with any form of life. Despite there being no sunlight, little oxygen and no food, the Devil's Worm (*Halicephalobus mephisto*) is still able to survive on the trace particles of water that are present (Borgonie *et al.* 2011).

There is however one place on Earth where there is extremely little, if any, life: the Atacama Desert in Chile where average rainfall is just 15 mm per year. Parts of this desert have not received rainfall in decades and some of its weather stations have never recorded rain. This lack of water is the principal reason for the almost total absence of life, and its extreme aridity may represent the 'dry limit' of microbial life (Navarro-González *et al.* 2003).

2.1.3 Distinguishing between Hydrology and Water Resources

Where there is a stock or supply of freshwater that can be drawn on by individuals, companies or water authorities, that water is termed a 'water resource'. The term 'resource' implies use, that is use to maintain life and to support standards of living, as well as use through agriculture and other activities. Water resources are sustained by a series of hydrological processes that can be collectively represented in the water cycle depicted in Figure 2.3. The hydrological processes that drive the water cycle are relatively straightforward and well defined:

- Water precipitates from the sky as rain or snow.
- On reaching the Earth's surface, some of this water soaks into the ground and is entrained within biota (from where it may evapotranspirate or become embedded within the organic matter). Some of the water continues to infiltrate into the soil and on into groundwater.
- The rest of the water either evaporates or runs off the ground surface, forming streams, rivers and lakes before ultimately reaching the sea.
- Water can evaporate at any of the stages occurring at the ground surface to eventually be precipitated in a new location.

Hydrological processes such as precipitation, evapotranspiration and runoff are not constrained by geopolitics or socioeconomic factors. In practice, however, the resultant

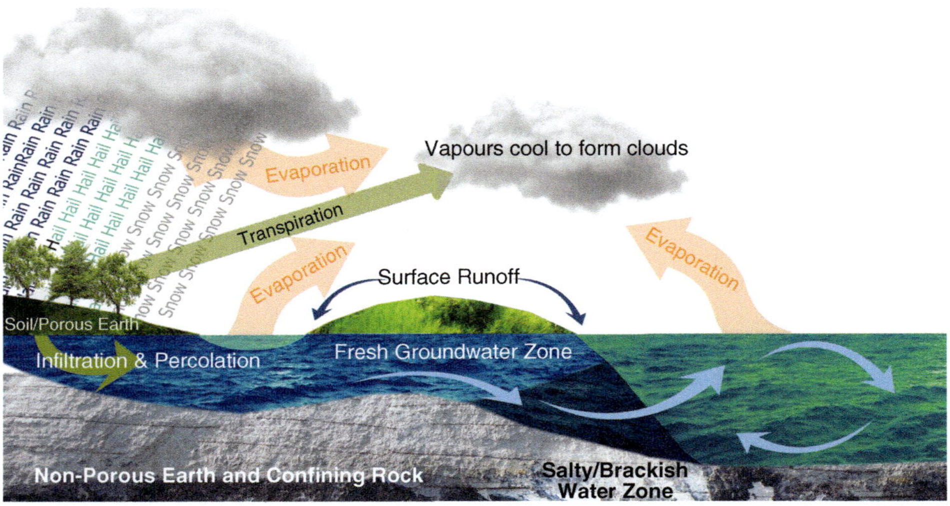

Figure 2.3 Traditional view of the hydrological water cycle (without human influence).

water resources are very much owned and governed (with varying levels of success; see Chapter 7). The Glossary at the start of this book sets out a number of terms used to describe water resources and water resource concepts based on definitions published by the United Nations Food and Agriculture Organisation (FAO 2003). Figure 2.4 illustrates how some of these concepts are defined by application of geopolitical and socio-economic factors to hydrological concepts.

2.2 Evolution of Water Resource Systems

So far Chapter 2 has established that water is a vital resource, is constantly in transition, moving through the water cycle, changing state and moving around the planet.

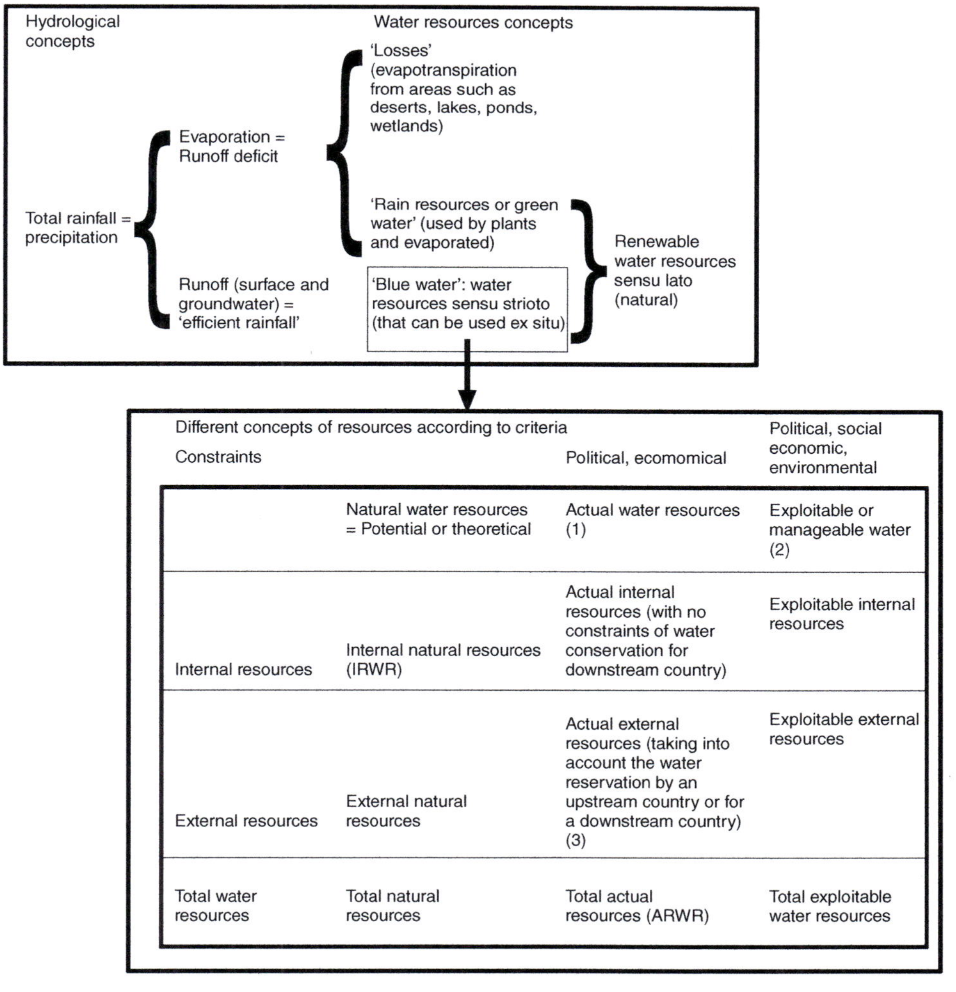

Figure 2.4 Distinguishing between hydrological and water resource concepts. *Source:* FAO (2003).

This section explores how 'hydroclimates' (the physical and chemical characteristics that define a particular aquatic habitat) create a diverse baseline of natural water resource systems, and how these systems have been manipulated by human activity into anthropogenic water cycles to increase the available water resource.

Let's start with those basic water resource systems and how they have helped to shape population and society.

2.2.1 Hydroclimates and Water Resources

Patterns of precipitation and temperature vary tremendously on a global scale. Detailed discussion of climate and climate types is beyond the scope of this book, but it is important to recognise the influence that climatic variability imparts on the distribution of water resources and on the types of water resource infrastructure and management systems that have developed. This is especially important as we consider the implications of population change, migration and how the global climate itself is evolving. The less predictable weather patterns that may accompany future climates will have major repercussions on our ability to manage resources and secure water supplies.

Wladimir Köppen's original (1900) and revised (1918) climate classification system sought to map the different terrestrial climatic systems (defined by temperature and aridity) in relation to (and impact on) the different vegetation zones that were also being mapped for the first time. Since then, climatologists have continued to refine the model and the classifications; the basic model identifies five main climatic types with 14 individual hydroclimates. Here we define the main climatic characteristics of each and the implications on water resources.

- **Type A: warm climates**. Low latitudes, high temperatures generally constant throughout the year (diurnal range in temperature is more significant than any monthly range). The fluctuations of the trade winds, the intertropical convergence zone (ITCZ) and the Asian monsoon exert the most seasonal influence.
  There are three main Type A climates:
  1) **Wet equatorial climate**: High rainfall volumes across the year with a summer peak. Air temperature is constant at around 30°C. High levels of runoff occur throughout the year tempered by land use and vegetative cover. *Examples include the Amazon rainforest.*
  2) **Tropical monsoon and trade-wind littoral climate**: Extremely high annual rainfall volumes occurring predominantly during summer. Air temperature is relatively constant at around 30°C. Monsoon rain can be unpredictable. Under natural circumstances, much of the runoff cannot be harvested; with the exception of the temporal surface water river systems, the monsoon climate does not naturally produce any significant water resource.

     More than half of the world's population lives in areas where water resources are supplied via the wet summer monsoon system (WCRP undated). Lifestyles have evolved around the cyclical nature of the monsoon and the main land use and economic activity is agriculture. Most places within monsoon-dominated climates do not have significant water storage facilities, except shallow aquifers. Artificial additional storage and irrigation systems are rare. A dependence on rainfall makes these regions highly vulnerable if the monsoon fails. *Examples include southeast Asia.*

3) **Tropical wet-dry climate**: High rainfall volumes in winter with very low rainfall in summer. Air temperature maintains a relatively flat profile of 20°C throughout the year. Depending on geology, this climate is able to facilitate winter surface water storage to support dry summer seasons. However, these supplies are highly vulnerable to dry winter scenarios. *Examples include central Brazil.*

- **Type B: arid and semi-arid climates**. Predominantly 15–30° latitude (in both hemispheres), low precipitation (although this varies greatly from year to year), low relative humidity, high evaporation rates (when water is available), clear skies and intense solar radiation.

  There are three main Type B climates:

  4) **Tropical and subtropical desert climate**
  5) **Mid-latitude steppe and desert climate**
  6) **Tropical and subtropical steppe climate**

  Desert and Steppe: Very low annual and monthly rainfall volumes (similar to the Arctic climate) but with an inverse air temperature pattern with lows of 10°C and highs exceeding 30°C. The limited rainfall generates very few natural water resources. Groundwater aquifers may be present but are vulnerable to over-abstraction.

  In some desert regions such as Peru and California water supply is supported by glacial meltwater, dependent on the ability of the population to invest in major storage infrastructure. *Examples include deserts in North Africa, Saudi Arabia and Central Australia.*

- **Types C and D: climates dominated by seasonal air masses**. Present at middle and high latitudes, seasonal variations in location and intensity of upper-level and mid-latitude westerly winds.

  Within this overall group there are five climates:

  7) **Humid subtropical climate (Type C)**: Moderately high annual rainfall across the year peaking in summer. Air temperature ranges from lows of 10°C in winter to 30°C in summer. Evaporation rates can suppress runoff volumes, although these remain relatively high due to the moderately high rainfall. Geology will affect the contribution of groundwater resources but surface water systems typically dominate. *Examples include regions of China.*
  8) **Mediterranean climate (Type C)**: Low annual rainfall with very low rainfall volumes in summer. A moderate annual range in temperature exists: between 10°C in winter up to 30°C in summer. There is typically limited runoff and few natural surface water storage systems. Groundwater (where available) augments water supply during dry periods when surface water is constrained. *Examples include Spain, southern France, Italy, the Balkans and Turkey.*
  9) **Marine west coast climate (Type C)**: Temperate latitudes tend to receive more stable and reliable rainfall driven by frontal weather systems, often influenced by topography. In the UK and across much of northern and western Europe, natural surface water resources and groundwater reserves are augmented by additional reservoir systems that capture and store excess winter rainfall runoff for use during the drier summer months. *Examples include northwest Europe and New Zealand.*
  10) **Humid continental climate (Type D)**: Moderate annual rainfall with a constant (flat) annual profile. Minimum winter air temperature of −10°C increasing to a maximum summer temperature of 20°C. Moderate but consistent rainfall can generate substantial resources in a humid continental climate due to limited evaporation, although the relatively low annual quantities are vulnerable to depletion from

over-abstraction. Depending on topography high runoff can occur, forming natural lake reservoirs and riverine water resources.

Extended droughts are rare in humid continental regions; however, periods of low precipitation and water scarcity are not uncommon (Juniata County 2009). As a result, human populations in this hydroclimate are subject to stable albeit fragile water resources. The extent of that fragility depends on the availability and accessibility of groundwater. Despite the regularity of rainfall, recharge can be limited by the relatively small absolute precipitation volumes, especially if the surficial and bedrock geology promotes runoff rather than infiltration. *Examples include the central and eastern USA, Eastern Europe and northern Japan.*

11) **Continental subarctic climate (boreal or taiga) (Type D)**: Very low annual rainfall with a slight summer peak. A wide range of temperatures from 0°C in winter to 25°C in summer. Due to the very limited rainfall, hydrology is dictated by topography, geology and land use that can limit available energy for evaporation and snowmelt. Water supplies are typically sourced from spring meltwater, followed by the concentrated summer rainfall. Water resource systems and management is strongly influenced by the seasonality of snowfall, melt, peak summer rain, small-scale surface waters (rivers), surface water reservoirs (where funding is available) and limited groundwater. *Examples include central and eastern Russia and the Canadian boreal zone.*

- **Type E: polar and arctic**. Very low temperatures with a steep gradient in annual air temperature from –50°C in winter to 0°C in summer. Exceptionally low annual rainfall controlled by air masses at high latitudes. Köppen identified two Type E climates – tundra and snow – but climatologists have added a third (Type H) climate specific to highlands.

12) **Tundra climate**
13) **Snow and ice climate**
14) **Highland climate**

In these climates water resources are dominated by snow and ice reserves with potentially large freshwater aquifers present beneath the ice. Water supplies provided by meltwaters will be under increasing threat as climate change accelerates ice loss and glacial retreat in many regions. Typically mountainous terrain and inhospitable interiors concentrate relative small human populations in coastal areas. Water supplies are typically subsistence based, accessed locally without significant water infrastructure. *Examples include Greenland, northern Canada and Siberia.*

Where there is very little rainfall (for example in arctic and desert regions), there is also very little opportunity for water resources to form. The result is that stocks of water are limited to either fossil groundwater or highly intermittent and unpredictable surface flows (such as wadi systems in deserts). Historically, arctic and desert regions supported small, scattered populations able to survive by accessing water within ice or shallow groundwater on an *ad hoc* basis. In more recent times however, driven by socioeconomic factors, larger, more densely populated communities have begun to develop in desert regions, particularly in the oil-rich areas of the Middle East. The population growth rate in the Middle East is among the highest in the world and this brings major challenges to societies, not least in relation to accessing water resources. According to a 2013 report by the Institute of Chartered Accountants in England and Wales, the population of Middle Eastern countries grew by 52% between 1990 and 2010 (compared to an average population growth across the world over the same period of 29%) (ICAEW 2013). In Bahrain, Qatar, and the United Arab Emirates, the population has more than doubled (these are countries that receive less than 0.2 km^3 of RWR a year; FAO 2003).

Hydroclimates impart a major influence on the distribution of water resources in the form of accessible surface water and ice. Geology can also influence the relative dominance of groundwater or surface water systems. Figure 2.5 illustrates how these three types of freshwater (ice, surface water and groundwater) are distributed around the world.

Just under 70% of freshwater is locked up in ice-cap glaciers (the vast majority in Antarctica); with 30% of the rest of it underground, groundwater is by far the most dominant source of 'accessible' freshwater. At the continental scale, the largest volume of groundwater is found in Asia. Most of the largest aquifer systems are in Africa (fairly evenly distributed across the continent albeit with a slight dominance across North Africa) followed by Asia (predominantly Russia), and North and South America. The 37 largest aquifer systems in the world cover 36% of the land area of continents (Margat 2008). It should however be noted that groundwater mapping is inherently uncertain and that information on stored volumes and hydraulic properties is limited. Many groundwater systems remain unidentified and unexplored and, for these reasons, assessment of how groundwater is used offers more tangible insight into this resource.

Margat (2008) explains how around 90% of groundwater recharge is estimated to re-emerge at the surface as baseflow, supporting surface watercourses, while the remaining 10% is exploited from springs and abstraction boreholes or exits the system via submarine outflow. By far the largest absolute volume of groundwater abstraction occurs in Asia (India, Pakistan, Iran and Bangladesh are the major consumers) followed by North America; see Table 2.1 (van der Gun 2012). Considering that a high proportion of the world's largest aquifer systems are located in Africa, rates of groundwater abstraction on the continent are currently very low.

Of the global groundwater that is abstracted, the vast majority (everywhere except Europe) is used to irrigate agriculture. The reliability of groundwater is thought to make irrigation by groundwater at least 20% more efficient than irrigation by surface water (EASAC 2010).

Surface water systems, typically formed by rainfall or meltwaters over non-permeable or semi-permeable geologies, are usually characterised by rivers or streams which criss-cross a landscape, forming lakes in depressions and draining catchment areas *en route* to the sea. Rivers, streams and lakes offer abundant services to mankind including navigation, opportunities for commercial and recreational fishing, provision of aquatic and riparian habitat, and outlets for the natural assimilation of waste. These types of services (or benefits) provided by the environment are defined as 'ecosystem services,' a concept discussed implicitly for decades but formalised and popularised by the Millennium Ecosystem Assessment (2005).

The availability of water on the surface makes it easy to divert, impound or abstract for drinking water, irrigation or industrial uses. However, surface waters fed by rainfall or meltwater are highly vulnerable to climatic variations, and the impacts of such events can be felt very quickly, even within a single season. The quality of surface water is also highly vulnerable to impacts associated with land uses such as farming which generates pollutant and sediment runoff that can degrade water quality in streams and rivers.

Table 2.2 list the world's most water-rich countries in terms of the total water resource (as a total of internal and external surface and groundwater). It shows that 60% of the

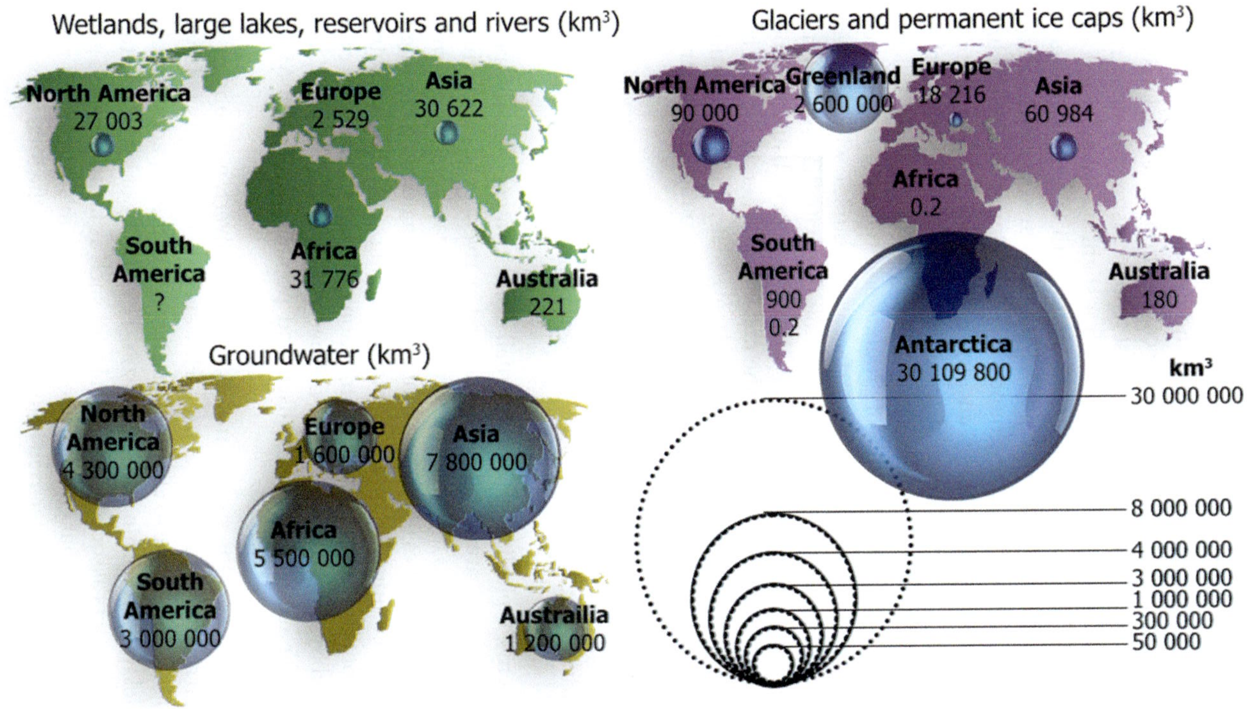

Figure 2.5 Global distribution of freshwater in ice, surface water and groundwater ('?' under 'South America' as per source).
Source: Adapted from graphic (Rekacewicz, P.) in UNEP/Grid Arendal (2008). Original *source:* IA Shiklomanov, State Hydrological Institute, St Petersburg; UNESCO, Paris; World Meterological Organisation; International Council of Scientific Unions; World Glacier Monitoring Services; United States Geological Survey, USA.

Table 2.1 Global groundwater abstraction in 2010.

Continent	Groundwater abstraction*				Compared to total water abstraction	
	Irrigation (km³/yr)	Domestic (km³/yr)	Industrial (km³/yr)	Total (km³/yr)	Total abstraction (km³/yr)**	Groundwater abstraction as% of total
North America	99	26	18	143	524	27
Central America and the Caribbean	5	7	2	14	149	9
South America	12	8	6	26	182	14
Europe (including Russian Federation)	23	37	16	76	497	15
Africa	27	15	2	44	196	23
Asia	497	116	63	676	2,257	30
Oceania	4	2	1	7	26	25
World	666	212	108	986	3,831	26

Source: van der Gun (2012).

* Estimated on the basis of IGRAC (2010), AQUASTAT (undated), EUROSTAT (undated), Margat (2008) and Siebert *et al.* (2010).

** Average of the 1995 and 2025 'business as usual scenario' estimates presented by Alcamo *et al.* (2003).

Table 2.2 Water-rich countries.

Country	Average precipitation (km³/yr) (1961–1990)	Total internal water resource (km³/yr)	Total external RWR (km³/yr)	Total RWR (km³/yr)	Internal water resource per inhabitant (m³/yr)
Brazil	15,236	5,418	2,815	8,233	31,795
Russian Federation	7,855	4,313	195	4,507	29,642
Canada	5,352	2,850	52	2,902	92,662
Indonesia	5,147	2,838	0	2,838	13,381
China (mainland)	5,995	2,812	17	2,830	2,245
Colombia	2,975	2,112	21	2,132	50,160
USA	5,800	2,000	71	2,071	7,153
Peru	1,919	1,616	297	1,913	62,973
India	3,559	1,261	636	1,897	1,249

Source: FAO (2003).

total water resource is spread across just nine countries. At the continental scale, the Americas have the largest share (45%) followed by Asia (28%), then Europe (15.5%). Africa has the lowest proportion with just 9% (see Figure 2.6).

Table 2.2 identifies Brazil as the country with the greatest RWR. This is largely due to the very high rainfall and resource available in Amazonia, a situation that means Brazil also has a high internal RWR per inhabitant. This view is overly simplistic however, because much of Brazil's water resource is located considerable distances from the urban areas that demand it most. As a result, the world's most water-rich country is struggling. In 2014, São Paulo, a megacity with a population of 23 million (one-fifth of all Brazilians), was gripped by its worst drought in 80 years. A slow response by the authorities meant that by the end of the 2014, São Paulo's reservoirs were at very low levels of storage and taps in some neighbourhoods were already dry.

Canada is arguably the world's most water-secure nation, with the highest internal RWR per inhabitant despite its lower levels of precipitation. This position of security largely reflects the country's sparse population density. However, even in Canada people can face water shortages. In fact, around 25% of Canadian municipalities have experienced past water shortages and have had to ask residents to voluntarily reduce their consumption as a result (Government of Canada 2013). The Canadian Prairies are particularly susceptible and suffered multi-season droughts in the 1930s and 1980s (Government of Canada 2013). Since then, there have been numerous single-season droughts threatening the agriculture and wetlands which dominate the landscape. Exacerbating these events, the region is now also subject to livestock intensification and oil sands extraction (University of Alberta 2007), processes which both alter patterns of water demand.

Water-poor countries are usually the smallest and most arid; however, there are important exceptions. Over 30 countries depend on other states for more than 50% of their RWR, and these include smaller nations (by geographic area) such as Bahrain, Israel and Benin but also larger countries such as Argentina, Azerbaijan, Bangladesh, Bolivia, Chad, Congo, Pakistan, Ukraine and Vietnam (FAO 2003).

2.2.2 Mechanisms of Human Interactions with Water Fluxes

Section 2.2.1 sets out the basic premise that the available water resource in a given location is a product of its hydroclimate, topography and geology. It also explains how human interventions have exploited natural systems, harnessing water in the environment to secure supplies for dependent populations. In most regions, this process of exploitation has occurred for centuries. In fact, the water management systems of the Romans (described in Section 2.3.1) continue to inform water supply planning to this day.

Many terms exist to describe the means by which people use water. For example, when water is taken from a natural source this is termed a *withdrawal* or an *abstraction*. The purpose for which the water is withdrawn is termed the *use*. The amount of water withdrawn which is taken up in the process of that use and not discharged back into the environment is termed *consumption*, and the amount which is surplus, or that which is rejected or wasted to a body of water, is termed a *return*. These terms are typically used when water is taken from what is referred to as a *blue water* source, a waterbody such as a river, lake or underground aquifer.

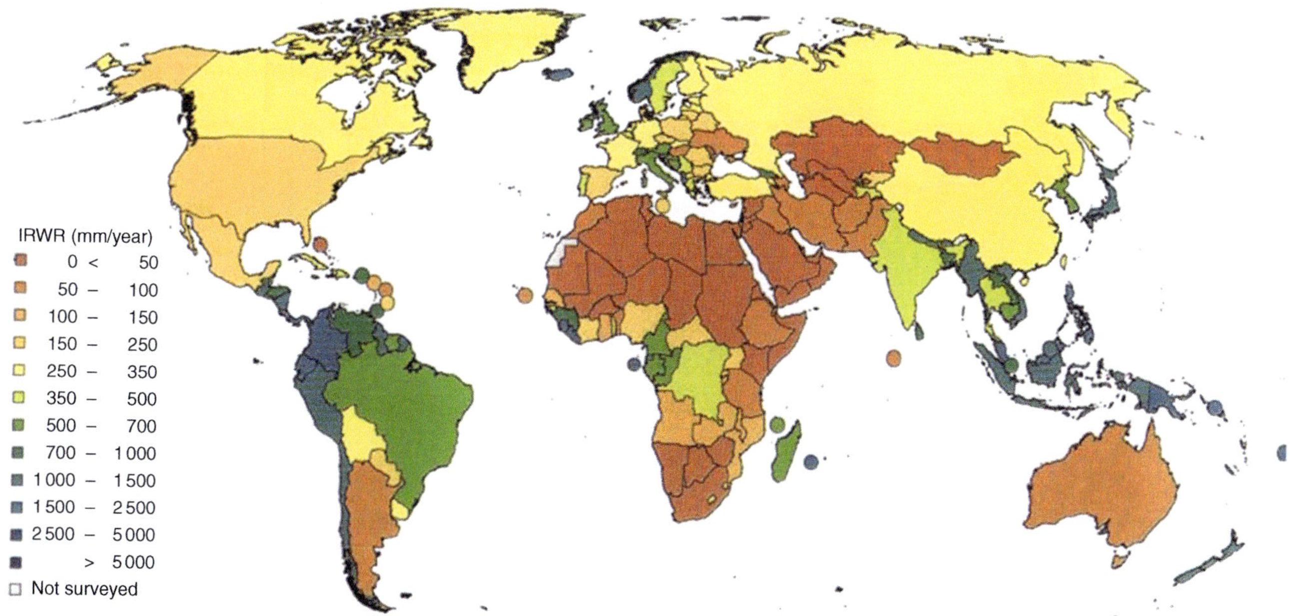

Figure 2.6 Global distribution of internal renewable water resources.
Source: FAO (2003).

The single largest use of water is in the growing of food. Where food is grown via rain-fed agriculture, the water withdrawal is from soil moisture, known as *green water*. The water is consumed in the process of evapotranspiration and, in this case, there is no return. An increasing proportion of food is however being grown through irrigated agriculture; because most irrigation systems deliver more water than is consumed by the crop, the excess is a return to the underlying aquifer or nearest river.

Table 2.3 summarises average annual withdrawals from blue water sources by continent. This shows that globally, 70% of blue water withdrawal is for irrigated agriculture, while only 11% and 19% is for domestic and industrial uses, respectively. Asia and Africa abstract (withdraw) the highest proportions for agriculture, almost twice the proportion in the Americas and almost three times that abstracted for agriculture in Europe. It should be remembered however that much of the water withdrawn for agriculture is consumptive, whereas much of the water withdrawn for domestic and industrial use is returned (via domestic wastewater and power generation cooling water). The consumptive use of agricultural water in Asia is immense, consuming 2010 km^3/yr.

2.2.3 Anthropogenic Influence: The Traditional Urban Water Cycle

Human interventions combine with the natural water cycle to create the anthropogenic water cycle (see Figure 2.7); the basis of many of the challenges facing today's water professionals.

In many places, the water that we see in the environment is not so much the product of the natural water cycle, but the result of highly engineered and regulated processes. Anecdotally, it has been said that in the UK over 90% of the water in rivers, streams and lakes is discharged from water infrastructure. It's not really important how accurate that figure is, but it certainly does reflect the high number and very high density of artificial influences that are found from headwaters down to river mouths.

Anthropogenic influences are discussed in more detail in Chapter 4, but here we introduce the following basic components:

- abstraction;
- storage;

Table 2.3 Blue water withdrawals by use (%) and by continent (km^3/yr).

Continent	Total withdrawal (%) by sector: Domestic	Industry	Agriculture	Total freshwater withdrawal (km^3/yr)	Freshwater withdrawal as % of IRWR*
Asia	9	9	82	2,451	20
Americas	16	35	49	790	4
Europe	16	55	29	374	6
Africa	10	4	86	215	5
Oceania	17	10	73	26	3
Global	11	19	70	3,856	9

Source: FAO (2011).

*Internal renewable water resources.

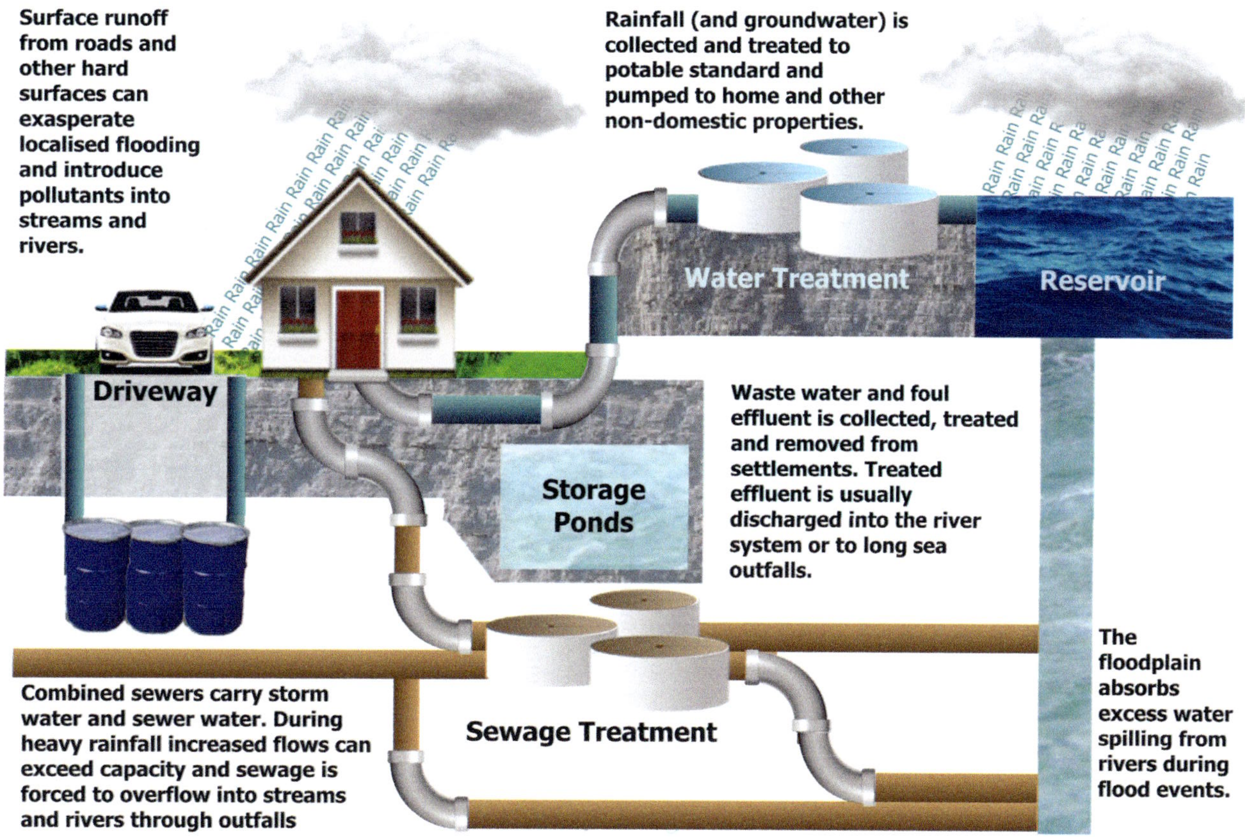

Figure 2.7 The water cycle in an urban context. Amended from Environment Agency (2007).

- water supply distribution systems;
- urban land use and stormwater runoff;
- sewerage systems; and
- wastewater treatment and discharge.

2.2.3.1 Abstraction

Abstracting water from the environment is the most basic form of human interference in the natural water cycle and has occurred for thousands of years. Water can be abstracted using different techniques: siphoning water from surface water resources; digging shallow wells to access groundwater from the water table; and drilling and installing boreholes which enable water to be abstracted from greater depths. As populations grow and demands for water increase, abstraction techniques have evolved to supply the larger volumes demanded.

In small volumes or when water availability is high, the impact of abstraction on the water cycle may be negligible (although it can often draw down environmental resources). Over-abstraction reflects a situation where the volume of water abstracted exceeds the natural capacity of the resource to recharge itself, and therefore typically leads to environmental and resource degradation.

Over-abstraction from surface waters often results in low river flows, reduced water quality and habitat degradation. Over-abstraction from groundwater often takes longer to translate into an observable impact. Similarly, groundwater resources often take much longer to recover from over-abstraction than surface water sources. The impacts of groundwater over-abstraction differ depending on the type of aquifer and its recharge rates, as well as its proximity to the sea. Aquifers that recharge very slowly, particularly those where the freshwater resource has been *in situ* for decades or centuries, are particularly vulnerable to abstraction impacts if volumes are not regulated in line with recharge. As well as indicating declining groundwater resources, a falling water table can have major impacts on surface water resources and even land stability.

Deteriorating water quality is also symptomatic of an over-abstracted water source, particularly in coastal aquifer systems where the change in hydrostatic pressure arising from over-abstraction causes water from the marine environment to move towards the freshwater aquifer.

2.2.3.2 Storage

Storage in lakes and groundwater is a natural part of the water cycle. Storage is also a major anthropogenic component. The purpose of creating additional storage is to reduce the pressure on other more limited and transient water sources (particularly rivers) and to increase resilience against acute water shortages by storing water during periods of high flow.

Reservoirs have the capacity to dramatically alter the natural water cycle. They are typically located in the upper reaches of river systems to enable stored water to later be released under gravity. Reservoirs may be recharged by direct inflows (where a river is dammed and the outflow managed via the dam structure) or diverted flows (where the reservoir is offline), including pumped storage whereby water is abstracted from a river and pumped into the reservoir.

2.2.3.3 Water Supply Distribution Systems

The anthropogenic water distribution system can be defined as the assets and processes that exist to manage water between its source and its point of use. Water in the distribution system is typically not subject to the natural processes of the water cycle (evaporation and infiltration, for example).

Water supply distribution systems can be extensive in size, superimposing a spatial layer that does not necessarily follow that of the natural catchment. Many water supply systems also include small service reservoirs which hold a specific number of days' water supply to increase flexibility and resilience within the network.

2.2.3.4 Urban Land Use and Stormwater Runoff

In the natural water cycle, rainwater either runs off the surface or infiltrates into the ground depending on the condition of the surficial geology and antecedent soil moisture.

Water is held within vegetation and soils and slowly permeates vertically or horizontally. In contrast, urban land use introduces different surface materials such as concrete and tarmac which are largely impermeable and therefore inhibit infiltration, replacing this process with runoff. It is this rapid movement of water through an urban area that can lead to stormwater (sometimes termed pluvial) flooding. Roads, pavements, buildings and hard-surfaced standing areas such as car parks all contribute to problems of water quality, water quantity, habitat and biological resource degradation, public health problems, and deterioration in environmental aesthetics. Hard surfacing increases flow volume, speed and channelling of water. As water flows through the urban environment it entrains contaminants such as hydrocarbons, metals and pathogens as well as nitrates, phosphates, synthetic organics and other materials which ultimately enter the natural water environment.

2.2.3.5 Sewerage Systems

Sewerage systems are a main water conduit through the urban water cycle. There are generally two main sewerage system models. One is the combined sewerage system, in which both sewage and urban stormwater drains within a single infrastructure network to a wastewater treatment works. This type of system is common in many older networks (the Victorian sewerage systems still in use across much of the UK, for example) and it often suffers from exceedances of capacity associated with intense rainfall. To prevent the water in the system backing up and flooding homes and other connected properties during heavy rainfall, combined systems have emergency outlets (termed combined sewer overflows) where the water escapes from the sewerage system into a natural waterbody, often introducing dangerous contaminants in the process.

The other sewerage system model keeps stormwater separate from sewage. This requires a dual drainage approach which is relatively straightforward to implement in new cities and developments, but more difficult to retrofit into existing urban areas that use the combined model. In separated sewerage systems, domestic (household) sewage may also be kept separate from industrial effluent.

2.2.3.6 Wastewater Treatment and Discharge

Wastewater treatment and discharge represent the last step in the anthropogenic water cycle. Water quality and the quantity of discharged water have a major impact on the receiving waters, downstream components of the water cycle and other users. In regulated systems, the quality of the water that can be discharged is often prescribed by way of permits. Water quality parameters and concentration limits are established in relation to the sensitivity of the receiving environment and the quality of influent into the treatment plant. Often, as environmental objectives have become more demanding (e.g. standards across European Union member states being driven up by the Water Framework Directive), permitted concentrations are reduced and treatment standards have to improve. In many places, volumetric demands for wastewater treatment continue to increase at the same time as environmental objectives are driving down permitted discharge concentrations. As a result, wastewater treatment capacity is emerging as a major constraint to urban development.

2.2.4 Anthropogenic Influence: Advancements in the Urban Water Cycle

Technologically advanced and often energy-intensive and expensive anthropogenic modifications to the water cycle, typically driven by water scarcity and the need to secure supplies, are summarised below and explored further in Section 4.3.4 (Alternative Approaches to Urban Water Management).

2.2.4.1 Desalination

Desalination refers to the removal of salts and minerals from saline or brackish water to create freshwater. Desalination is completely independent of rainfall; it therefore represents a major departure from the natural water cycle and is a climate-independent water supply. Several desalination technologies are available and can typically be grouped into either thermal or membrane-based alternatives.

Israel produces 40% of its domestic water from seawater desalination; however, the majority of the world's desalination plants are located in the Arabian Peninsula. The world's single largest plant, the Jebel Ali Desalination Plant in the United Arab Emirates, produces 640,000 m^3/day of treated water.

2.2.4.2 Reuse

Reuse refers to the recycling of treated effluent such as from municipal wastewater treatment works. The cascading reuse of water through human cycles is now proven as a technically viable water source in a range of applications, geographies and scales. Importantly, its availability is relatively unresponsive to climate change; like desalination, it therefore represents a source that could significantly enhance the resilience of water supply systems. It is however currently an energy-intensive supply option, particularly for systems producing water of drinking water quality. Treatment technologies are also often complex and expensive.

Arguably, one of the biggest constraints on the reuse of treated wastewater for urban purposes is negative perception, the 'Yuck Factor'. The power of this perception to completely derail technically robust reuse schemes is demonstrated in numerous examples of mothballed projects, such as the Toowoomba reuse project in Australia and the original version of the San Diego urban reuse system (which is now progressing after an overhaul, particularly in terms of messaging, communications and branding). Water management professionals must not forget that even the most technically proficient projects can fail unless public and stakeholder attitudes are taken seriously and the necessary actions implemented to secure support. Notwithstanding these constraints, improving technologies are gradually reducing the costs and energy requirements of reuse systems to the point where, in a growing number of locations, they now represent viable components of water supply portfolios.

2.2.4.3 Managed Aquifer Recharge

Managed aquifer recharge (MAR) is a generic term collectively describing means of artificially introducing water into an aquifer to support water resource management in the catchment or to protect against saline intrusion. Aquifer Storage and Recovery (ASR) and Shallow Aquifer Recharge (SAR) are two types of MAR.

- ASR works by injecting treated water directly into the saturated zone of an aquifer via wells, thereby artificially increasing the volume of stored water. It can also alter the level of the water table which can, in turn, affect surface hydrology where these are augmented by groundwater baseflows. The injected water may eventually be re-abstracted for other uses.
- In contrast to ASR, SAR recharges the unsaturated vadose zone near the surface and above the water table. The purpose of SAR is to augment the recharge that occurs via natural infiltration.

2.2.4.6 Water Transfers

Water transfers refer to the artificial transfer of water across the landscape. One way of balancing the uneven distribution of water sources is by artificial transfers. Such transfers typically take water from where it is available to where it is demanded, allowing populations to grow in locations that may otherwise be far from a water source. Water is heavy and so transfers (other than by gravity) use large amounts of energy to pump water. Depending on the local situation, artificial transfers can therefore be very expensive.

Anthropogenic water supply schemes do not necessarily operate in isolation. For reuse schemes, once water has been reclaimed, either from wastewater effluent or from the sea via desalination, it can be set aside for later use in MAR schemes, thereby providing a valuable buffer against potential future droughts. The Orange County groundwater replenishment scheme (described in Breakout Box 2.2) is one of the world's leading examples of a combined reuse and MAR scheme.

2.2.5 Anthropogenic Influence: Agriculture

Agricultural processes introduce highly significant disruptions to the natural water cycle. Land-use practices and irrigation are two of the most significant.

- **Irrigation:** The aim of irrigation is to increase crop yield above that which is achievable by purely rain-fed agriculture. Ideally, this is achieved by targeting watering at that stage of the crop growth cycle that benefits most from abundant water supply. In this section of the book, our interest is in the volume of water that is abstracted and ultimately lost from the local system as a result of evapotranspiration. More information on the relationship between water and agriculture is provided in Chapter 5.
- **Land-use practices:** Agricultural land-use practices can have a very significant influence on catchment hydrological processes. Use of pesticides and herbicides along with practices such as furrowing combine with rainfall and its subsequent runoff to contaminate waterbodies with chemicals and sediment, in turn degrading water quality

Breakout Box 2.2 Orange County groundwater replenishment scheme

Orange County is a semi-arid region in California that receives on average 325 mm of rainfall a year. The Orange County Water District serves 2.4 million residents, largely via abstractions from a large groundwater basin holding some 49 km^3 of water with an annual yield of nearly 370 million m^3. Historically, the main source of water for natural basin replenishment had been infiltration from the Santa Ana River; however, water flows fluctuate from year to year and natural recharge has not been able to replenish the annual volumes withdrawn from this aquifer since the 1940s. Rather than solely relying on importing water over long distances (an energy-intensive and therefore expensive process), the Orange County Water District responded by launching a groundwater replenishment program.

The groundwater replenishment scheme is the world's largest advanced water purification system for potable reuse. It takes treated wastewater that would otherwise be discharged to the Pacific Ocean and purifies it using a three-step process consisting of microfiltration, reverse osmosis and ultraviolet disinfection. This purification process produces water of a quality that exceeds all state and federal drinking water standards.

Operational since January 2008, this project can produce up to 265,000 m^3 of high-quality water every day, enough to meet the needs of nearly 600,000 residents in north and central Orange County. Half of this water is injected underground to act as a seawater barrier that prevents the intrusion of saline water into the freshwater aquifer, while the other half is pumped to infiltration basins where it filters through sand and gravel to recharge the groundwater to support potable supply. Source: Orange County Water District (Undated).

and reducing its suitability for abstraction. Entrained sediments in river flow can also interfere with the performance of water treatment facilities and sedimentation reduces the volumes of water stored in reservoirs.

Ultimately there are limits to our abilities to optimise the quantity of water that is available for human activity. Physical, social and economic constraints define the limits of what can be achieved, for example: how much water can be abstracted from a river; how much storage can be created; how much water can be transferred from one place to another; how much water can be diverted from the environment; how much treatment we can apply to wastewater; and how much water can be discharged back into the environment at a given time. These limits are in part dictated by what tradeoffs we are prepared to accept and how much we value different aspects of those tradeoffs (e.g. the carbon emissions associated with advanced wastewater treatment, pumping and distribution, and desalination). Anthropogenic modifications to the natural water cycle inevitably disrupt baseline hydrological processes; these disruptions can be negative, positive or neutral, depending on their scale and combined effects.

2.3 Water, Society and the Biosphere

2.3.1 Water and Civilisation

Our earliest societies naturally established themselves in areas with reliable sources of water, either adjacent to rivers or close to groundwater springs. Where water was scarce, populations were limited to sparsely distributed communities; larger rivers and groundwater supplies supported more densely populated centres. Community growth and decline was therefore intrinsically coupled to the availability of water.

Many past civilisations have risen and catastrophically collapsed as water resources have dried up, shifted in location, or became contaminated:

- **The Mayan civilisation:** This civilisation established itself in the Yucatan area of what is now modern Mexico, a region that was essentially a seasonal desert, completely reliant on rainfall for water supply. It is thought that between the eighth and ninth centuries AD, a prolonged series of droughts caused by rapid climate change depleted water sources to the extent that water and food supplies were exhausted, causing a collapse in population and ultimately the demise of the entire civilisation.
- **The bronze-age megacities of the Indus Valley:** It is thought that a dramatic increase in the frequency and intensity of drought led to large declines in the population of this once-prosperous region (of up to 100,000 people) which then never fully recovered. It is likely that, as large community groups became untenable, populations were forced into smaller and more disparate clans. Evidence of droughts persisting for more than 200 years can be seen in sediment deposits in the Arabian Sea, the Gulf of Oman, and in stalactites in caves in northeast India and southern Arabia.
- **The Roman Empire:** While earlier civilisations made noteworthy strides forward in the development of systems to capture and distribute drinking water, it was the Romans who perfected the art of urban water management, examples of which can still be seen today. It can be argued that Rome's claim as the first great city derives from it being the first to have a truly integrated water infrastructure system of multiple water sources and water transfers, piped water supplies, drainage systems, sanitation and disposal. Interestingly, the great feats of water engineering weren't driven by drinking needs alone, but also by the needs of Rome's magnificent bath houses, public fountains, gardens and public toilets.

The water infrastructure of Rome was perhaps the first example of a publicly owned water supply system. The aqueduct systems were owned and funded by the state, but only the wealthy citizens who had connections to their houses paid for water. The average citizen was provided with drinking water free of charge. The concepts of ownership and related topics are explored further in Chapter 8.

2.3.2 The Human Right to Water

On 28 July 2010 through Resolution 64/292, the United Nations (UN) General Assembly explicitly recognised the human right to water and sanitation and furthermore acknowledged that clean drinking water and sanitation are essential to the realisation of all human rights. Despite the recognition of this right, around 700 million people continue to lack access to safe drinking water.

Figure 2.8 shows those regions and countries which met the Millennium Development Goals drinking water target. It is interesting to note that the arid states of the USA, Australia, and the wealthier nations of North Africa and the Middle East have near-universal access to water supplies despite their limited natural water resources.

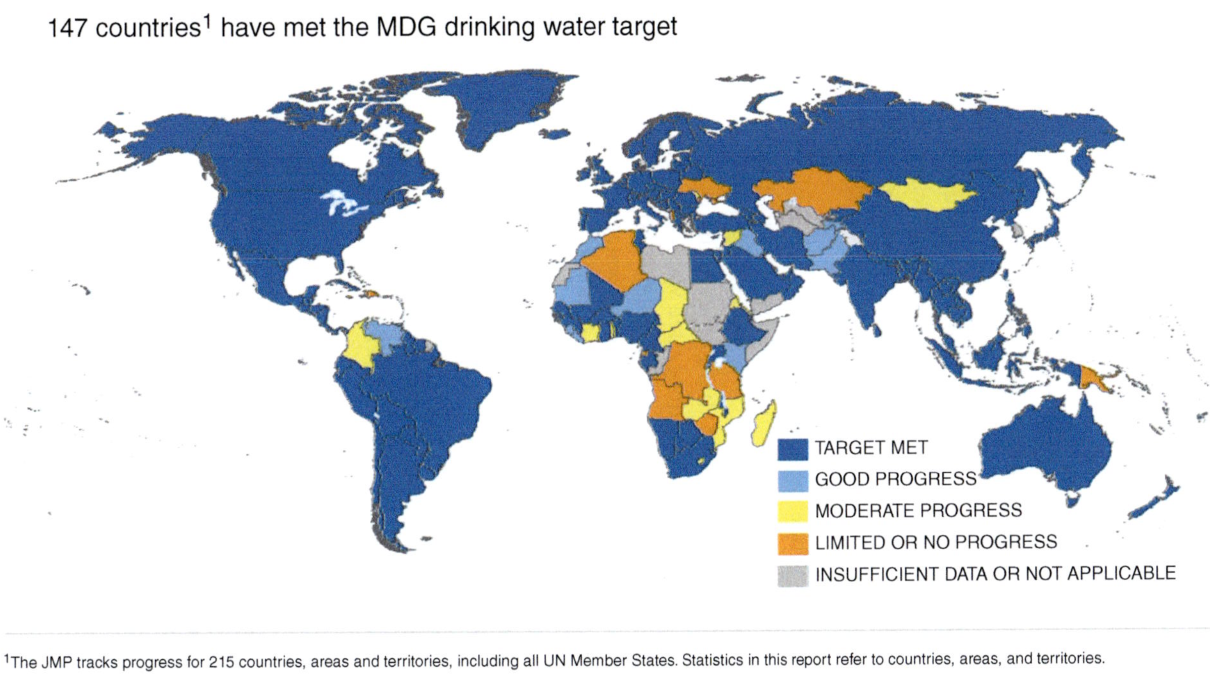

Figure 2.8 Progress of nations to the Millennium Development Goals for Drinking Water (2015 data).
Source: UNICEF/WHO (2015).

2.3.3 Population Growth and Mobility

While the human right to water is now recognised by the UN, the challenge of ensuring its delivery to all remains vast. The global population has trebled in the space of 60 years and reached 7.4 billion by mid 2016 (Population Reference Bureau 2016). Should this rate of growth be sustained, the planet's available resources will have little chance of sustainably meeting the demands imposed upon them. Fortunately, the rapid rates of recent population growth are unlikely to be maintained. Most developed countries have now progressed to the latter stages of the demographic transition model (see Breakout Box 2.3) and are now witnessing a tailing off in population growth and, in some instances, population decline. Mid-range projections by the UN predict a global population of 8.1 billion by 2025, 9.6 billion by 2050 and 10.9 billion by the end of the century (UN 2013). These growth rates are however uncertain and depend to a large extent on trends in Africa and Asia. Table 2.4 shows that in Africa, in particular, population growth is predicted to remain robust.

Population growth is only one factor influencing the demand for water. Individual direct per capita consumption typically increases as populations move from rural to urban lifestyles, and also in line with economic prosperity. These issues are examined further in Chapter 4.

Breakout Box 2.3 The demographic transition model

The demographic transition model describes how birth and death rates (and therefore population rates) change as a country moves through a simplified process of economic development. The stages in this development process are shown below.

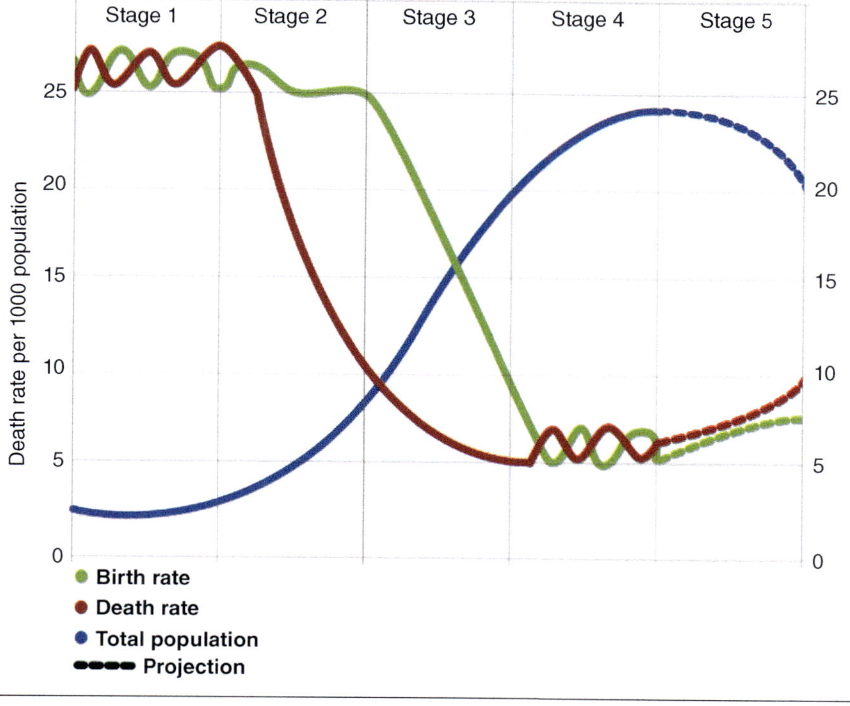

Table 2.4 Population growth to 2050.

Region	Population (millions)		Percentage change
	2013	2050	
Africa	1,111	2,393	+115%
Asia	4,299	5,164	+20%
Europe	742	709	−4.7%
Latin America and the Caribbean	617	782	+27%
Northern America	355	446	+26%
Oceania	38	64	+68%
Total	7,162	9,551	+33%

Source: UN (2013).

2.3.4 Disparity between Water Resources and Population

Section 2.2 describes the reasons behind the globally uneven distribution of water resources and explains how hydroclimates have influenced the types of man-made water infrastructure that now exist. Here we consider the problems that arise when disparities between water resources and populations exist. Section 2.3.5 goes on to explore the challenge of 'positive feedback loops', how a water system can destabilise if natural 'checks' are breached or overridden.

Historically, the ability to access secure supplies of water was one of the primary factors governing the location and success of human communities. However, other socioeconomic factors now exert a far greater influence on an individuals' migration or habitation decisions. As a result, we now find booming communities in locations where their populations are disproportionate to the volumes of water available.

Around 35% of the world's population lives in Asia, a continent which has just 28% of the world's water resources. India's and China's populations are vastly disproportionate to the volume of water resource within their territories. Conversely, even though Russia and Canada have low internal RWRs, these are shared between relatively small populations.

We have already discussed how anthropogenic modifications disrupt the natural hydrological processes of water resource systems. The FAO (2003) concluded that the threat and impact of anthropogenic disruption on water regimes, the vulnerability of chronically overutilised resources, saline intrusion and the disappearance of water sources combine to intensify water scarcity and problems of disparity in access to water. Transboundary water-sharing issues add further complexity, and already exist between numerous countries (for example in the Balkans and Nile river basin). Transboundary water politics are explored further in Section 6.5.2.

2.3.5 Ability to Access Local Water Resources

Economic prosperity affords nations the ability to invest in infrastructure to access otherwise hard-to-reach water resources. As demands for water have grown, traditional supply options such as reservoirs, boreholes, river abstractions and their associated water distribution networks needed to be supplemented. The often energy-intensive and expensive new processes described in Section 2.2.4 (water transfers, desalination, MAR and treated effluent reuse) are often perceived to provide more resilience against the changing water environment.

It is clear that a community's ability to access and, more importantly, to stabilise and secure water resources increases opportunities for its social and economic prosperity, in turn stimulating population and industrial growth and, as a result, reinforcing the need for more and more sophisticated water management solutions. This reinforcing cycle, while potentially supporting significant socioeconomic gains, also has the potential to encourage water exploitation, to exceed safe levels, and trigger positive feedback loops (see Breakout Box 2.4 and Figure 2.9 and more on systems thinking in Chapter 9).

Breakout Box 2.4 Feedback in water resource systems

Feedback occurs when the output of a system also serves as one of its inputs, thereby leading to a change in the state of the system (Botkin and Keller 2000). Negative feedback is stabilising, and the system's response is in the opposite direction to the output (e.g. when a city's water supply capacity is exceeded, the output is a slight reduction in quality of life, the system response is a slight outward migration or reduced immigration).

Positive feedback on the other hand is destabilising and very dangerous in terms of water resource management (e.g. a city's water supply capacity is exceeded so additional infrastructure is introduced to increase capacity. The output is an increase in water supply capacity and maintained quality of life. The system response is further migration into the city and possibly increased per capita demand for water).

Destabilising positive feedback loops could create situations where populations continue to expand unabated, further exacerbating the strain on resources, and increasing demands for technological solutions until the system ultimately fails.

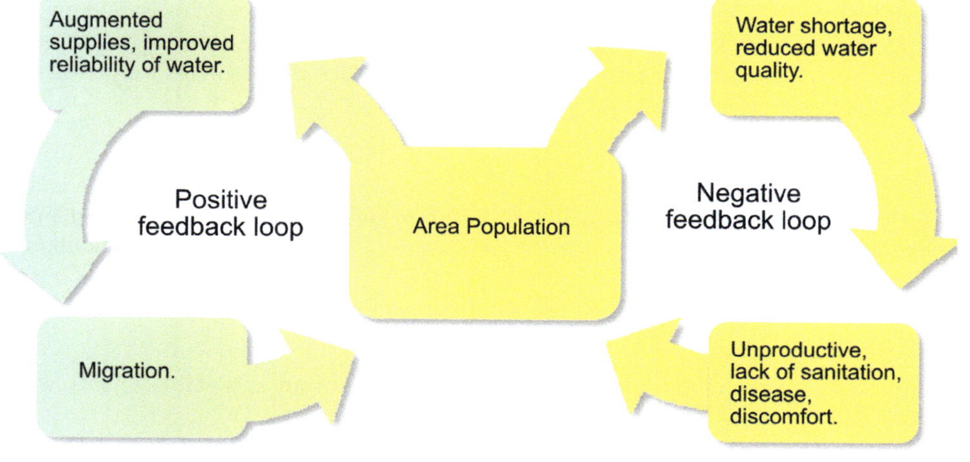

Figure 2.9 Diagram of feedback loops in water resource systems.
Source: Adapted from Botkin and Keller (2000).

In contrast to cases where physical water availability is insufficient to meet demand, in many parts of the world people often lack the finances necessary to invest in water supply infrastructure or the fees required to secure water connections, despite there being sufficient water available in the environment to meet demand.

The set of eight Millennium Development Goals (MDGs) intended to be met by 2015 included a target to halve the proportion of the global population without sustainable access to safe drinking water and basic sanitation. A 2014 report claimed that the basic sanitation component of this MDG target was one of the furthest MDGs from being achieved; in 2012, 2.5 billion people did not use an improved sanitation facility and 1 billion people still resorted to open defecation (UN 2014).

While a 2015 report stated that 4.2 billion people have access to piped water supplies and that 147 countries had met the MDG drinking water target, only 95 countries had met the MDG sanitation target and only 77 countries had met both targets. Sanitation levels are increasing; since 1990 2.1 billion people have gained access to improved sanitation and the proportion of people practising open defecation has fallen almost by half. However, there are still huge gaps between rich and poor, rural and urban people. About 50% of people living in rural areas in developing countries still lack improved sanitation facilities, compared to only 18% of people in urban areas (UN 2015).

The World Health Organization has identified government inability to translate what may otherwise be good strategies into effective implementation programs as a core constraint to progress in water supply and sanitation (WHO 2014). Limited investment, especially in rural areas, and insufficient monitoring and evaluation of current programmes are also problems. Achieving further progress in providing access to safe water and sanitation will be a key facilitator to achieving the Sustainable Development Goals (SDGs), those targets that build on the MDGs and set the post-2015 development agenda.

Privatising water supply and sanitation service provision is one mechanism that has the potential to improve water resource management outcomes. Section 8.1 explores this issue in depth; however, in many examples privatisation has led to worsening water access, particularly for those communities in developing nations that live in peri-urban regions and slums. Karmarker (2012) examined the impact of water privatisation on the urban poor in Mumbai, and found a system that excludes large sections of the population due to lack of basic infrastructure in slums and a lack of incentives for privatised companies to resolve the situation.

2.3.6 Different Types of Water Scarcity

The concept of *water stress* is defined in this book as the effects felt when the quantities and qualities of water available in a given location are insufficient to meet the demands placed upon them.

Water scarcity occurs when the demand for water is greater than the available resource, and can be defined in three ways (FAO 2013):

- *Physical water scarcity*: a physical shortage of water of an acceptable quality with respect to aggregate demand;
- *Infrastructural water scarcity*: where even though the actual RWR may be sufficient to meet demand, there is inadequate infrastructure to get the water to where it is needed; and

- *Institutional water scarcity*: where institutions, legislation and/or regulation fail to ensure that water is supplied in an affordable and equitable manner. This type of water scarcity is sometimes referred to as *economic water scarcity*.

The FAO (2013) also suggests that physical water scarcity is experienced when water withdrawals exceed 20% of RWR. While this benchmark is a useful guide for the degree of water stress experienced in a country or river basin, it is only an indication. The impacts of water scarcity will be heavily conditioned by local and site-specific factors such as the effectiveness of coping mechanisms. It should also be remembered that a portion of the RWR will be required to sustain aquatic ecosystems, often referred to as the environmental flow.

As population and water demands have expanded, even regions with high absolute RWRs are finding themselves exposed to the impacts that a lack of water can cause. Globally, around 1.1 billion people experience chronic, long-term water scarcity and therefore water stress, in one form or another, while a further 2.7 billion regularly experience instances of acute water scarcity (for at least one month per year during a monitoring period from 1996–2005; Hoekstra *et al.* 2012).

2.3.7 Ability to Access Distant Water Resources

Disparities in water scarcity are not just conditioned by the ability to access local water resources. In many regions, complex distribution systems also allow distant water resources to be tapped by communities that may be many hundreds of kilometres away. Controversial dam schemes (such as Ethiopia's Grand Renaissance Dam on the Blue Nile) or river transfers (such as the diversion from the Caspian Sea through Iran to the over-abstracted Lake Urmia or the Chinese South-to-North Water Transfer Project) illustrate how engineering solutions can relieve the pressure on nations or populations that can afford the investment. Note however that the relief may only be temporary if positive feedback systems are triggered, or the root causes of the water scarcity not addressed.

There is however another, much less visible, but arguably much more significant activity which enables nations and populations to access the water resources of others while simultaneously reducing the pressure on their own resources. The trade in virtual water, 'hidden water' embedded within crops and other products, is vast and largely overlooked in current water management policy and planning. Section 3.3 explores the complexities of the trade in virtual water in more detail. As an indication of the significance of virtual water trade, Figure 2.10 shows the flux of embedded water traded in products between different regions. It shows that South America has overtaken North America as the biggest exporter of water, serving the largely food-based needs of Asia and Europe. The thickness of the arrows indicates the relative volume of the trade in virtual water. The largest flows are indicated in black numbers (as measured in cubic kilometres).

Whereas water management in the past has typically been national or regional in focus, the globalisation of food supplies and other products has radically changed the landscape over which water management must be considered. The trade in virtual water is a global issue affecting almost all nations, and cannot be ignored in national water management planning.

2.3.8 Modern Water Politics

The issues addressed in the preceding subsections – the human right to water, the different types of water scarcity, the exploitation of local and distant water supplies, and

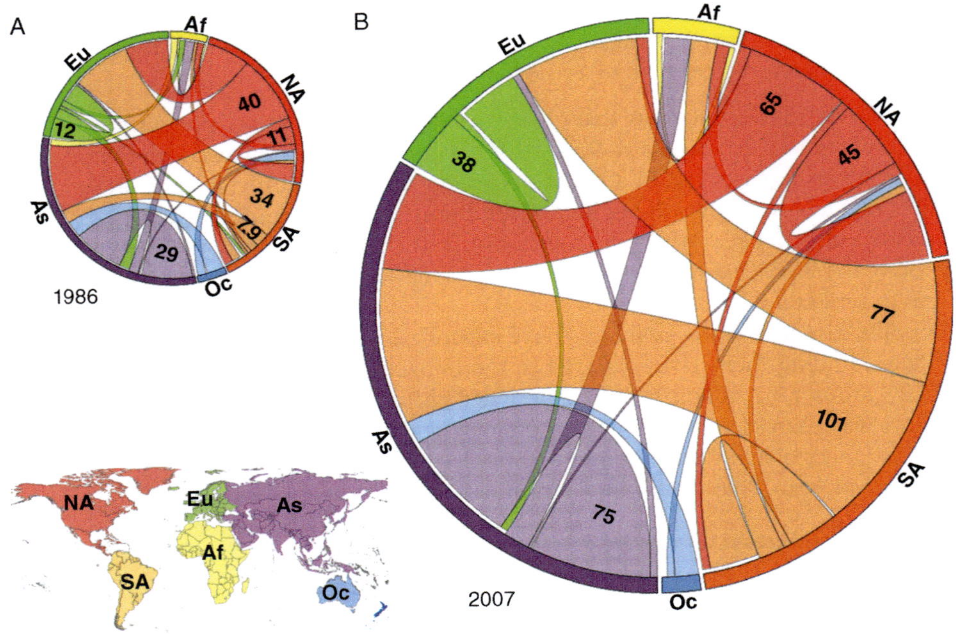

Figure 2.10 Virtual water trade.
Source: Dalin *et al.* (2012). Reproduced with permission of National Academy of Sciences.

the growing importance of virtual water – all mean that the management of water in the 21st century requires the consideration of a large number of socioeconomic processes. As a result, water management has never been a more highly politicised issue. In some cases, political decisions have led to tensions both within, and increasingly between, different nations.

In the Middle East and Africa there are a number of examples of potential conflict over water. Possibly one of the most serious is the dispute between Israel and Palestine. After the Second World War, the relocation of hundreds of thousands of people to Israel increased populations on the margins of some of the driest deserts in the world. Pressure on limited water resources has been further increased by decisions to develop water-intensive commercial agriculture. Following the 1967 war, Israel gained exclusive control of water resources in the West Bank (from a mountain aquifer) and the Sea of Galilee and now obtains 60% of its annual freshwater from these sources (approximately 1 km^3/yr). Water consumption in both Israel and Palestine often exceeds available supply and this occasionally leads to requirements for rationing, clearly a situation which is not conducive to peace.

The importance of modern-day water politics cannot be overstated, although there is debate between water professionals on the likelihood of water conflicts. While there is no shortage of tension over water, and notwithstanding that climate change will likely alter and enhance the role of water as a stress multiplier in transboundary geo-political issues (Connell 2013), because of the complexities of achieving effective water management in a dynamic and globalised world, physical conflict is unlikely to help any country or group achieve its water security aims (Dunn 2013).

In 2014, Peter Gleick and Matthew Heberger published a contrasting and sobering analysis of water-related conflicts. They found that the number of reported water-related conflicts had increased rapidly in the last decade from an average of about two a year until 2000 to between 10 and 18 in 2011. The authors' belief is that the potential for within-country conflict and local violence related to water now represents a greater threat than international conflict. They argue that equitable access to water, strategies for sharing during shortages and water contamination are some of the most important issues to be addressed (Gleick and Heberger 2014).

Water transfers or impoundments are a particular source of tension. Ethiopia's intention to dam the River Nile to secure its water resources has renewed tensions with Egypt. Similar controversy has surrounded Turkey's long-held plan to expand the Atatürk Dams, a plan that has fuelled tension with those downstream countries that are dependent on flows from the Tigris and Euphrates (see Breakout Box 2.5).

Regardless of whether water is a direct, indirect or unrelated factor behind conflict, when wars and disputes do occur, water and sanitation supplies are almost always one of the services hardest hit. In early 2013 for example, the third year of the Syrian conflict, access to water and sanitation was considered 'severely limited' in over 90% of the country (see Figure 2.11). Water availability had dropped from 75 litres to 25 litres per person per day, and the proportion of the population able to access wastewater treatment facilities had fallen from 70% to 35% (UNICEF 2013). In 2014 it was reported that while in 2011 approximately 85% of the population in Syria had access to safe drinking water this had dropped to just 40% (Syria Recovery Trust Fund 2014). Updates have been more limited since then, but in January 2016 UNICEF began reporting that water supply to major parts of Aleppo were continuing to be entirely cut off (UNICEF 2016).

Poor political planning has the potential to result in unintended consequences for water resources that subsequently stir tensions in local and regional communities. California experiences huge challenges maintaining water supply to its population under increasing

Breakout Box 2.5 Shared resource conflicts on the Nile and Euphrates

Ethiopia is a country that has been ravaged by drought many times. In an attempt to stabilise its water security, the country has initiated an ambitious engineering project that involves constructing the Grand Renaissance Dam on the Blue Nile. Egypt is downstream of Ethiopia on the River Nile and, in response to the plan, Egypt's president demanded that construction stop and vowed to protect the nation's historical rights to the river 'at any cost' (Gleick and Heberger 2014).

Turkey is a country with geographically uneven water resources, a growing population with increasing levels of per capita water demand, and ambitions to support economic growth and social prosperity through large-scale water resource engineering projects. The most contentious of these plans is the expansion of the Atatürk Dams in the southeast of the country (the Southeastern Anatolia Project), and the source of 90% of the flow in the Euphrates River. The existing dams have reduced flow in the Euphrates, supplying Syria and Iraq, by a third. Average annual rainfall in Iraq is 154 mm and the Shatt Al-Arab basin, the only river basin in the country, is formed by the confluence of the Euphrates and Tigris. Tensions between Iraq and Turkey, and other neighbouring countries, intensified when Turkey began the Southeastern Anatolia Project, and its plans to expand the program have raised tensions further (Shamout and Lahn 2015).

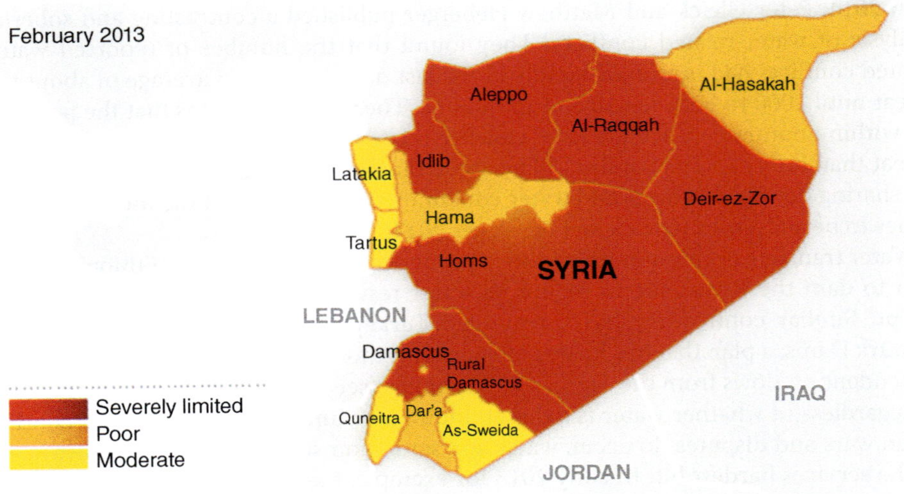

Figure 2.11 Access to water and sanitation in Syria. *Source:* UNICEF (2013).

pressures of climate change and population growth. These water-related pressures are nothing new. The 'California Water Wars' of the early 20th century were fuelled by a decision to construct a 320-km-long aqueduct to divert water from the agricultural lands of the Owens Valley to the increasingly thirsty city of Los Angeles, a decision which led to the almost total desertification of the Owens Valley (Forstenzer 1992). What's more, Los Angeles' rapid growth meant that, by the 1940s, the city needed yet more water supplies. This time, water was diverted from Mono Lake and the resultant damage to the ecosystem, important for migrating birds, led to a 15-year litigation battle culminating in a 1994 law which protects Mono Lake water rights (Mono Basin Clearinghouse Undated). Los Angeles has since had to invest in alternative schemes to secure its water supply, including state-funded water conservation and recycling projects.

Water contamination is another source of national and within-country tension. The world is peppered with examples of water supplies being contaminated, often as a result of industrial mismanagement or accident:

- In 2000, heavy rains in Romania led to failures in the tailing dams of the Baia Hare gold mine. Approximately 100,000 m^3 of cyanide-contaminated liquid poisoned the drinking water supplies of over 2 million people in Hungary. An investigation concluded that the accident was caused by the inappropriately designed tailings dams, inadequate monitoring of the construction and operation of those dams, and by severe, though not exceptional, weather conditions (WISE Uranium Project 2001).
- In 1999, 700,000 tonnes of cyanide tailings spilled from a damaged concrete pipe in the Philippines, burying 17 homes and inundating 51 hectares of rice land (WISE Uranium Project 2016).
- In 1988 the water supply of a town in Cornwall, UK was poisoned with 20 tonnes of aluminium sulphate when the chemical was accidentally tipped into a public water supply manhole instead of the intended chemical tank. It is thought that up to 20,000

local people and 10,000 holidaymakers drank the contaminated water in the hours and days after the spillage. This accident has been the subject of retrospective investigations into the long-term health implications on people who drank the water (Rowland *et al.* 1990).

- The failure of a copper main dam embankment in Canada in 2012 led to a non-consumption order on the local water supply (WISE Uranium Project 2016).
- The problems of mining and in particular acid mine drainage on water resources are not confined to developing countries. The United States National Wildlife Federation has issued several reports listing its concerns over the risks of acid mine drainage affecting the public water supplies of major cities including Michigan (National Wildlife Federation Undated).

International and national water policies and treaties on the equitable distribution of water are important and necessary steps in the process of improving the governance of water. However, it is at the local scale where the impacts of these decisions will be borne out and therefore where decision-making must arise. Concerted and collective community action is required to manage the competing water demands of towns and cities, farmers, industry and the environment. Such a collaborative approach is necessary to ensure that all requirements for water are identified and appropriately balanced.

Politics and governance is also a core element of the water risk theory that is increasingly being recognised by businesses and investors as a threat to productivity and profitability. Stabilising local, regional, national and international water-related tensions is examined in greater detail in Chapter 6.

References

Alcamo, J., Döll, P., Henrichs, T., Kaspar, F., Lehner, B., Rösch, T. and Siebert, S. 2003. Global estimates of water withdrawals and availability under current and future 'business-as-usual' conditions. *Hydrological Sciences Journal*, 48(3), 339–348.

AQUASTAT. Undated. On-line database of FAO. Available at: http://www.fao.org/nr/water/aquastat/data/query/ (accessed February 2017).

Bass, M. S., Finer, M., Jenkins, C. N., Kreft, H., Cisneros-Heredia, D.F., McCracken, S.F., Pitman, N.C.A., English, P.H., Swing, K., Villa, G., Di Fiore, A., Voigt, C.C. and Kunz, T.H. 2010. Global conservation significance of Ecuador's Yasuní National Park. *Public Library of Science*, 5(1), doi: 10.1371/journal.pone.0008767.

Belluscio, A. 2010. Hostile volcanic lake teems with life. Available at: http://www.nature.com/news/2010/100402/full/news.2010.161.html (accessed January 2017).

Borgonie, G., García-Moyano, A., Litthauer, D., Bert, W., Bester, A., van Heerden, E., Möller, C., Erasmus, M. and Onstott, T. C. 2011. Nematoda from the terrestrial deep subsurface of South Africa. *Nature*, 474(7349), 79–82.

Botkin, D.B. and Keller, E.A. 2000. Environmental Science: Earth as a Living Planet, Eighth Edition. Wiley, London, UK.

Connell, D. 2013. Water wars, maybe, but who is the enemy? Available at: http://www.globalwaterforum.org/2013/04/10/water-wars-maybe-but-who-is-the-enemy/ (accessed January 2017).

Dalin, C., Konar, M., Hanasaki, N., Rinaldo, A. and Rodriguez-Iturbea, I. 2012. Evolution of the global virtual water trade network. *Proceedings of the National Academy of Sciences of the United States of America*, 109(16), 5989–5994.

Dunn, G. 2013. Water Wars: A Surprisingly Rare Source of Conflict. Available at: http://hir.harvard.edu/water-wars-a-surprisingly-rare-source-of-conflict/ (accessed February 2017).

EASAC. 2010. Groundwater in the Southern Member States of the European Union: Country Reports France, Greece, Italy, Portugal and Spain. Available at: http://www.easac.eu/home/reports-and-statements/detail-view/article/groundwater.html (accessed January 2017).

Environment Agency. 2007. Water Services Infrastructure Guide: A Planning Framework. Available at: https://www.gov.uk/government/uploads/system/uploads/attachment_data/file/292913/gean0107blln-e-e.pdf (accessed February 2017).

EUROSTAT. Undated. On-line database of the European Commission. European Commission, EUROSTAT. Available at: http://ec.europa.eu/eurostat (accessed February 2017).

Food and Agriculture Organisation of the United Nations (FAO). 2003. Review of World Water Resources by Country. Available at: ftp://ftp.fao.org/agl/aglw/docs/wr23e.pdf (accessed January 2017).

Food and Agriculture Organisation of the United Nations (FAO). 2011. The State of the Worlds Land and Water Resources for Food and Agriculture. Available at: http://www.fao.org/nr/solaw/solaw-home/en/ (accessed January 2017).

Food and Agriculture Organisation of the United Nations (FAO). 2013. Topics: Water Scarcity. Available at: http://www.fao.org/nr/water/topics_scarcity.html (accessed January 2017).

Forstenzer, M. 1992. Dust to Dust. Available at: http://articles.latimes.com/1992-04-10/news/mn-179_1_owens-lake (accessed January 2017).

Gleick, P.H. (ed.) 1993. Water in Crisis: A Guide to the World's Fresh Water Resources. Oxford University Press, New York.

Gleick, P.H. and Heberger, M. 2014. Water Brief 3: Water and conflict. Events, trends, and analysis (2011–2012). In Gleick, P. *et al.* (ed) The World's Water Volume 8: The Biennial Report on Freshwater Resources. Island Press, USA.

Government of Canada. 2013. Environment Canada. Threats to Water Availability in Canada. Available at: https://www.ec.gc.ca/inre-nwri/default.asp?lang=En&n=0CD66675-1&offset=4&toc=hide (accessed February 2017).

Hoekstra A.Y., Mekonnen M.M., Chapagain A.K., Mathews R.E. and Richter B.D. 2012. Global monthly water scarcity: blue water footprints versus blue water availability. *PLoS ONE*, 7(2), doi: 10.1371/journal.pone.0032688.

IGRAC. 2010. Global Groundwater Information System (GGIS). Delft, the Netherlands, IGRAC. Available at: https://www.un-igrac.org/global-groundwater-information-system-ggis (accessed February 2017).

Institute of Chartered Accountants in England and Wales (ICAEW). 2013. Economic Insight: Middle East. ICAEW, London, UK.

Juniata County. 2009. Pennsylvania Comprehensive Plan. Available at: http://www.co.juniata.pa.us/departments/planning/comprehensive-plan (accessed January 2017).

Karmarkar, D. 2012. Impact of water privatisation on urban poor: a case of Mumbai. *International Inter-Disciplinary Research*, 2(1), 63–73.

Margat, J. 2008. Les Eaux Souterraines dans le Monde. UNESCO, Paris, France.

Millennium Ecosystem Assessment. 2005. Ecosystems and Human Well-being: General Synthesis. World Resources Institute, Washington DC, USA.

Mono Basin Clearinghouse. Undated. Decision 1631 Background. Available at: http://www.monobasinresearch.org/timelines/d1631.php (accessed January 2017).

National Wildlife Federation. Undated. Acid mine drainage: A threat to Michigan's water quality. Available at: https://www.nwf.org/pdf/Great-Lakes/Acid_mine_drainage_factsheet.pdf (accessed January 2017).

Navarro-González, R., Rainey, F.A., Molina, P., Bagaley, D.R, Hollen, B.J., de la Rosa, J., Small, A.M., Quinn, R.C., Grunthaner, F.J., Cáceres, L. and Gomez-Silva, B. 2003. Mars-like soils in the Atacama Desert, Chile, and the dry limit of microbial life. *Science*, 7(5647), 1018–1021.

Orange County Water District. Undated. Groundwater Replenishment System. Technical Brochure. Available at: http://www.ocwd.com/media/4267/gwrs-technical-brochure-r.pdf (accessed January 2017).

Population Institute. 2010. Talking Points and Facts. Available at: https://www.populationinstitute.org/programs/gpso/gpso/ (accessed January 2017).

Population Reference Bureau. 2016. 2016 World Population Data Sheet. Available at: http://www.prb.org/Publications/DataSheets/2016/2016-world-population-data-sheet.aspx (accessed February 2017).

Rowland, A., Grainger, R., Smith, R.S., Hicks, N. and Hughes, A. 1990. Water contamination in north Cornwall: a retrospective cohort study into the acute and short-term effects of the aluminium sulphate incident in July 1988. *Journal of the Royal Society of Health*, 110, 166–172.

Shamout, M.N. and Lahn, G. 2015. The Euphrates in Crisis Channels of Cooperation for a Threatened River. Chatham House. Available at: https://www.chathamhouse.org/sites/files/chathamhouse/field/field_document/20150413Euphrates_0.pdf (accessed January 2017).

Shiklomanov, I. and Rodda, J.C. 2003. World Water Resources at the Beginning of the Twenty-First Century. Cambridge University Press, International Hydrology Series, Cambridge, UK.

Siebert, S., Burke, J., Faures, J., Frenken, K., Hoogeveen, J., Döll, P. and Portmann, T. 2010. Groundwater use for irrigation – a global inventory. *Hydrology and Earth System Sciences*, 14, 1863–1880.

Syria Recovery Trust Fund. 2014. Rehabilitation of a Water Supply Network in a City in Idleb Governorate – Phase II. Available at: http://www.srtfund.org/articles/34_rehabilitation-of-a-water-supply-network-in-a-city-in-idleb-governorate-phase-ii (accessed January 2017).

UNEP/Grid Arendal. 2008. Vital Water Graphics - 2nd Edition. Available at: http://www.unep.org/dewa/vitalwater/rubrique14.html (accessed January 2017).

UNEP. 2014. Towards integrated water resources management: International experience in development of river basin organisations. UNEP, Khatoum, Sudan.

UNICEF. 2013. Running dry: Water and sanitation crisis threatens Syrian children. Available at: http://www.unicef.org/mena/Syria_Crisis_WASH-Syria-Feb-2013-En.pdf (accessed January 2017).

UNICEF. 2016. UNICEF Syria Crisis Situation Report, January 2016. Available at: http://reliefweb.int/report/syrian-arab-republic/unicef-syria-crisis-situation-report-january-2016 (accessed January 2017).

UNICEF/WHO. 2015. Progress on sanitation and drinking water – 2015 update and MDG assessment. Available at: https://www.wssinfo.org/fileadmin/user_upload/resources/JMP-Update-report-2015_English.pdf (accessed February 2017).

United Nations. 2013. The World's Population Prospects: the 2012 revision, highlights and advance tables. United Nations, New York. Available at: http://esa.un.org/unpd/wpp/publications/Files/WPP2012_HIGHLIGHTS.pdf (accessed January 2017).

United Nations. 2014. The Millennium Development Goals Report 2014. United Nations, Ney Work, USA.

United Nations. 2015. The Millennium Development Goals Report 2015. United Nations, Ney Work, USA.

United States Geological Society (USGS). 2014. How much water is there on, in, and above the Earth? Available at: http://water.usgs.gov/edu/earthhowmuch.html (accessed January 2017).

University of Alberta. 2007. Running Out of Steam? Oil Sands Development and Water Use in the Athabasca River-Watershed: Science and Market Based Solutions. University of Alberta, Alberta, USA.

van der Gun, J. 2012. Groundwater and Global Change: Trends, Opportunities and Challenges. UNESCO, Paris, France.

WISE Uranium Project. 2001. The Aurul S.A. tailings dam failure (Baia Mare, Romania). Available at: http://www.wise-uranium.org/mdafbm.html (accessed January 2017).

WISE Uranium Project. 2016. Chronology of major tailings dam failures. Available at: http://www.wise-uranium.org/mdaf.html (accessed January 2017).

World Climate Research Programme (WCRP). Undated. The Global Monsoon Systems. Available at: http://www.wcrp-climate.org/documents/monsoon_factsheet.pdf (accessed January 2017).

World Health Organisation (WHO). 2014. Urban population growth. Available at: http://www.who.int/gho/urban_health/situation_trends/urban_population_growth_text/en/ (accessed January 2017).

Part II

Stresses and Strains

3

Key Concepts

In his 2010 BBC Reith Lectures referred to in Chapter 1, Astronomer Royal Professor Sir Martin Rees was alluding to the stresses and strains that humanity's activities have imposed on the biosphere's natural resources, and to the critical point we have reached in our evolution. He postulates that we must act within the relatively short time frame of this century if those stresses and strains are to be managed within sustainable limits.

Part II of the book (Chapters 3–6) explores how human activities have resulted in stresses and strains on water in the biosphere. For this purpose, human activities are classified into three groups based on our needs.

- **Live:** the basic water requirements for human life as well as the water used to generate power and support social and economic health.
- **Eat:** the demand we place on the water environment to provide our food.
- **Consume:** the role of water in the production and processing of other major consumer goods.

In order to establish a framework against which the stresses and strains imposed by these activities can be described, it is necessary to define a number of key concepts:

- that of water fluxes: the rainfall, runoff and evaporation on which we rely to live, eat and consume;
- the mechanisms by which people access and influence these fluxes, namely water withdrawal (also known as abstraction), water consumption and water return (discharges back into the environment);
- the impact of these mechanisms on the water fluxes through water stress and water scarcity; and
- the impacts that the water footprint of individuals, communities or businesses can have on distant locations via the global trade in *virtual water*.

3.1 Water Fluxes in Space and Time

Chapter 2 describes the broad distribution and total volume of water across the globe. At that scale the numbers are hard to grasp, so water planners working at regional scales apply several concepts to define the water resource. The building blocks of these

Water Resources: A New Water Architecture, First Edition.
Alexander Lane, Michael Norton and Sandra Ryan.

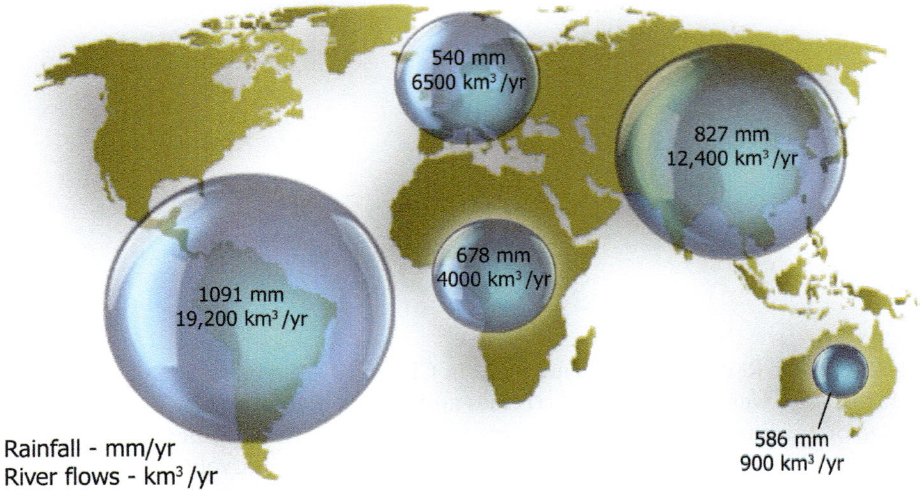

Figure 3.1 Water fluxes by continent.

concepts are what we term *fluxes of water*: rainfall, runoff, evaporation and evapotranspiration, for example. In an average year, 110,000 km^3 of rainfall falls on the global land surface and results in around 43,000 km^3 of runoff into the oceans (see Figure 3.1). The balance returns to the atmosphere through evaporation and evapotranspiration. It is the runoff figure (km^3/yr) that is often termed the renewable water resource (RWR) (see Glossary).

Fluxes of water vary considerably between nations and continents, depending primarily on the prevailing climate and topography. For example, the annual RWR of Brazil is 8,200 km^3 whereas the comparable figure for Kuwait is just 0.02 km^3 (Hoekstra and Chapagain 2007). Fluxes also vary significantly over time, and both short-term and long-term trends and cycles can be observed (see Breakout Box 3.1).

Climatic processes are a critical influence on water fluxes and how these processes change over time will in turn impart huge influence on water stress and scarcity. Projections of these impacts have been made for more than 20 years, but the Fourth Assessment of the Intergovernmental Panel on Climate Change (IPCC), released in 2007, was perhaps the first to gain widespread acceptance of its findings. The Fifth Assessment was finalised in 2014, and concluded that the projections made in 2007 were conservative and that future impacts are likely to be even more severe. The Fifth Assessment also stressed that the unpredictability of weather will increase as a result of climate change. The evidence of the past few years would seem to add weight to this argument, from extreme rainfall events (northeast USA 2012, southwest UK 2014, Kashmir 2014, Philippines 2014), to drought (England 2012, California 2014–15, Sudan 2011–12, Africa 2015–16). Runoff decreases are projected in the Mediterranean, Middle East, Central and South America, and parts of Australia, while increasing precipitation and runoff is expected in some northern latitudes. Runoff increases are also projected for parts of the Arabian Peninsula, the Horn of Africa and the Indian subcontinent (Haddeland *et al.* 2014).

Breakout Box 3.1 Example of extreme alternating water fluxes

Extremes in water fluxes are increasingly common to many countries and regions. The English drought between 2010 and 2012 is one such example. Over 24 months rainfall was persistently below average, with 14 months receiving less than 70% of average rainfall. The drought was then broken in spectacular fashion, with record rainfalls in the late spring and early summer of 2012 quickly transforming a problem of drought to one of flooding (Marsh *et al.* 2013).

To formulate predictions, the IPCC and other similar research organisations rely on complex global and regional climate models (GCMs and RCMs), which in turn make a number of assumptions about the future climate, not least the extent to which society chooses to implement measures to reduce the concentrations of CO_2 and other greenhouse gases in the atmosphere. While the projections these models derive are highly sensitive to their inherent assumptions, by comparing a number of different simulations, the following trends emerge:

- more extreme rainfall events everywhere;
- increased annual rainfall volumes in higher latitudes and parts of the tropics;
- reduced annual rainfall volumes in the mid-latitudes and some subtropical regions;
- higher seawater and freshwater temperatures everywhere; and
- net decline in overall snowpack volumes.

Notwithstanding these broad trends, there remains huge uncertainty around climate change at the regional and local scales on which most water management impacts are felt. While climate change may cause some locations to experience net benefits, for example through agriculture benefitting from increased rainfall, the overall prognosis at a global scale is that the impacts of climate change are most likely to reduce the availability and reliability of those water fluxes on which we otherwise rely to live, eat and consume.

3.2 Mechanisms of Human Interaction with Water Fluxes

When describing the means by which people use and interact with water, there is widespread and often inaccurate use of many terms. This book uses the definitions listed in the Glossary to describe how humans use water to live, eat and consume.

When water is taken from a body of water, this is termed a *withdrawal* or an abstraction. The purpose for which the water is withdrawn is termed the *use*. The amount of the water withdrawn that is taken up in the process of that use and not discharged back into the environment is termed *consumption*, and the amount which is surplus or that which is rejected or wasted to a body of water is termed a *return*. These terms are typically used when water is taken from what is referred to as a *blue water* source, a waterbody such as a river, lake or underground aquifer.

Humanity's single largest use of water is in the growing of food and the majority of global crops are sustained by soil moisture from rainfall, referred to as *green water*. The green water is consumed by the crop via evapotranspiration and, as a result, there is no return flow of water. For the increasing proportion of mankind's food being grown through irrigated agricultural means, blue water is withdrawn from rivers and groundwater sources. Because most irrigation systems deliver more water to the crop than is actually consumed, the excess is returned to the underlying aquifer or via overland flow to the nearest surface watercourse.

Table 3.1 summarises average annual withdrawals of blue water by continent. It shows that 70% of blue water withdrawal is for irrigated agriculture, with 11% and 19% used for domestic and industrial purposes, respectively. Importantly, while much of the water use for agriculture is consumptive and therefore subsequently unavailable for other local uses, many domestic and industrial water uses are non-consumptive and result in water returns (for example as domestic wastewater or as power generation cooling water).

Table 3.1 also shows that the global annual average water withdrawal is close to 4,000 km^3 compared to the global RWR of 43,000 km^3. This then begs the question, what is the fuss about? Despite withdrawal of what is a low proportion of the global average RWR, water stress is repeatedly and increasingly experienced in a variety of locations all across the globe. The reasons for this situation are primarily related to the uneven spatial and temporal distribution of water resources and people (see Section 2.3). At regional and local scales, the necessary infrastructure to store water in wet years to meet demand when RWR is below average is often lacking. Furthermore, much of the RWR occurs in remote unpopulated locations such as northern Canada and northern Russia, or is concentrated in particular locations or waterbodies such as the Amazon River. Finally, it must be remembered that the RWR must fulfil not only our human demands to live, eat and consume, but also, crucially, the environmental flow required to sustain the aquatic and terrestrial environments on which ultimately, albeit often indirectly, all human activities rely.

Table 3.1 Blue water withdrawals by use and continent.

	Water use					
Water use category	**Oceania**	**Africa**	**Asia**	**Europe**	**Americas**	**Global**
Domestic (%)	17	10	9	16	16	11
Industry (%)	10	4	9	55	35	19
Agriculture (%)	73	86	82	29	49	70
Total volume (km^3/yr)	26	215	2,451	374	790	3,856

Source: FAO (2011).

3.3 Water Stress and Water Scarcity

The concept of *water stress* is defined in this book as the effects felt when the volumes and qualities of water available in a given location are insufficient to meet the demands placed upon them. The concept of *water scarcity* refers to the availability of water resources and is defined by the FAO in three ways.

- **Physical Water Scarcity**: where the RWR is insufficient to meet demands.
- **Infrastructural Water Scarcity**: where even though the RWR may be sufficient to meet demand, there is inadequate infrastructure to get the water to where it is needed.
- **Institutional Water Scarcity**: where institutions, legislation and/or regulation fail to ensure that water is supplied in an affordable and equitable manner. This type of water scarcity is sometimes referred to as economic water scarcity.

Chronic, long-term physical water scarcity is experienced by around 1.2 billion people globally, while a further 1.6 billion experience infrastructural and institutional instances of water scarcity (UN Water 2015). Water scarcity is also not an issue confined solely to poorer nations in hot, arid climates. It represents an increasingly real threat to human and business livelihoods around the world (as illustrated in Figure 3.2).

The FAO defines physical water scarcity as being experienced when water withdrawals reach approximately 20% of RWR. While this rule of thumb is useful in providing an indication of the degree of water stress likely to be experienced in a given location, more sophisticated assessment methods, such as the European Commission's Water Exploitation Index, are needed to inform decision making on water management. Any assessment of water stress in any country, river basin or sub-basin requires a sound understanding of local water resources and water demands, both human and environmental.

In research published in 2012, basin-scale water-use information from around the globe was used by Hoekstra *et al.* (2012) to calculate the number of months when consumptive demands for water exceeded the available RWR after required environmental flows were removed. Figure 3.2 shows global water stress calculated in this way for the largest river basins and therefore represents a measure of physical water scarcity. It does not however show where there may be areas of infrastructural and/or institutional water scarcity. For example, some regions of Sub-Saharan Africa have ample RWR to meet demand but are characterised by widespread infrastructural and institutional failings.

To illustrate how water stresses can vary at a national and local scale, Figure 3.3 depicts waterbodies at risk of stress in England and Wales, mapped by the UK's Environment Agency (EA 2013). The Environment Agency identifies serious water stress as occurring when:

- the household demand for water is a high proportion of the effective rainfall available to address that demand; or
- the future household demand for water is likely to be a high proportion of the effective rainfall available to address that demand.

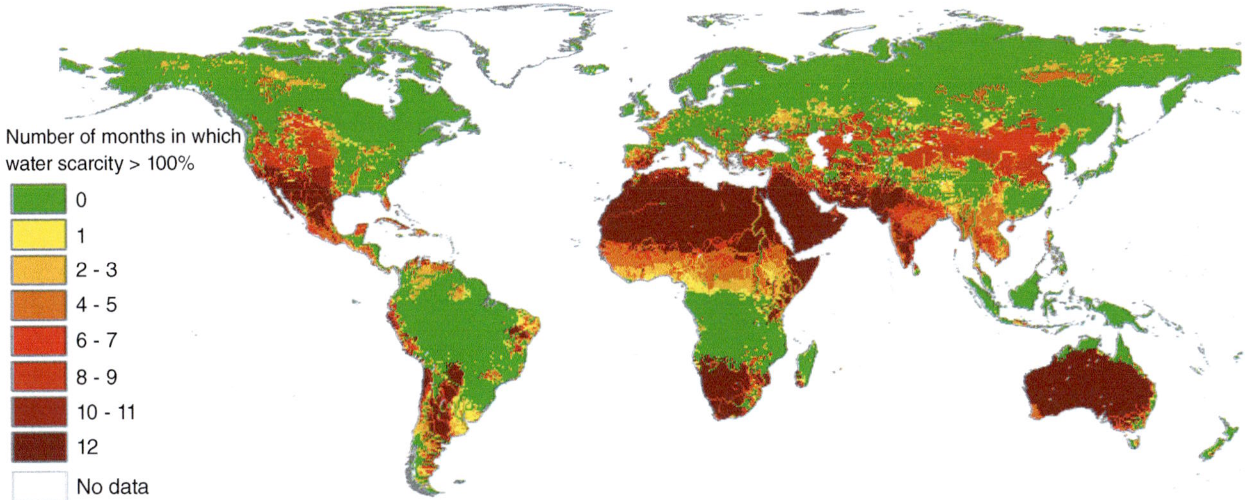

Figure 3.2 Global water scarcity assessment from water footprint data.
Source: Mekkonen and Hoekstra (2016). Reproduced with permission of Arjen Hoekstra.

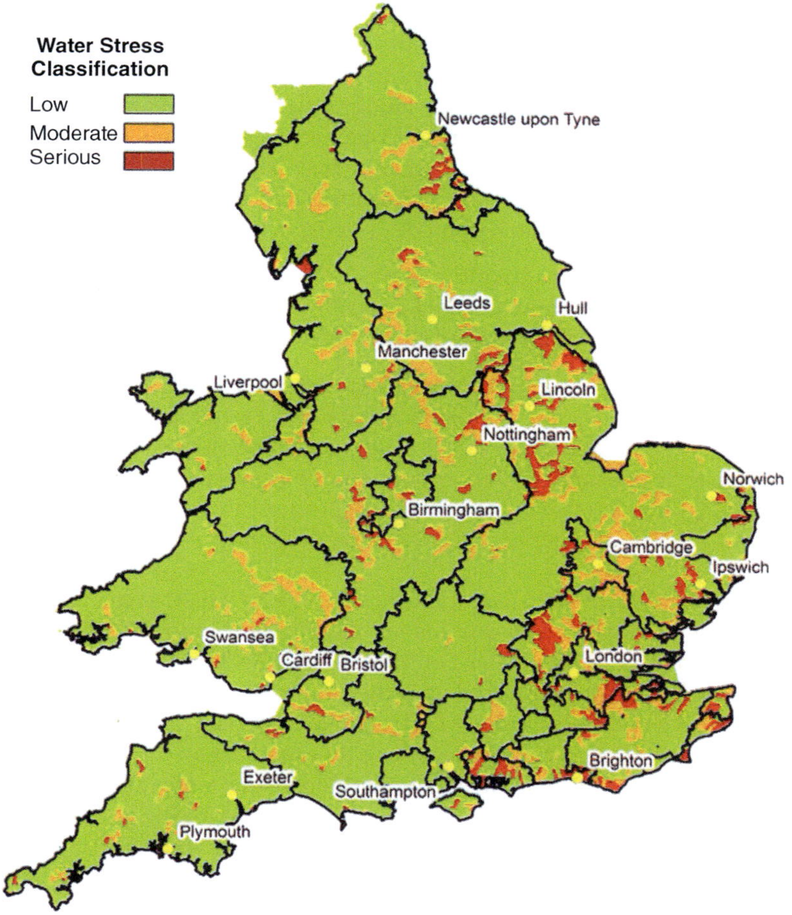

Figure 3.3 Water stress classification for waterbodies in England and Wales. It should be noted that this map indicates stress levels for waterbodies only, not the entirety of land area.
Source: Environment Agency (2013).

Figure 3.3 shows the waterbodies at risk of stress within the areas of responsibility of individual water companies. The Environment Agency concluded that of the 24 water company areas considered, only one was experiencing a 'low' stress level. Nine water companies were found to be experiencing a 'serious' stress level.

3.4 Virtual Water and the Water Footprint

The term *virtual water* emerged in the early 1990s through the research of Professor Tony Allan of King's College London. He studied the responses of Middle Eastern countries to water scarcity and used the term 'virtual water' to describe how nations with low RWR accessed water in other nations through trade in food and other goods. The concept of virtual water is therefore linked closely to that of the *water footprint*, the volume of water consumed in the production cycle of a given product.

Often, the basic water footprint concept is refined to identify the relative proportions of blue and green water consumed, as well as to include *grey water*[1], the volume of water needed to assimilate the pollutants of any returns of water to the environment.

Figure 3.4 depicts a theoretical example of a country or river basin exporting food and other products through agricultural and industrial activities supported by both blue and green water consumption. The virtual water content of its exports therefore comprise:

- **green water**: consumed by rain-fed agriculture;
- **blue water**: consumed by industry and irrigated agriculture; and
- **grey water**: the volume of blue water required to assimilate returns of used water.

Academics and researchers have found the water footprint concept to be a useful and easy-to-understand indicator of the size of a demand for water. By linking this concept to that of virtual water, the small size of our direct water consumption in comparison to the water we each indirectly consume in the production of our everyday products can also be effectively highlighted. Furthermore, both concepts are valuable in identifying countries that rely on the RWR of other nations to sustain the ability of their citizens to live, eat and consume.

In 2007 Hoekstra and Chapagain published the *Globalization of Water*, a seminal text in recent water literature that included an assessment of the water footprint of nations. That work was updated in 2011 by the Water Footprint Network (WFN; Hoekstra and Mekonnen 2011a, b) to include more granular data at the river basin scale (see Figure 3.2). These two publications triggered an explosion of research into national- and catchment-scale comparisons of virtual water trade. Some researchers expanded on the application of the virtual water and water footprint concepts to develop virtual water accounts for particular nations and basins. Figure 3.5 illustrates one such account for the UK in 2011, taken from WFN data. The figure shows the internal and external elements of virtual water as well as the components related to water consumption and production. The re-export element accounts for adding value to a product that has been imported.

In 2007, Hoekstra and Chapagain identified 6,500 international trade routes collectively moving around 567 km^3 of virtual water around the globe each year, a doubling of the flow calculated for 1986. That's roughly equivalent to 25% of the total global consumptive use of blue water, although it does of course include green and grey water.

This international trade in goods, especially food, has meant that many nations have been able to support populations and industries that have expanded far beyond the size that could otherwise have been sustained by domestic water sources alone. In this way, water scarcity has been averted in some nations, but may have been inadvertently exacerbated in others. Figure 3.6, developed by University of Twente and WFN in 2011 (Hoekstra and Mekonnen 2011a, b), depicts average annual virtual water flows between nations between 1997 and 2008. It shows how regions such as North and South America act as virtual water exporters, supplying virtual water importers such as Western Europe and Japan.

---

[1] This should not be confused with greywater, that water used in household kitchens, showers and bathroom hand basins and which represents a potential source for local water recycling.

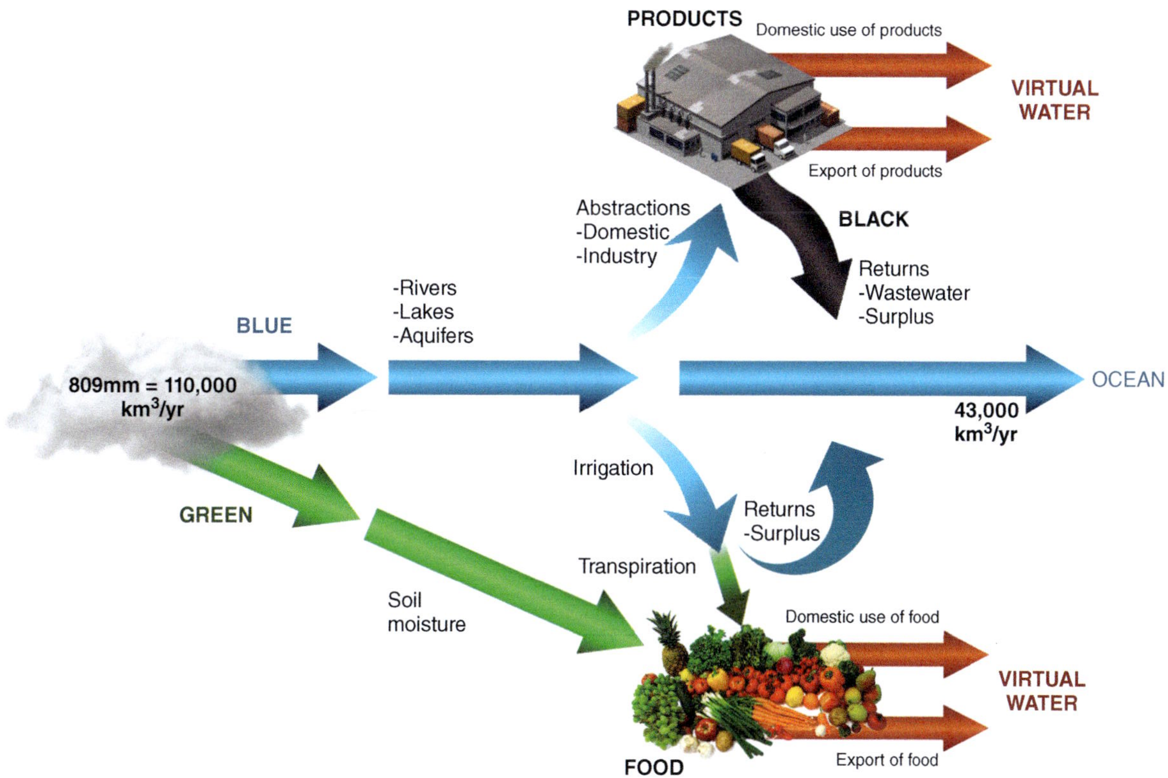

Figure 3.4 Virtual water in relation to blue and green water use in food and products.

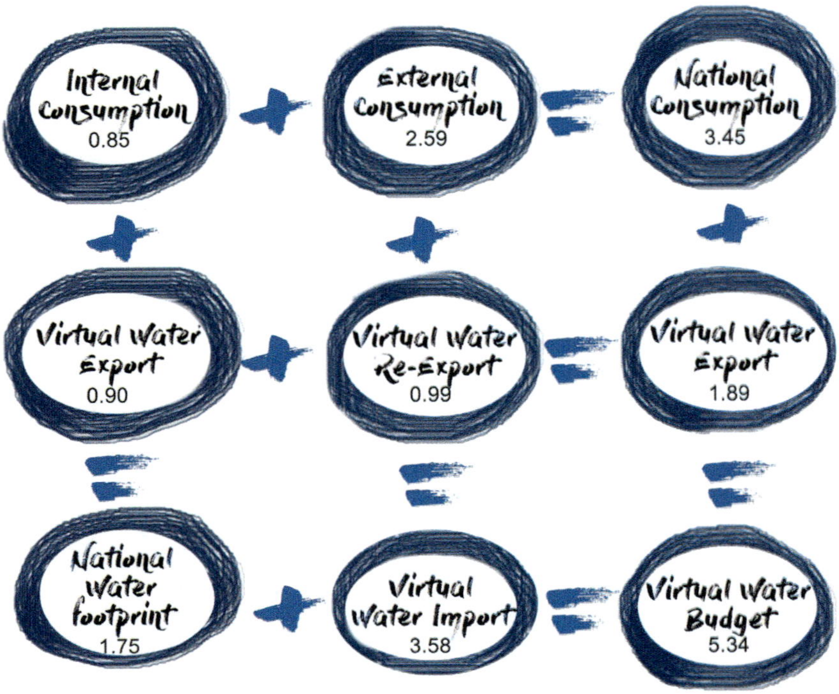

Figure 3.5 Virtual water account for United Kingdom, m^3 per capita per day.

Figure 3.7 shows the water footprint of consumption per capita over the same period as Figure 3.6; some interesting observations emerge when the two figures are compared. For example, while North America has been a substantial exporter of virtual water, its citizens also have a high water footprint of consumption; this is a consequence of the USA and Canada having high RWRs and using these to sustain a high standard of living and a food export industry.

The data used to develop Figure 3.7 can be presented and analysed in a number of different ways. For example, Figure 3.8 focuses on comparing the per capita internal (national) and external (international) water footprint of consumption of select countries. Typically, high external water footprint is correlated with low RWR and high imports of food. This general trend can however be influenced by political decision making, for example to support agricultural production with subsidies or through the imposition of export tariffs. Figure 3.8 shows that:

- citizens of the USA have double the water footprint of citizens of the UK and China, but they are not dependent to a high degree on water in other nations;
- citizens of the UK and Ethiopia have similar water footprints, but while the UK is highly dependent on water in other nations, Ethiopia is virtually self-sufficient; and
- citizens of Australia and Israel have similar water footprints, but Israel is highly dependent on water in other nations.

In Figure 3.9, the diagram identifies the different uses of freshwater in the top 10 freshwater-consuming countries, while Figure 3.10 identifies the largest net exporters

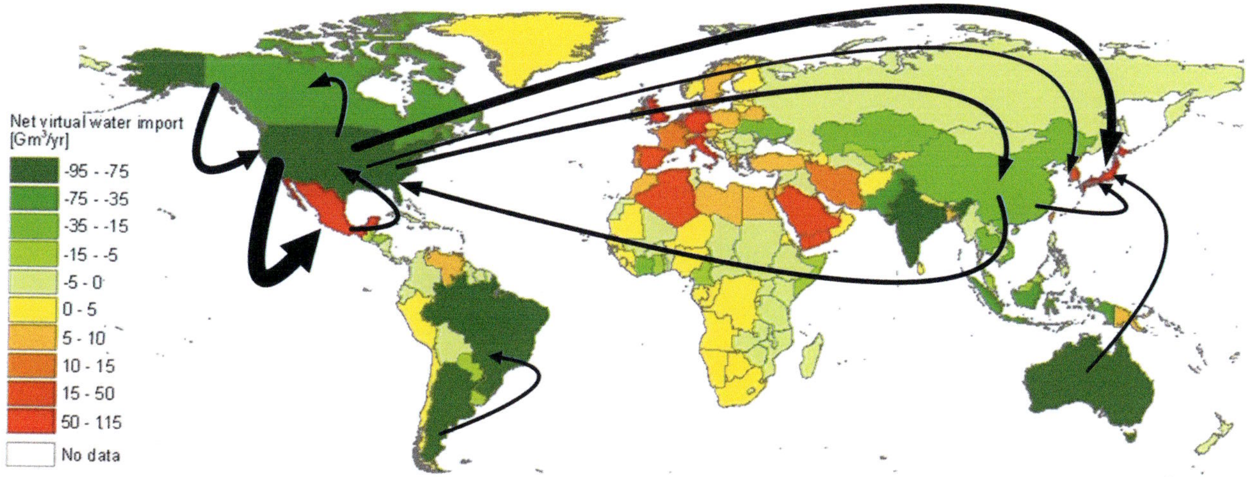

Figure 3.6 Global virtual water trade 1997 to 2008, average km^3/year.
Source: Hoekstra and Mekonnen (2012). Reproduced with permission of Arjen Hoekstra.

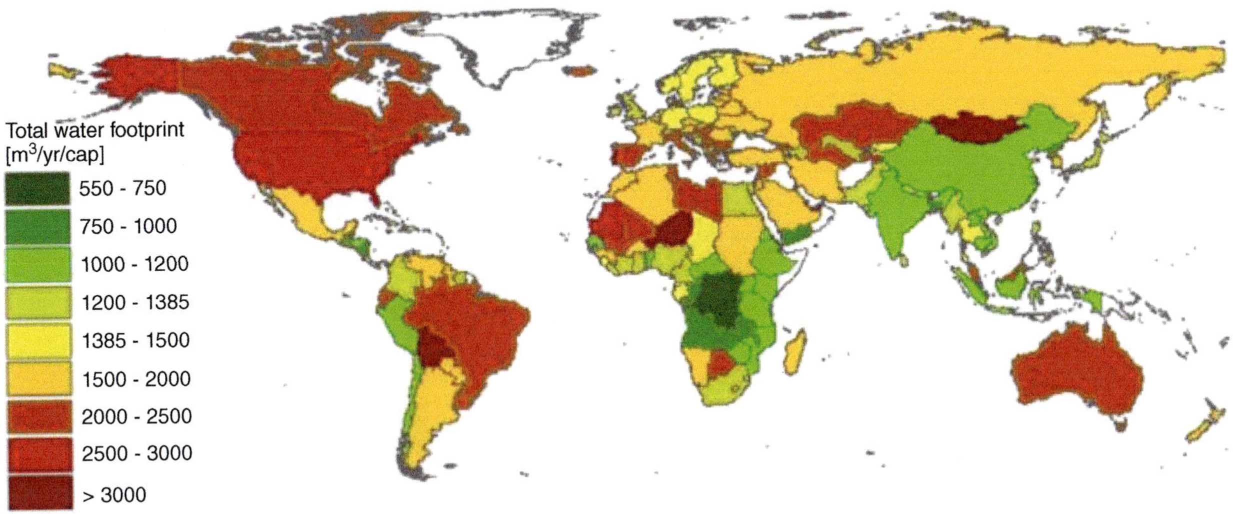

Figure 3.7 Average water footprint of national consumption in m^3 per year per capita in the period 1996–2005. Countries shown in green have a water footprint that is smaller than the global average; countries shown in yellow or red have a water footprint larger than the global average.
Source: Hoekstra and Mekonnen (2011a). Reproduced with permission of Arjen Hoekstra.

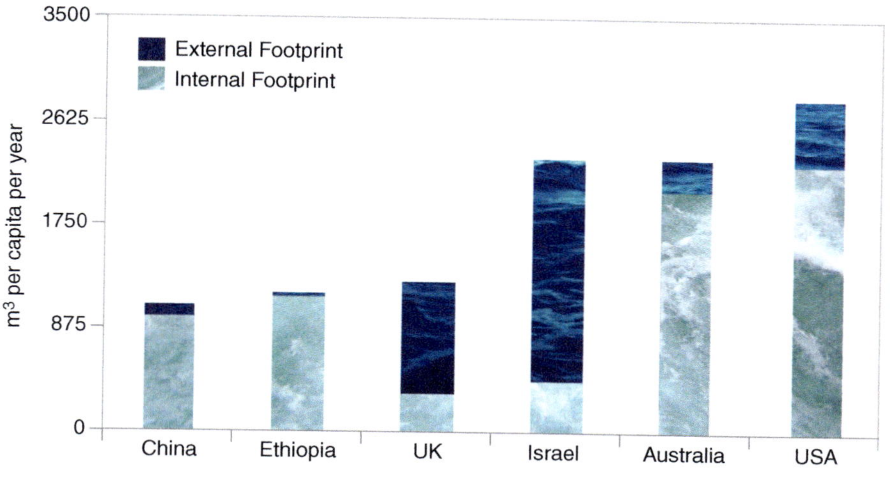

Figure 3.8 Water footprint of selected nations, m^3 per capita per year.

and importers of virtual water. By analysing Figure 3.9, one can see that China, India, the USA and Brazil have the largest footprints of consumption as a consequence of high population (1,369 km^3, 1,144 km^3, 821 km^3 and 355 km^3, respectively). Brazil and USA also have very large RWRs per capita (45,157 m^3/yr and 9,589 m^3/yr, respectively; see Table 2.2), which both supports a high level of water consumption and has enabled them to become leading virtual water exporters.

The reasons for these export and import patterns lie with governance and, in particular, with how the past and present policies of particular nations sought and seek to exploit their natural resources for economic comparative advantage. It is perhaps only now, as the growing competition for water pushes more and more countries into positions of water stress, that virtual water trade will become an explicit consideration of water and foreign trade policy.

- In the **USA,** much of the mid-west and southwest are now heavily water stressed from growth in urban demands, but more so from increased demands for irrigation in agriculture, a major source of US export and interstate revenue. Does the USA gain more from crop exports than it loses in the impacts associated with water stress? And can those impacts be sustainably maintained into the future?
- In **India,** water stress is manifest in the extreme overdraft of aquifers by unregulated pumping for irrigated agriculture (also often supported by subsidies on fuel). While at a national level India has significant RWR, most of this is located in the north and east, distant from the nation's major agricultural areas. Major projects to transfer water between states have been proposed but are hugely expensive.
- In **Australia,** the tension between water stress and economic stability is focused on the Murray Darling River basin in the nation's southeast. The government has attempted to allocate the limited RWR between competing demands using sophisticated water-trading frameworks that include government 'buy-back' of water to maintain environmental flows. The intent has been to expose agri-business to the true economic value of water and thereby to encourage it to adopt more innovative approaches to water use and conservation.

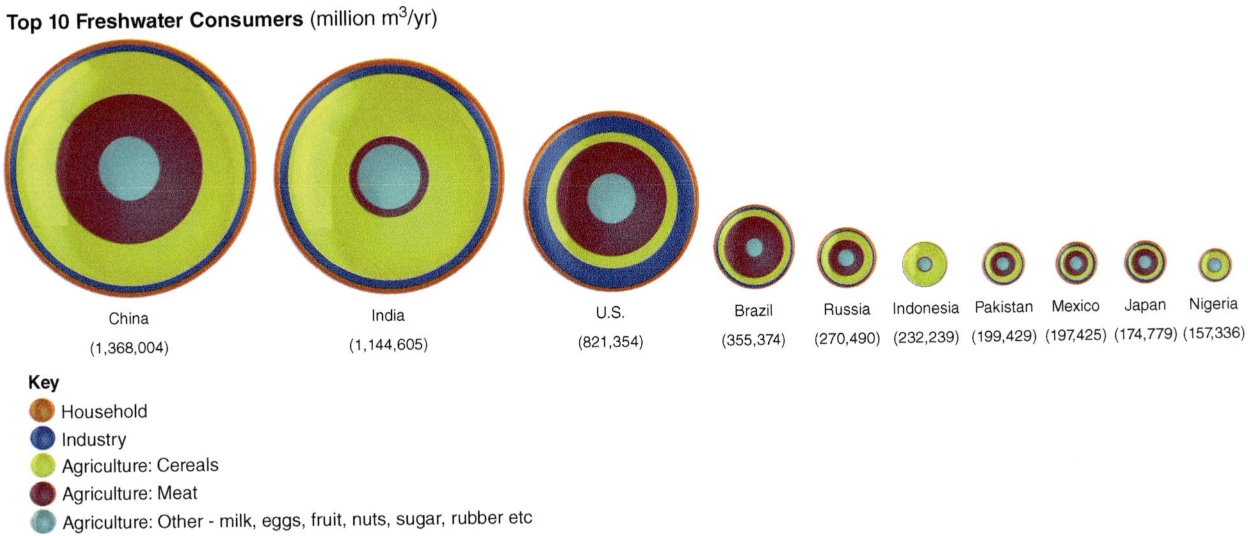

Figure 3.9 Water footprint of the top 10 most water-consumptive nations. Adapted from Hoekstra and Mekonnen (2012).

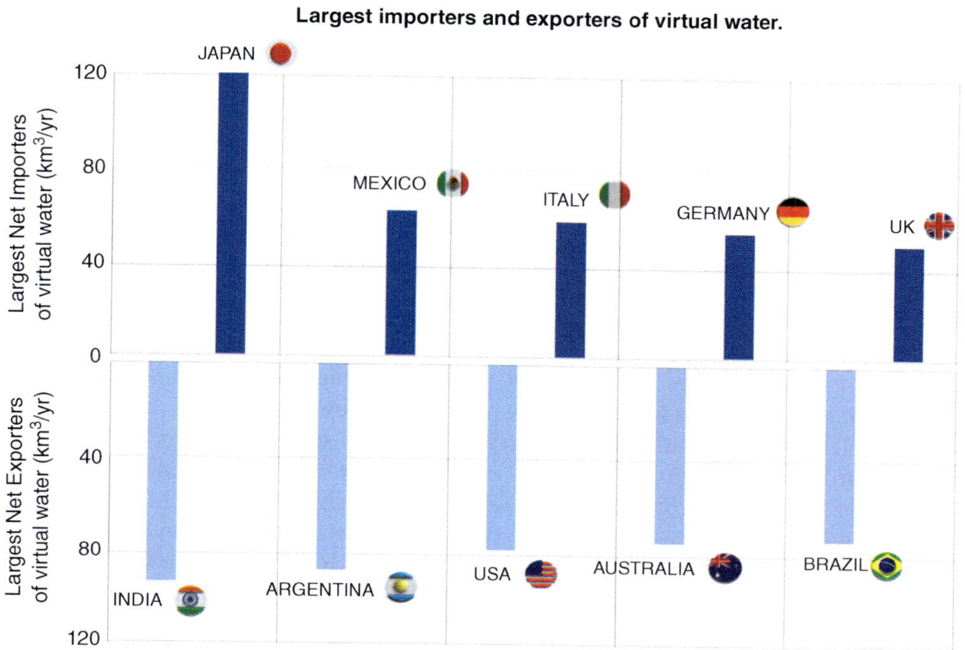

Figure 3.10 The largest importers and exporters of virtual water. Adapted from Hoekstra and Mekonnen (2012).

Returning to Figure 3.10 and turning now to the top virtual water importers, we again see countries that seem out of place. For example, while the list includes countries you would expect, such as those with low or moderate RWR and high population (Japan and Mexico), it also includes three European nations whose RWR is relatively high in comparison to their respective populations. The RWRs of the UK, Germany and Italy are higher than their water footprints; however, their governments have chosen to import virtual water and, in turn, free up domestic freshwater sources for other activities, principally those associated with business, amenity and aquatic ecosystem protection.

In the case of the UK, Hoekstra and Mekkonen (2016) argue that importing >50% of required blue water is unsustainable, and they point to four possible mitigation strategies which include growing more food in UK. Interestingly, a comparison is also drawn with the supply chain risk of businesses importing raw materials from other nations.

So what impacts do these trades in virtual water have on global, regional and local water resources? And to what extent are these trades, vulnerable as they are to the vagaries of commodity prices and the actions of other governments to distort markets, responsible for the increasing water stresses now being experienced in an ever-increasing number of countries? To consider these questions, researchers have used network theory in an attempt to unravel the highly complex trading pathways, and recent studies have reached a number of conclusions.

- 'In trading goods across national borders, we are effectively trading the services of water' (Reimer 2012).

- 'International food trade has led to enhanced savings in global water resources over time. Overall, less water-efficient countries have been increasingly importing from more efficient countries' (Konar *et al.* 2012).
- 'Virtual water flows tend to be driven by GDP and social development status rather than water scarcity' (D'Odorico *et al.* 2012).
- 'Long-distance transport of food weakens the resilience of the coupled natural-human system' (D'Odorico *et al.* 2012).
- 'There is a tendency for developing countries to source water-intensive commodities from abroad while protecting their own water resources' (Lenzen *et al.* 2012).

The latter is an important observation because it might imply that national policies that encourage food imports are knowingly or otherwise exacerbating water stress in other nations. Put another way, national policy could be protecting national water resources at the expense of water resource protection elsewhere.

The proliferation of virtual water research has also been accompanied by lively debate about the utility of the 'virtual water' and 'water footprint' concepts. While it is important to recognise that the concepts do make a number of significant simplifications and that they ignore a number of important factors influencing water management, for example the impact of water use in the host location, their simplicity does make them highly powerful tools for communicating with the public. For decision makers, because of their inherent assumptions, the concepts of 'water footprint' and 'virtual water' should be recognised as high-level and indicative aids, and applied principally to identify those issues requiring more detailed study.

While much of the existing research has been focused on evaluating volumes of virtual water, there are fewer examples of research which attempt to assess the impact of virtual water trade on water stress. In a discussion paper published by the University of Bonn (Lenzen *et al.* 2012), the researchers looked at degrees of national water scarcity alongside economic input–output analysis of the virtual water trade network.

Figure 3.11 is taken from data in that paper and shows the leading ten importers and exporters of 'scarce' water. What this shows is that the five leading importers of virtual water – Japan, Mexico, Germany, Italy and UK – are all also leading importers of 'scarce' water from nations experiencing water scarcity. In the case of the UK, 'scarce' water accounts for some 25–30% of the total (green and blue) virtual water imports. The research also shows that several nations who are not net importers of virtual water are however importing virtual water from water-scarce nations, for example France and the USA.

3.5 Live, Eat, Consume: The Conceptual Framework of Water Stress and Virtual Water

Chapters 4, 5 and 6 consider our water-consuming activities against the framework of concepts described in this chapter. The framework is one rooted in the hydrological principles of rainfall and runoff, but widened to emphasise the importance of accounting for the green, blue and grey components of water withdrawal and consumption while also acknowledging the critical role that virtual water trade now plays in influencing locations of water stress.

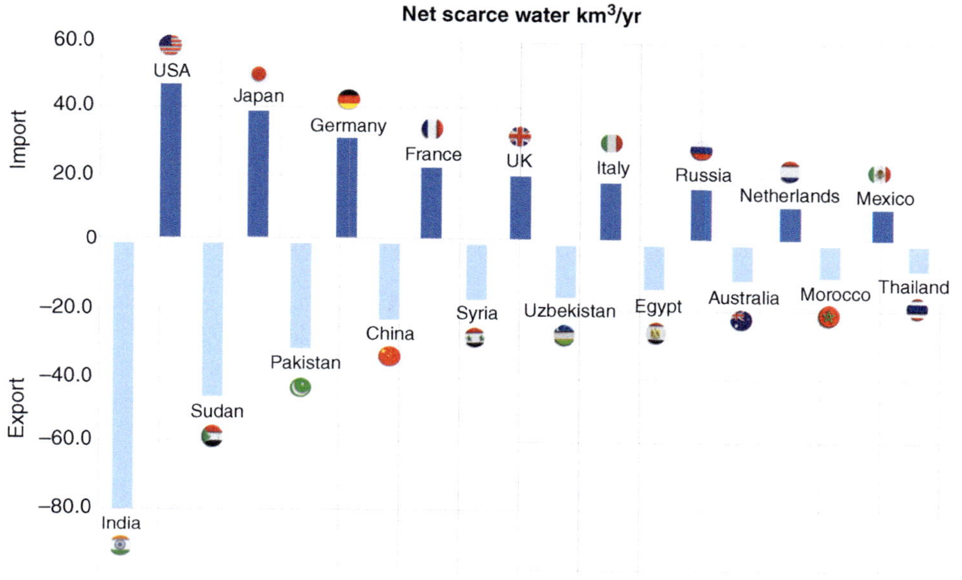

Figure 3.11 Net scarce virtual water exporters and importers.

Analysis of virtual water flows highlights that water stress is no longer solely the result of local communities acting in isolation. Indeed, it is now widely postulated that some nations have, unwittingly maybe, alleviated their own water scarcity by promoting agricultural and other trade policy that results in, or worsens, water stress in other nations (Lenzen *et al.* 2012).

To fully appreciate the role that water plays in allowing societies to live, eat and consume, it is first important to acknowledge that the use of water has both essential and discretionary components. Every day, we each need to drink between 2 and 4 litres of clean water and use 20 to 30 litres of clean water for sanitation. Beyond this volume, however, water use largely becomes a matter of choice. Personal and corporate desires dictate how much additional water we use; because water contributes to the success or otherwise of so many human activities (e.g. agriculture and energy production), immediate social and financial gain often increases with increasing discretionary water use. The tendency for human populations to overexploit water is therefore strong and this is one of the primary causes of water stress.

To avoid this overexploitation and the resulting eventual collapse of our natural resource pools, those responsible for water allocation and supply need to quantify the size of the essential component of water use and then manage the supply of the essential and discretionary volumes differently. In many countries however, this distinction is absent. In order to appease their populations, or to encourage economic growth, governments may provide all water for free or at a heavily subsidised price, even in countries or regions where water is scarce. At the other extreme, where government provision of the essential water component is inadequate, opportunistic private enterprise may fill the gap, often supplying populations at high prices; in turn, this introduces issues of economic water scarcity for those too poor to afford the service.

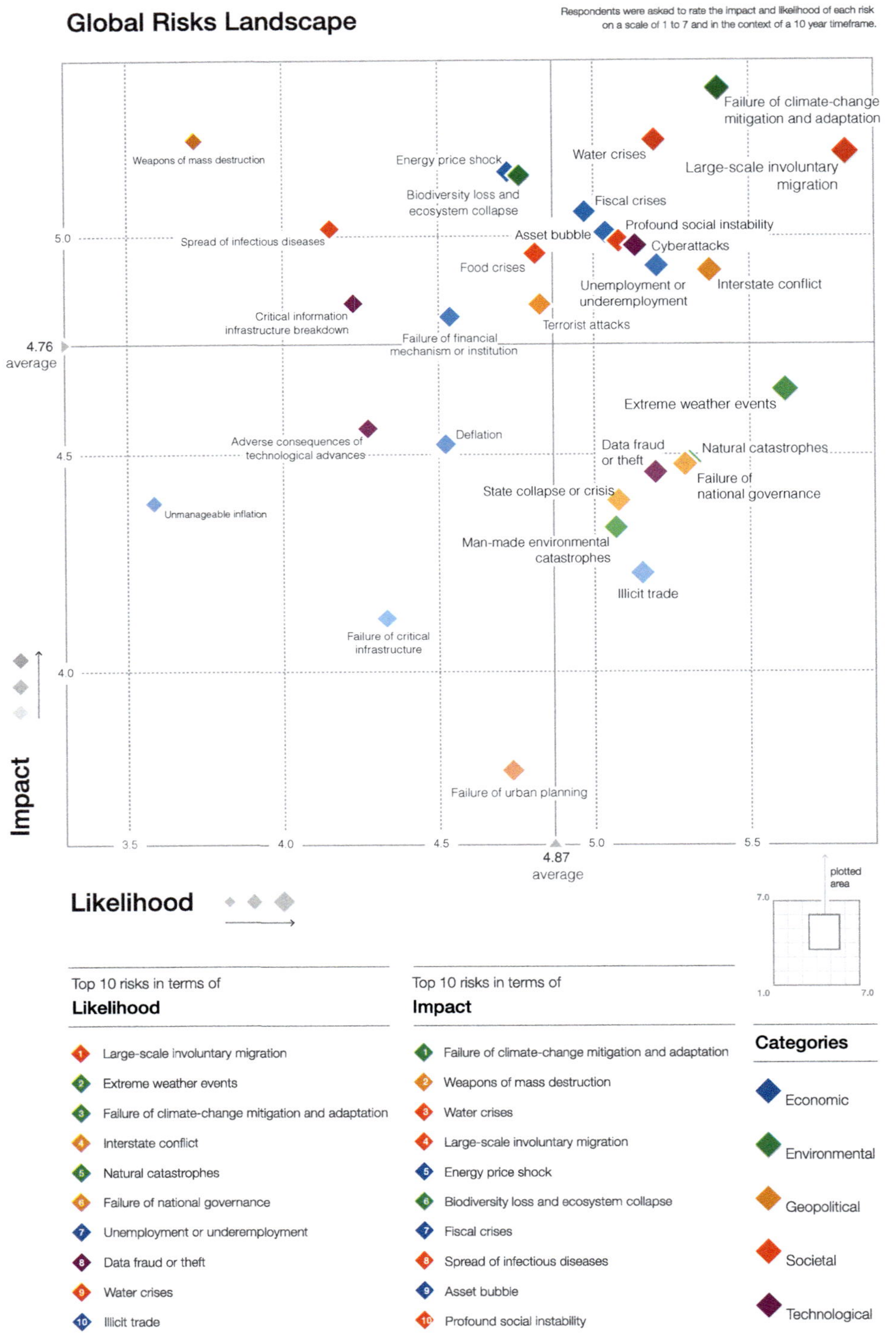

Figure 3.12 World Economic Forum global risks 2016. Reproduced with permission of World Economic Forum.

Failures of water governance such as these are heavily influenced by our own perceptions of water. Section 2.3.2 discusses the human right to water, a vitally important recognition of the essential component of water use. In many societies however, historic social and political norms have conditioned a perception that all water is a right that should be provided free of charge, regardless of the difficulties of supplying and managing the resource. This perception therefore acts as a powerful social brake on changing the water supply *status quo* to one which advocates a more sustainable approach.

Notwithstanding these constraints, appreciation of the true value of water is emerging, particularly among businesses and industries. As more and more companies feel the financial impacts of water stress, the need for water risks to guide business planning is increasingly acknowledged. Each year, the World Economic Forum (WEF) publishes its Review of Global Risks, a survey of more than 1000 experts on a spectrum of 50 societal, economic, environmental, technological and geopolitical risks. Since 2007, 'water supply crises' and 'food shortage crises' have risen steadily in their perceived likelihood and impact. In the 2016 Global Risk report (World Economic Forum 2016), and as illustrated in Figure 3.12, 'water supply crises' were listed as the ninth-most likely global risk to occur and the third-most significant in terms of its potential impact. Note also in Figure 3.12 that food crises, energy price shocks, ecosystem collapse, and failure to mitigate and adapt to climate change, all inextricably linked to water, also have high likelihood and impact scores.

While national governments and intergovernmental entities continue to debate 'water' in a proliferation of national and international think tanks, focus groups, cooperation networks and the like, the WEF's analysis is one founded on the opinions of those who are directly in touch with the economic impacts of water scarcity. The analysis is evidence that action to address water management is needed right now. If not, humanity will be severely compromised in its ability to live, eat and consume into the future.

References

D'Odorico, P., Carr, J.A., Laio, F. and Ridolfi, L. 2012. Spatial organization and drivers of the virtual water trade: a community-structure analysis. *Environmental Research Letters*, 7.

Environment Agency. 2013. Water Stressed Areas – Final Classification. Environment Agency, Bristol UK.

Food and Agriculture Organisation of the United Nations. 2011. The State of the World's Land and Water Resources for Food and Agriculture. FAO, Rome.

Haddeland, I., Heinke, J., Biemans, H., Eisner, S., Flörke, M., Hanasaki, N., Konzmann, M., Ludwig, F., Masaki, Y., Schewe, J., Stacke, T., Tessler, Z.D., Wada, Y. and Wisser, D. 2014. Global water resources affected by human interventions and climate change. *Proceedings of the National Academy of Sciences*, 111(9), 3251–3256.

Hoekstra, A.Y. and Chapagain, A.K. 2007. Globalization of Water: Sharing the Planet's Freshwater Resources. John Wiley and Sons, London, UK.

Hoekstra, A.Y. and Mekonnen, M.M. 2011a. National Water Footprint Accounts: The green, blue and grey water footprint of production and consumption. UNESCO-IHE, Delft, Netherlands.

Hoekstra, A.Y. and Mekonnen, M.M. 2011b. Global water scarcity: monthly blue water footprint compared to blue water availability for the world's major river basins. Value of Water Research Report Series No. 53, UNESCO-IHF, Delft, Netherlands.

Hoekstra, A.Y. and Mekonnen, M.M. 2012. The Water Footprint of Humanity. National Academy of Science, USA.

Hoekstra, A.Y. and Mekonnen, M.M. 2016. Imported Water Risk: the case of the UK. *Environmental Research Letters*, 11, 5.

Hoekstra, A.Y., Mekonnen, M.M., Chapagain, A.K., Mathews, R.E. and Richter, B.D. 2012. Global monthly water scarcity: blue water footprints versus blue water availability. *PLoS ONE*, 7.

Konar, M., Dalin, C. Hanasaki, N. Rinaldo, A. and Rodriguez-Iturbe, I. 2012. Temporal dynamics of blue and green virtual water trade networks. *Water Resources Research*, 48, 1–11.

Lenzen, M., Bhaduri, A., Moran, D., Kenmoto, K., Bekchanov, G. A. and Foran, B. 2012. The role of scarcity in global virtual water flows. ZEF – Discussion Papers on Development Policy No. 169. Centre for Development Research, University of Bonn.

Marsh, T.J., Parry, S., Kendon, M.C. and Hannaford, J. 2013. The 2010-12 drought and subsequent extensive flooding. Centre for Ecology & Hydrology, Wallingford, UK.

Mekonnen, M.M. and Hoekstra, A.Y. 2016. Four billion people facing severe water scarcity. *Science Advances*, 2, 2.

Reimer, J.J. 2012. On the economics of virtual water trade. *Ecological Economics*, 75, 135–139.

UN Water. 2015. The World Water Development Report 2015, Water for a Sustainable World. UNESCO, Geneva, Switzerland.

World Economic Forum. 2016. Global Risks 2016, 11th Edition. WEF, Geneva, Switzerland.

4

Live

4.1 Introduction

Access to sufficient quantities and appropriate qualities of water underpins all human activity. In this chapter, the first of three that explore some of the most complex and challenging water management issues, we consider how water interacts with our need for energy and its relationships with the phenomenon of urbanisation, one of the key demographic trends of our time. A series of case studies are used to illustrate how human ingenuity and innovation has optimised, and can continue to optimise, our use of all resources. For example, there remains enormous untapped potential to better harness our increasingly vast stores of data in the search of synergies between different users of a given resource. The lessons we can learn from how nature meets its water and energy needs will also be invaluable, as will be the ability of political and corporate leaders to develop appropriately integrated and astute resource policy. The ability of our increasingly interconnected society to recognise the need for change and then galvanise action will be no less vital.

Cities provide an excellent setting within which to observe the complexities of water management. They are centres of resource consumption, promoting economic growth that attracts migrants in a self-perpetuating cycle of immigration and growth. The difficulties of achieving equitable provision of water in such a dynamic environment are mirrored by the challenges associated with the provision of energy and other key resources. By exploring the relationships between water and energy and by considering the provision of these resources in urban settings, this chapter highlights how the management of all resources can be optimised to improve social, environmental and economic outcomes.

4.2 Water and Energy

4.2.1 The Nexus of Water and Energy

No economic activity of note can occur in isolation from the interacting inputs of energy and water. The transport and treatment of water and wastewater requires energy while the production of almost every known energy source requires water, either during its extraction and refining, during fuel production or in order to cool power plants. Indeed, the relationships between these two resources persist at the

Water Resources: A New Water Architecture, First Edition.
Alexander Lane, Michael Norton and Sandra Ryan.

most basic levels of human survival. It is the same proportion of the global population who lack access to safe water and sanitation that are also most likely to lack access to electricity (UNWWAP 2014).

The direct and indirect links between the management of water and energy abound, and this situation creates both risks and opportunities for planners. Despite this dual reliance however, the management of energy and water typically continues to occur in discrete silos; the policies related to one are often formulated with little or no regard to their impacts on the other. This situation can in part be explained by the perceived economic value of each resource.

The economic value of energy is well known. Energy sources, particularly fossil fuels, are highly monetised assets, traded across global markets that ensure management decisions are largely based on economics. In contrast, the management of water is often viewed as a public health and welfare issue and a socio-political prerogative (UNWWAP 2014). These differences in perception mean that integrated water and energy policy is rare. In locations where water is abundant, the impact of this oversight is hidden. However, with population and climate change leading to increased instances of physical and economic water scarcity, examples of energy supply systems being forced to shut down as a result of water constraints and of water management being limited by energy availability are increasing in their number and severity (see Breakout Box 4.1).

The differing scales at which water and energy are managed also act as an important impediment to integrated decision making. While energy is often supplied via a national grid, water resources are typically exploited at much smaller geographic scales with the impacts of inappropriate water management expressed as relatively localised events.

Breakout Box 4.1 Water shortages and energy outages

There exist numerous examples of the interrelationships between water and energy, and of the associated risks to human interests:

- during a summer drought in the USA in 1988, high temperatures and low river levels forced Commonwealth Edison power station to reduce power output by 30% while the Dresden and Quad Cities plants in Illinois were forced to shut down completely;
- between 2000 and 2010, concerns over water availability halted power plant construction or operation in no less than 14 US states (Grubert and Kitasei 2010);
- in 2003, heat waves in France forced nuclear power plants to reduce electricity production due to a lack of water for cooling;
- in 2007, the Australian National Electricity Market experienced severe water shortages that saw generation capacity curtailed and a threefold increase in the wholesale price of electricity (Bildstein 2007; Hussey and Pittock 2012);
- residents of the Palestinian Territory of Gaza have repeatedly experienced power outages which have caused the city's sewage pumps to fail and its sewers to overflow (IRIN 2013). As a result, the population is exposed to the health risks associated with water-borne disease; and
- it has been estimated that, due to drought and environmental warming induced by climate change, power production could fall by between 4% and 16% in the USA and by between 6% and 19% in Europe between 2031 and 2060 (van Vliet *et al.* 2012).

Decisions over energy provision therefore need to be made with greater regard to regional and local constraints and opportunities, whereas the provision and management of water needs to begin at the scale of the catchment and, in many cases, also consider regional factors to identify efficiencies in water supply and wastewater management.

4.2.2 Energy Use in Water Management

Developed countries have, for the most part, adopted a linear approach to harnessing their water resources. Freshwater is extracted from its natural source, be it a surface waterbody or a groundwater store, and then piped to individual users via a water treatment plant which is operated to achieve a desired water quality. Any wastewater generated by a user is then piped back to a natural water store; this is not necessarily the same store it was extracted from, but typically via a wastewater treatment plant in an attempt to limit the impact of the entrained pollutants on the receiving environment.

This approach to water management is highly effective at ensuring that water supply systems achieve their primary goal: that of maintaining public health. Historically, by implementing this linear water management approach, industrialising countries have been able to greatly reduce instances of waterborne disease and, as a consequence, swell the workforce. This in turn supported economic growth that raised standards of living in a reinforcing cycle.

The opposite socioeconomic trends are true of those nations that lack safe water and sanitation. In such situations, mortality rates, especially among the young, are high and economic development suffers as a result. Furthermore, in order to access the few safe sources of water available, families typically have to travel long distances; this task is usually assigned to women which, in turn, prevents girls attending school and locks in subsistence livelihoods and gender inequality.

While the linear approach to water management has greatly improved public health, it is not devoid of constraints. The take, use, dispose ethos is inherently single-minded and therefore wasteful of many of the embedded resources present in so-called 'wastewater'. In addition, it ignores other factors that may influence or be influenced by the water management system and so foregoes otherwise valuable and mutually beneficial outcomes. In this respect, energy represents perhaps the greatest oversight. Energy is required to extract and distribute water, to purify it, to gather wastewater and to treat wastewater before it is returned to the environment. Furthermore, the way water is used by the customer may also consume significant volumes of energy, particularly when water is heated to warm homes and businesses.

4.2.2.1 Energy Demands from Water Management

Water has several characteristics that make its provision energy intensive. Water is heavy. One cubic metre of pure water weighs one tonne, around a third more than one cubic metre of gasoline. Water also has a high heat capacity meaning that a relatively large amount of energy is required to raise its temperature. For example, to heat a given volume of water by 1°C requires double the amount of energy needed to heat the same volume of oil by 1°C.

Considered at national scales, the water sector is typically only a minority user of energy. At regional and local scales, however, its influence can be much more significant. For example, while the US water and wastewater sector is responsible for between

3% and 4% of national energy use (USEPA 2013), in California the sector uses almost 20% of all electricity (Copeland 2014). Furthermore, the subsequent heating of water by customers consumes more than 30% of the state's natural gas (Cooley and Donnelly 2013). Globally, the water and wastewater sector is responsible for around 3% of all greenhouse gas emissions (World Bank 2010).

For the public or private body responsible for the provision of water and wastewater services, energy represents a significant component of their operating expenditure. In the US, this figure is around 40% (USEPA 2013) and second only to labour (Copeland 2014). The amount of energy required to operate a given water supply system is controlled by three key factors:

1) the antecedent quality of the source water in comparison to its intended end use;
2) the choice of water treatment technologies adopted; and
3) the amount of pumping required by the conveyance system.

Antecedent Water Quality Both the antecedent quality of a natural water source and the required water quality of the end use will vary from location to location and from customer to customer. For a given water treatment system and type of water source, energy requirements will be lowest when the difference in quality between the source water and the required final water quality is smallest. In some instances, water treatment may not be required at all; however, because most systems function by centralising water treatment at a small number of plants, and because most of these plants feed a single pipeline network supplying all customers, all water must be treated to meet the most stringent demand (invariably potable use). Over-treatment is therefore common and, as such, energy efficiencies might therefore be achieved if water was instead supplied on a fit-for-purpose basis. In developed countries where this goal could only be achieved through the retrofitting of complex pipeline networks, it is likely to be uneconomic. In contrast, however, through careful infrastructure planning, developing countries have the chance to learn lessons from their already developed counterparts and avoid becoming locked in to energy-intensive water management practices.

Maintaining high water quality in natural water sources is also an important means of limiting the amount of energy required during treatment. Watershed protection is a useful concept in this respect, and one that is capable of providing a number of mutually beneficial outcomes. In an example from the UK, improved soil and fertiliser management in upland catchments of the southwest of England was able to solve water quality problems at around one-sixth of the cost of the centralised treatment alternative, and with a substantially reduced carbon footprint (Wessex Water 2011). Furthermore, watershed management approaches such as these also act to slow the passage of water across the landscape, reducing the likelihood and impact of downstream flooding and limiting the erosion of soil and nutrients from the land.

Method of Water Treatment Once extracted from a natural source, water is typically screened and then treated to render it suitable for potable use. The energy demanded by these processes, although notable, is typically small compared to that associated with conveying freshwater to the treatment plant and with distributing treated water to the customer. Water treatment processes are responsible for less than 10% of total energy demand, mostly resulting from the need to continually mix stored water within the treatment plant to prevent stagnation. Caffour (2008) estimates that the treatment of 1 m^3 of water for potable use in the UK requires around 0.59 kWh of energy, and that 0.63 kWh is required to treat every 1 m^3 of wastewater prior to its disposal. It should also be noted that groundwater usually requires less treatment prior to use than surface water (ESMAP 2012).

As the number of water-stressed countries and regions has grown, attention has increasingly turned to technologies that treat saline water to reduce its salt content and thereby make it suitable for a broader range of uses. Globally, there are more than 16,000 desalination plants in operation, with almost half of these located in the Middle East. Global desalination capacity increased by 65% between 2008 and 2011 (IDA 2011), and is predicted to double by 2020 (UNWWAP 2014). A variety of desalination technologies exist (see Breakout Box 4.2); however, they all have significant energy demands. In fact, as highlighted by Figure 4.1, desalination is widely regarded as the most energy-intensive water treatment technology (UNWWAP 2014).

The technologies employed at wastewater treatment plants can be broadly grouped into primary, secondary and tertiary (or advanced) categories (see Table 4.1 and Figure 4.2). In addition, an important component of the overall energy demand of wastewater management relates to the further treatment and disposal of the separated solid sludge.

Breakout Box 4.2 Desalination basics

The first desalination facilities were developed in the Gulf States in the 1950s and typically relied on thermal or electrical energy to heat seawater, causing the pure water component to evaporate before its subsequent condensation and collection. These technologies were extremely energy intensive, consuming around 10 kWh for each cubic metre of potable water produced (NWC 2008). Although still energy demanding, modern installations allow thermal desalination technologies to integrate drinking water and electricity production, with the waste heat from a power plant used as the heat source for the desalination process. Because demands for energy and water typically vary over different timescales however, a tradeoff between optimising either water or energy efficiency is often required.

Many modern desalination plants use reverse osmosis (RO) as their technology. RO requires the application of hydraulic pressure to force saline water to pass through a membrane that rejects salts and thus generates a purer water stream. RO desalination consumes less energy than thermal alternatives, and recent advances in technology mean that seawater RO plants can now perform with energy requirements of between 3 kWh/m^3 and 4 kWh/m^3 (Elimelech and Phillip 2011).

In addition to energy demand, an important consideration when planning desalination schemes is management of the waste from the desalination process. The salts in the input water are concentrated into a highly saline brine that must be carefully managed in order to prevent environmental damage. Options include ocean disposal, underground injection, crystallisation of salt for beneficial use and landfill. Each of these alternatives has its own management constraints ranging from the protection of marine ecology to identifying suitable geologies, securing markets for salt products and selecting appropriate landfill liner materials. In addition, whenever brine is pumped to a disposal point or heated to cause evaporation, further energy is consumed.

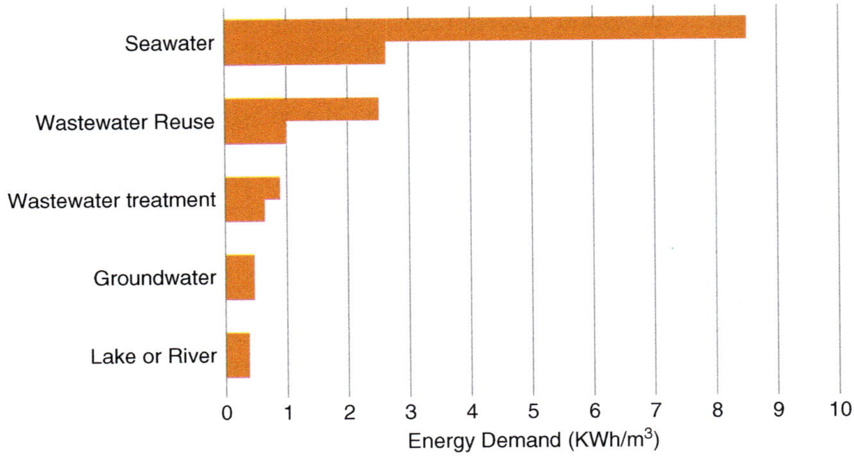

Figure 4.1 Energy required to supply 1 m^3 of potable water.
Note: split bars indicate a range of energy demand for the source in question.
Source: Adapted from UNWWAP (2014).

Table 4.1 Wastewater treatment technologies and their energy constraints.

Treatment	Description	Energy constraints
Primary	Primary treatment consists of removal of solids removal via sedimentation in settling basins with chemical dosing to enhance the process where required. The removed solids are either treated and reused as fertilisers, incinerated, or disposed of in landfills.	Kennedy *et al.* (2008) estimate that the energy intensity of primary wastewater treatment in Australia is around 0.22 kWh/m^3.
Secondary	Secondary treatment removes organic matter and suspended solids from the water via biological treatment. The activated sludge method, which relies on aerobic microorganisms to digest and mineralise organic matter, is the most widely applied technology in many developed countries (SWIE 2013). Other methods include trickling filter (where water is passed over a medium to which bacteria are attached) and anaerobic digestion (where digestion by microorganisms occurs in the absence of oxygen).	ESMAP (2012) calculates an average energy use for activated sludge systems at large wastewater treatment plants in the USA of 0.27 kWh/m^3. For trickling filter systems, the energy use was estimated at 0.18 kWh/m^3, with the difference reflecting the need for aeration via mixing in the activated sludge process. Anaerobic digestion, while not requiring aeration, may require energy to maintain optimum temperatures. This demand can however be offset by reusing the biogas produced during digestion (USEPA 2013).
Tertiary/ advanced	Tertiary or advanced treatment technologies include chlorination, ozone and ultraviolet (UV) light exposure. These technologies reduce the concentrations of contaminants such as nutrients, pesticides, pharmaceuticals and dissolved solids.	Ozone and UV light exposure are the most energy intensive of the tertiary treatments (SWIE 2013); however, energy consumption can be highly variable between plants. UNWWAP (2014) suggest that the UV tertiary treatment process can consume as little as 0.04 kWh/m^3.

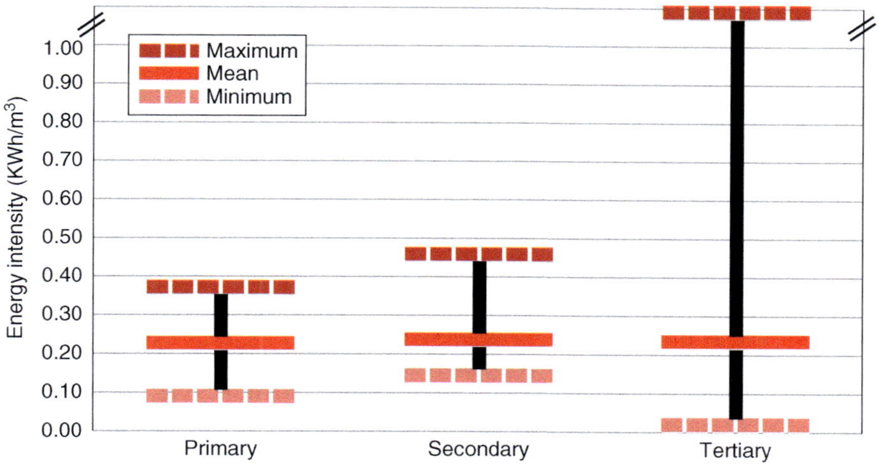

Figure 4.2 Energy intensity of wastewater treatment in Australia.
Source: Adapted from Kennedy *et al.* (2008).

Caffour (2008) estimates that for wastewater treatment plants in the UK, sludge treatment and disposal can consume 40% of the total energy requirements of the plant.

Over time, as sanitation systems have improved, primary treatment systems have gradually been coupled with secondary and tertiary modules. Secondary treatment is now mandated for municipal wastewater treatment plants in the USA, and in northern and central Europe more than 70% of the population is connected to a wastewater treatment plant employing tertiary treatment methods (ESTAT 2013). In less-developed regions however, secondary and tertiary systems are much less common. Even in south-eastern Europe, only 9% of the population has a tertiary water treatment connection (ESTAT 2013). Most communities in developing African countries have no sewage treatment facilities at all (WHO/UNICEF 2008).

Pumping The operation of pumps in pipelines and water treatment plants is responsible for the vast majority of the energy consumed by the water sector (see Figure 4.3). To meet water demand in southern California, water is pumped through 4,800 km of pipelines,

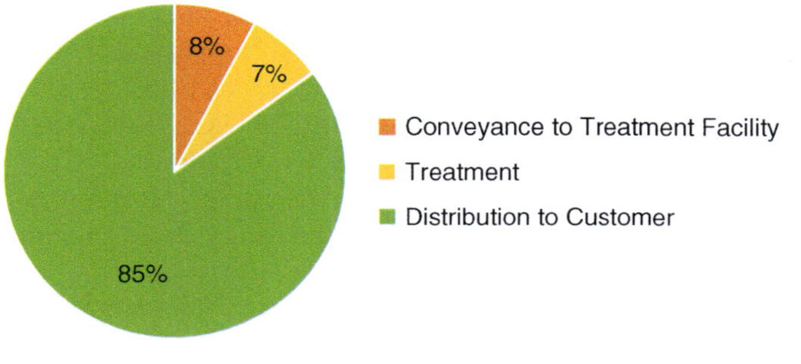

Figure 4.3 Energy intensity of water treatment processes.
Source: Adapted from SWIE (2013).

tunnels and canals with the result that 2.4 kWh of energy is required to import just 1 m^3 of water to San Diego (SWIE 2013). By contrast, in New York the energy intensity of water supply is around 0.7 kWh/m^3 (Copeland 2014). To reduce these energy costs, transmission infrastructure should be carefully located to take advantage of natural land gradients. Minimising the distance between the water source and the ultimate consumer is also an obvious advantage.

In cases where groundwater is the primary water source, pumping costs associated with water extraction can be a significant burden. Aware of the constraints imposed by these costs and in an effort to promote industrial growth, the governments of many developing countries have subsidised the fossil fuels used to pump groundwater. While making the extraction process economically viable for the immediate user, these subsidies can have a devastating effect on groundwater levels. Cheap energy prices supported by subsidies mean that, for example, in India one million new groundwater tube wells are drilled each year (UNWWAP 2014). Between 1950 and 1990 in Beijing, China it is estimated that the groundwater table fell by some 45 m (Kennedy *et al.* 2007).

4.2.2.2 Energy Consumption by the Customer

Once water reaches its intended user, the extent to which that water is heated acts as the primary determinant on energy use. In fact, heating of water represents the dominant energy demand on the entire water management system. Hussey and Pittock (2012) cite studies which estimate that water heating is responsible for one-quarter of residential energy demand in Australia. In the USA, three-quarters of residential electricity is used to heat water, equivalent to around 37% of the nation's entire energy generation (Sanders and Webber 2012). In Sweden the amount of energy used to heat water is 100 times greater than that used during its prior treatment (UNWWAP 2014).

4.2.2.3 Reducing Energy Demands in the Water Sector

The most obvious means of reducing the amount of energy consumed by the water sector is to use less water and to heat less of what is left. By reducing water use, less energy is required to extract, distribute and treat water. The associated infrastructure is also likely to require less maintenance and repair, reducing the energy inputs associated with these activities.

In addition to improved water efficiency, the energy demands of the water sector can also be moderated by reducing leaks within distribution systems and by improving the efficiency of pumping. In 2012/2013, UK water company Thames Water lost 646,000 m^3 of water every day to leaks in its pipelines, equivalent to 25% of its total supply and enough to meet the demand of more than 4 million people (Thames Water 2014a, b).

Lost water is also wasted energy. Byers (2012) calculated that in the UK, a leakage rate of 25% equated to lost energy that could otherwise have been used to power 400,000 homes. Of course, reducing leaks is only economical up to a certain point. However, by accounting for the associated energy savings as well as the water savings, this economic level of leakage can be more accurately calculated.

Given that the energy demands of the water sector are dominated by the pumping of water, energy efficiency initiatives focused on optimising this process are likely to yield the greatest benefit. Potential measures include:

- ensuring that pumps are not oversized for their purpose or of an age beyond their normal design life;
- using specialised, high-efficiency motors and drives; and
- improving monitoring and control systems and using data more effectively in order to optimise operation of the water supply network.

Sustainable groundwater management also has the potential to indirectly reduce the energy intensity of the water sector. By controlling rates of abstraction and thereby maintaining groundwater levels, pumped extraction systems are not required to work as hard to achieve the same rates of water withdrawal.

A number of different innovations show great promise for reducing the energy intensity of the water sector and a selection of some of the most interesting are presented in Table 4.2.

Table 4.2 Opportunities for improved energy and water management.

Harnessing the power of data

Ever-increasing volumes of data are collected during the processes of water extraction, treatment and distribution. Increasingly, flow rates and water demands are continuously monitored, energy consumption rates recorded and chemical inputs tracked in precise detail. By carefully analysing this data for synergies and trends, and then by operating water infrastructure in a precise and responsive manner, so-called smart grids can achieve significant savings in energy consumption, infrastructure maintenance and labour requirements. As analytical techniques improve and as technologies enable increasingly precise control of infrastructure from remote locations, the potential benefits of smart grids will continue to grow.

Efficiencies in desalination

Aforementioned advances in membrane technologies have already achieved substantial reductions in the energy demands of desalination systems. Research and development is now increasingly focused on attempting to harness renewable energies to power the desalination process. Concentrated solar power (CSP), whereby solar energy is captured and redirected by mirrors to heat fluids, is one alternative. The application of CSP is currently significantly more expensive than fossil-fuel-based alternatives; however, initiatives in the Middle East and North Africa are working to improve its economic viability. This region's solar energy potential is estimated to be 1,000 times greater than that of its other renewable energy sources combined (UNWWAP 2014).

New processes of desalination also show promise. One such group of technologies takes the opposite approach to the more common approach of reverse osmosis (RO), utilising rather than working against the natural process of osmosis. In theory, osmosis can be harnessed to concentrate a wastewater stream while simultaneously diluting a higher-strength solution. If the higher-strength solution is seawater, its gradual dilution means that subsequently reclaiming the water by traditional RO now requires less hydraulic pressure and therefore less energy. While osmotically driven membrane systems remain largely unproven at commercial scales, testing does highlight their potential water and energy co-benefits (see Hancock *et al.* 2013 for example).

Natural treatment systems

Natural ecosystems such as wetlands clean water using very little energy. Constructed wetlands aim to replicate the natural treatment process to support, or in some cases replace, stages of the modern wastewater treatment process. While the process of installing a constructed wetland requires energy, once operational its ongoing demands are comparatively small. Furthermore, the biomass of the wetland acts as a carbon sink meaning that constructed wetlands can have net negative carbon emissions (Kalbar *et al.* 2013) while also supporting local biodiversity.

Table 4.2 (Continued)

Energy recovery

The flow of wastewater through pipes represents a potential energy source. SWIE (2013) estimates that future in-pipe hydroelectricity generation from municipal and irrigation pipe networks in the USA could reach 1,100 GWh a year, around 10% of the monthly electricity generation by New York State (USEIA 2014). The city of Vienna already generates a small proportion of its electricity (around 0.2%) from turbines within its water pipeline network (Wien Energie 2014).

The burning of biogas produced when wastewater is digested in anaerobic conditions also represents a potential resource. By burning the methane gas, heat and power can be generated that can in turn contribute to meeting the treatment plant's heating and electricity demands (UNWWAP 2014). Having an on-site power source separate from the grid can also increase the reliability of the treatment plant, especially in countries or regions where power cuts are otherwise common.

4.2.3 Water Use in Energy Production

Energy consumption is a fundamental driver of economic growth and one which underpins the lifestyles of people around the world, particularly in developed countries. Global average per capita energy use is 1.9 tonnes of oil equivalent (toe) a year; however, the national disparities are vast (see Figure 4.4), ranging from more than 17 toe in Iceland and Qatar to less than 0.2 toe in Eritrea (World Bank 2014).

Access to electricity is a particularly crucial metric for human development and is often closely linked to the availability of clean water and sanitation. Globally, per capita electricity use in 2011 was around 3,000 kWh (World Bank 2014). Iceland is the highest per capita electricity consumer at more than 50,000 kWh a year. Haitians use the least, at just 32 kWh per person per year (World Bank 2014). The International Energy Agency (IEA) estimates that in 2009, one in five people lacked access to electricity, almost all of whom lived in developing countries (IEA 2013a).

With a growing global population and a burgeoning middle class, energy demand is expected to increase by more than 30% between 2012 and 2035 (IEA 2013a, BP 2014). Virtually all of this growth is expected to occur in countries outside the Organisation for

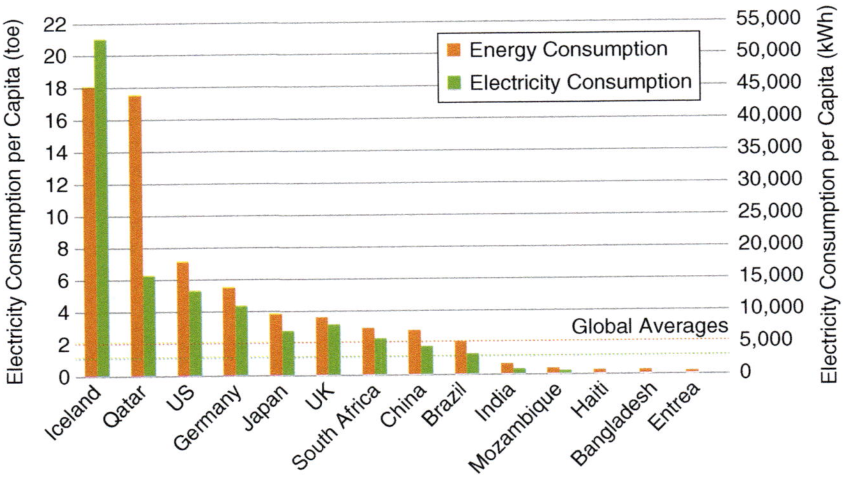

Figure 4.4 Per capita energy consumption in 2011.
Source: Adapted from World Bank (2014).

Economic Co-operation and Development (OECD). Demands from China and India will be particularly important, and by 2050, electricity generation in Africa is expected to have grown sevenfold (BP 2014). In contrast, the World Energy Council (WEC) (2010) predicts that by 2050, European energy consumption will have fallen by 5% and that only marginal growth will have been experienced in North America.

The means by which this increase in global energy demand is met will have profound implications for the planet's environment. The burning of fossil fuels releases gases that absorb solar radiation, retaining heat in the atmosphere and, in turn, influencing global climate patterns. Numerous rounds of international negotiations have focused on reducing the rate at which these gases are emitted into the atmosphere, and almost every country on Earth acknowledges the need to bring these emissions under control. By contrast, the important role played by water in supporting energy production often goes unrecognised. As a consequence, the impacts of energy policies and portfolios on local and regional hydrological cycles are largely ignored, thereby jeopardising not just future water resources but also the security of future energy supply should water management systems fail.

Water is vital to the production of energy from virtually all known sources. From the extraction of primary energies to the generation of final energy types (see Breakout Box 4.3), water quantity and quality are key controls on process efficiency. Water is required to extract oil and natural gas from the Earth, to mine coal and uranium, and to grow crops for biofuels. More water is then required during the generation of electricity in thermal power stations (as a cooling medium), during the production of transport fuels in oil refineries and bioreactors, and during the generation of hydroelectric power (via evaporation from the associated reservoir). Finally, used water is often returned to the environment at a temperature and quality different to that of the receiving water, and frequently in a different location to that from which it was extracted.

These water demands mean that the energy sector is often responsible for a significant proportion of national water withdrawals; in the USA and Europe, this proportion exceeds 40% (Rodriguez *et al.* 2013; Perrone and Hornberger 2014). Globally, some 583 billion m^3 of water is withdrawn from the environment to support energy production and around 11% of this is consumed (evaporated) (IEA 2013a).

4.2.3.1 Water Use in Primary Energy Supply

The volumes of water required to support the extraction processes of our various primary energy sources vary widely. The water demands of a given primary energy source will also typically change depending on its physical characteristics, and on the hydrological and climatic characteristics of the region from which it is being extracted. Figure 4.5 illustrates the relative importance of the various primary energy sources to global primary energy supply in 2011, alongside an indication of the relative consumptive water demands of each.

Breakout Box 4.3 Primary energy sources and final energy types

A *primary energy source* is an energy source that has not undergone any transformative process. Primary energy sources can be renewable, such as wind and solar, or non-renewable such as coal, oil and natural gas.

Final energy is that energy available to a user once a primary energy source has undergone conversion. Final energy types include electricity, gasoline and diesel oil.

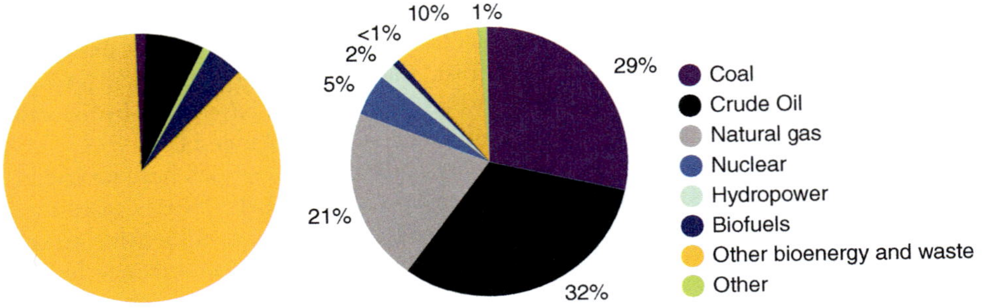

Figure 4.5 Primary energy supply in 2011 (left) and relative water consumption (right).
Source: Adapted from IEA (2013b) and WEC (2010).

Figure 4.5 shows that while the fossil fuels of coal, oil and natural gas made up more than 80% of global primary energy supply in 2011, they were responsible for only a minor component of total water consumption. In contrast, biomass and biofuels provided some 10% of primary energy supply but were responsible for the vast majority of total water consumption. It should be noted that water consumption data related to energy production can vary significantly depending on regional climate, the adopted energy production processes and the method of water accounting (Granit 2011). For example, the WEC (2010) found that water consumption associated with biofuel production varied six-fold depending on location, from 24,000 m^3/1000 GJ in the Netherlands to 143,000 m^3/1000 GJ in Zimbabwe.

Once portrayed as a panacea for reducing human dependency on fossil fuels, production of first-generation biofuels is now increasingly perceived as too water intensive to warrant this status. Dominguez-Faus *et al.* (2009) estimate that in the USA, between 500 litres and 4,000 litres of water are required to produce 1 litre of bioethanol, the most common form of biofuel. Furthermore, in many developing and emerging economies, biofuel cropping is associated with either direct forest loss or with indirect land clearance through the displacement of existing land uses to marginal areas.

Notwithstanding the fact that the fossil-fuel-dominated energy sector is responsible for more than 60% of global greenhouse gas emissions (IEA 2013a), coal, oil and natural gas exhibit relatively low rates of water consumption. The absolute volumes are still notable however, and the water demands of the techniques required to extract these fossil fuels are sufficient to cause impacts to energy production where water is scarce. For example, one technique used to enhance the extraction of oil is to flood the oil reservoir by pumping in water and forcing the crude oil towards production wells where it can be extracted (Lenzen *et al.* 2012). Once used for this purpose, water that returns to the surface must be carefully treated to prevent environmental harm. WEC (2010) estimates that in 2005, global coal, crude oil and natural gas supplies were responsible for the consumption of some 210 billion m^3 of water, greater than the total water consumption of the UK. In fact, Matichich *et al.* (2012) claim that the petroleum industry handles a larger volume of water than it does oil.

4.2.3.2 Water Use in Final Energy Consumption

Of the 13,113 million toe of primary energy supplied globally in 2011, around one-third was used to generate electricity and a further one-third used to produce petroleum products at oil refineries (IEA 2013b). Oil refineries are used to produce a variety of products (light distillates such as gasoline, middle distillates such as diesel and heavy distillates such as fuel oils) and require water for cooling (the dominant use), as boiler feed and as process water. Wu *et al.* (2008) estimate that approximately 43 m^3 of water is required to generate 1,000 GJ of energy in the form of gasoline. Bioethanol production also requires water for grinding, liquefaction, fermentation, separation and dehydration. Wu *et al.* (2008) found the process to be more water-intensive than crude oil refining, estimating that the production of 1,000 GJ of energy consumed 142 m^3 of water.

The volumes of water consumed in oil refineries and bioreactors are small in comparison to those associated with the generation of electricity. Most electricity in most regions of the world (80% when considered globally) is generated by burning a primary energy source, usually a fossil fuel, to generate heat that turns water into steam. That steam then spins a turbine, which in turn drives an electric generator. After passing through the generator, the steam is cooled and condensed before commencing the cycle again. In almost all thermal power plants, the cooling of steam is achieved by heat exchange with water withdrawn from the environment, some of which subsequently evaporates and is therefore consumed. The remainder of the water is typically discharged back into the environment, potentially (if poorly managed) at an elevated temperature that can alter oxygen levels in the receiving waters and cause ecological impacts.

The efficiency of a power plant is typically the primary determinant of direct water demand. The more efficient the power plant, the less heat to be dissipated and therefore the less cooling water required. The resource being burnt and the type of cooling system also have important ramifications for water withdrawal and consumption however. According to the US National Energy Technology Laboratory (NETL), a typical nuclear plant requires an average of just under 3 m^3 of cooling water to produce a single megawatt of energy. In contrast, NETL estimate that coal and natural gas plants require an average of 1.9 m^3 and 0.7 m^3 of water, respectively, to produce the same amount of energy (Reardon 2012).

Breakout Box 4.4 explains how once-through, closed-loop and dry-cooling systems function and shows that the choice of cooling system is not simply constrained by a consideration of water availability. Rather, it requires an understanding of complex tradeoffs between plant efficiency, water withdrawal, water consumption and cost. Table 4.3 summarises this tradeoff and shows that no one cooling system performs best against all criteria.

Breakout Box 4.4 Power plant cooling systems

In *once-through* cooling systems, water passes through the plant in a linear pathway. Large volumes of water are therefore required; however, because only one exchange of heat occurs, most of the water is returned to the environment and only a small amount, around 1%, is evaporated (WEC 2010). Around 40% of thermal power plants in the USA use once-through cooling systems (USDoE 2009).

In contrast to once-through systems, *closed-loop* approaches use a re-circulating cooling cycle. The same water is used for several heat exchanges and so the required water withdrawals are significantly reduced. However, the greater number of heat exchanges means that a higher proportion of withdrawn water is consumed, typically around 90% and more in absolute volume compared to once-through systems. Closed-loop systems are also more expensive than once-through systems (WEC 2010). In the USA, just over half of thermal power plants use re-circulating cooling (USDoE 2009).

The third type of cooling system, *dry cooling*, does away with water altogether, instead using air as the heat transfer medium. Dry cooling systems can therefore decrease the water consumption of a thermal power plant by as much as 90% (Rodriguez *et al.* 2013); however, they are three to four times more expensive than once-through and closed-loop systems (WEC 2010). Dry cooling is also much less efficient than wet cooling and this reduces the overall efficiency of the plant, meaning that more primary energy supply (and any associated water used during its extraction) is required to produce the same volume of energy. The UN suggests that the cost of dry-cooling systems needs to fall by around 50% to make them economically competitive in more regions of the world (UNWWAP 2014). Dry-cooling thermal power plants are currently rare; less than 1% of plants in the USA employ this method (USDoE 2009).

Table 4.3 Thermal power plant cooling system tradeoffs.

Cooling System (fossil fuel thermal power plant)	Water Withdrawal (m^3/MWh)	Water Consumption (m^3/MWh)	Plant Efficiency (Rank)	Capital Cost (% of once through wet system)
Once Through Wet	142.5	0.38	1	100
Re-circulating Wet	4.5	4.2	2	140
Dry	0	0	3	420–560

Figure 4.6 presents the water consumption of various final energy types and indicates the relative importance of primary energy production and final energy generation to the respective totals. It shows that for fuels (from either oil or biomass), the vast majority of water consumption occurs during primary energy production. In contrast, for all forms of electricity, the power generation process is responsible for most water demand. Biofuels are by far the most water-intensive energy source. Only energy generated by wave, wind, geothermal or solar photovoltaic (PV) sources have negligible water demands.

4.2.3.3 Power Station Vulnerabilities Related to Water

Power stations use a variety of fuels: gas, coal, nuclear, wind, hydropower, biomass and oil. Large volumes of water are used during the power production process, mostly to prevent power stations from overheating; it is this need to cool, usually using water, that generates a number of vulnerabilities as follows:

- **low flows**: water for cooling cannot be abstracted when flows drop below the level of the water intake pipe. Lowering the pipe intake is an expensive option and increases the risk of pulling in sediment that can damage the cooling system. Depending on the water-abstraction permitting regime in the affected region, the amount that can be abstracted can be capped using flow-related licence conditions;
- **heat waves**: power stations operate more efficiently when coolant water is lower in temperature. Heat waves can increase the temperature of fresh or seawater to the point where it becomes unusable as a coolant; and
- **heated effluent**: inevitably, the outflow water will be warmer than the inflow water. However, if that inflow water is warmer due to the external temperature then the subsequent outflow temperature can be even higher, potentially exceeding environmental

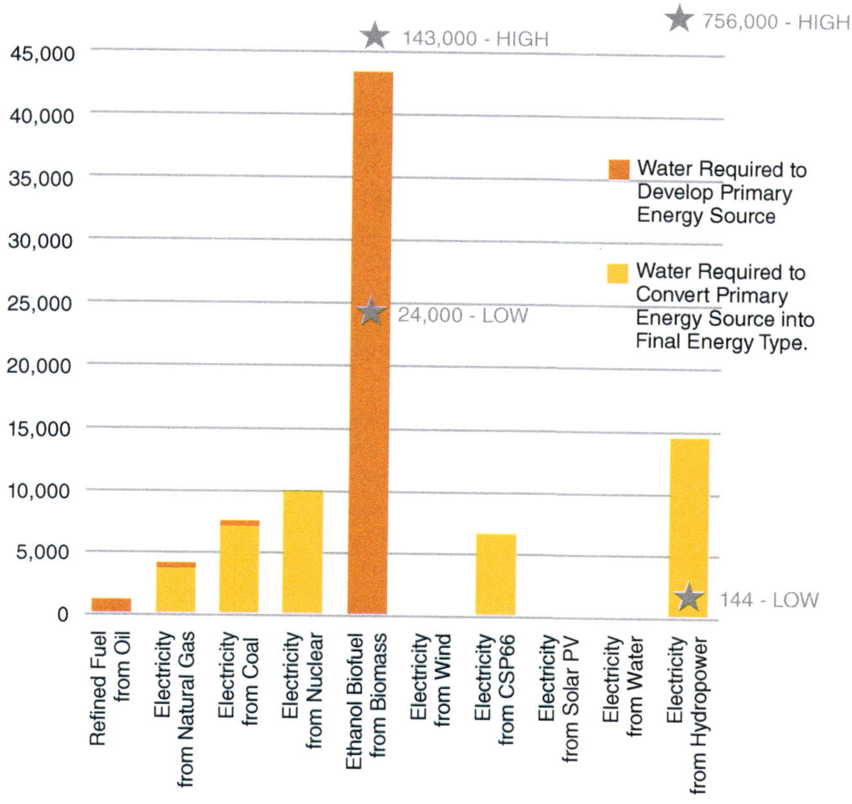

Figure 4.6 Water consumption of different energy types.
Note that in all examples shown wet closed-loop cooling systems are assumed.
Source: WEC (2010), except for water required to convert oil to refined fuel and ethanol biofuel from biomass for which data are sourced from Wu *et al.* (2008).

compliance levels and forcing the power station to shut down. Where present, regulations generally prohibit overheated discharges as these can trigger algal blooms, reduce dissolved oxygen and threaten aquatic life.

There have been many examples of major power stations either being forced to reduce their output, or being forced to shut down altogether, due a lack of water for cooling. As climate change leads to more frequent and more severe droughts and heat waves, and as populations continue to increase, the strain on power generation and the associated water resources will continue to become more acute.

4.2.3.4 Hydropower

Hydropower schemes are responsible for generating approximately 15% of global electricity (IEA 2013b) and some 86% of renewable electricity (WEC 2010). In 2009, 35 countries obtained more than half their electricity from hydropower; for seven countries, including Nepal and Paraguay, the proportion was virtually 100% (IEA 2012).

The majority of hydropower schemes work by damming a natural watercourse and controlling the release of the stored water to drive a turbine and electric generator.

Because this process can be initiated more rapidly than any other power generation mechanism, hydropower can build resilience and flexibility into electricity networks. Hydropower systems can also be operated as an energy store. By pumping water up to the dam reservoir, energy is effectively stored for later use as and when it is required. Because of this pumping, however, such systems are in fact net consumers of energy (UNWWAP 2014).

In a limited number of settings, where the prevailing hydrology permits, run-of-river hydropower schemes (with no dam but often a weir) may be viable. Such systems rely on natural river flow to drive the turbines and so do not require upstream storage. Globally, small hydropower schemes (which include run-of-river schemes) make up around 6% of total hydropower capacity (IREA 2012); however, because the schemes allow little control over the flow of water, electricity generation is less responsive to demand.

The damming of large bodies of water to support hydropower generation promotes water consumption through evaporation. The rate of evaporation varies depending on climate and reservoir geometry and can range from 0.04 m^3/MWh to as high as 210 m^3/MWh (WEC 2010). Hydropower schemes therefore have the potential to remove significant volumes of water from the local hydrological cycle; however, it should also be remembered that hydropower reservoirs may support a number of other important societal functions such as irrigated agriculture, navigation, recreation and flood risk management.

4.2.3.5 Emerging Primary Energy Sources

The last two centuries of economic growth achieved by many now-developed countries relied heavily on the availability of cheap energy generated by the burning and refining of fossil fuels. Current efforts to decouple the consumption of oil, coal and gas from economic growth are driven by two key factors: the scarce and ultimately finite nature of fossil fuels; and the pressing need to limit human-induced climate change. In addition, governments are also required to juggle a number of other energy policy objectives that include ensuring energy security for their citizens, supporting the growth of key industries and maintaining energy affordability. Unfortunately, sound stewardship of water resources is rarely included in these objectives and the impact of emerging energy sources on water resources is therefore often ignored. The following subsections consider a few of the most pressing potential issues.

Unconventional Fossil Fuels Breaking the link between economic growth and fossil fuel consumption requires strong political will, business leadership, time and significant investment. These attributes are rare to find in combination, meaning that global reliance on fossil fuels persists and is slow to change. Furthermore, as fossil fuel resources become more scarce, their market prices increase, in turn making unconventional resources (those types that are harder to extract) more profitable. Recent years have witnessed significant growth in the exploitation of unconventional fossil fuels. Shale gas is perhaps the most well-known; however, a broad variety exist. One characteristic common to most is their increased reliance on water for extraction in comparison to conventional fossil fuels (see Table 4.4).

Extracting unconventional oil and gas relies on complex but proven processes that typically both require and produce significant volumes of water. Often, a high-pressure

Table 4.4 Unconventional fossil fuels and their water demands.

Fuel	Description	Water demands
Shale gas	Natural gas found in low-permeability shale formations. Resources are globally widespread; however, significant commercial development is underway in only a few countries.	Shale gas extraction relies on hydraulic fracturing (pumping water, sand and a small proportion of chemicals underground at high pressure in order to force the target formation to fracture). The volume of water used varies significantly between wells depending on geology. Between 60% and 80% of the injected fluid returns to the surface, where it may require treatment before reuse or disposal.
Coal bed methane	Natural gas held in coal seams by water pressure. Several large CBM projects are in operation in Australia. Globally, the largest CBM reserves are found in the former Soviet Union, Canada, China, Australia and the USA.	CBM extraction involves pumping water out of the coal seam to the surface in order to allow the gas to flow. Not all CBM wells require hydraulic fracturing. CBM production therefore requires less water than for shale gas, but generates significant volumes of water that may require treatment before reuse or disposal. On average, each CBM well in Queensland, Australia produces 20,000 litres of water per day (CSIRO 2012).
Tight gas	Natural gas found in low-permeability formations other than shale. Large reserves are found in Russia, the USA, Australia and Venezuela.	The process of extracting tight gas is very similar to that required for shale gas. Hydraulic fracturing is necessary and a proportion of the fracturing fluid returns to the surface.
Oil Sands	Oil sands comprise mixtures of sand, clay, water and a dense and viscous form of petroleum known as bitumen. Canada has approximately three-quarters of known global deposits and is currently the only country commercially exploiting these resources.	A common means of extracting the oil is to pump steam into horizontal wells drilled into the oil sands. The US Bureau of Land Reclamation (2012) estimates that the production of one litre of oil from oil sands uses between 2.3 litres and 5.8 litres of water.
Oil shale	A sedimentary rock containing kerogen that has not undergone enough geologic pressure or heat to become conventional oil. The vast majority of global resources are located in the USA.	Oil shale can either be extracted and processed at the surface or processed underground. Mining of oil shale requires water and the US Bureau of Land Reclamation (2012) estimates that one litre of oil produced from oil shale uses 2.6–4 litres of water.
Shale oil	Crude oil trapped within tight, impervious shale formations.	Shale oil is found in similar situations to shale gas; the required extractive processes and associated water demands are therefore comparable.
Methane hydrates	A crystalline solid consisting of a methane molecule surrounded by a cage of interlocking water molecules. Methane hydrates are formed at high pressures and low temperatures with resources found in seafloor sediments such as those off the coasts of Japan and Canada.	Little is known about how to extract methane hydrates at commercially viable rates and test drilling has taken place in only a few countries. Equally little is known about the potential environmental risks, particularly associated with the potential release of methane during the extraction process.

mixture of water, sand and a minor component of chemicals is forced into underground rock formations via wells to cause it to fracture and, in turn, to release its gas or oil. This process of hydraulic fracturing has been practised for many years; however, public understanding of the process and its associated risks is limited. This lack of understanding often promotes polarised opinion and conflict between stakeholders.

In instances where produced water from unconventional petroleum development is well managed, it can provide a valuable resource to the local community. For example, produced water associated with the coal bed methane (CBM) industry in Queensland, Australia is treated and then used to irrigate crops, as processed water in industrial applications, and is applied to suppress dust on unpaved roads. Produced water associated with shale gas operations in the USA is often treated and then reused to hydraulically fracture the next well. Where surface water management is poorly performed however, leaks and spills can occur, in turn jeopardising environmental assets and human interests.

It should be remembered that the volumes of water required to extract unconventional fossil fuels are relatively small when considered at a national or regional scale. For example, the annual water consumption of the shale gas industry in Pennsylvania represents less than 0.25% of the state's total water consumption (Accenture 2012). However, when considered on a local scale and over short timeframes, the relative volumes of water used can be significant; conflicts with other land uses, notably agriculture, are therefore not uncommon. Shale gas developments in Europe as well as CBM projects in Australia and oil sands schemes in Canada are all characterised by water resource constraints that require careful management.

Biofuels Between 2000 and 2011, global production of biofuels increased more than six-fold to almost 230 million litres a day (Perrone and Hornberger 2014). This growth was largely driven by government biofuel mandates that existed in at least 60 countries. Given that the water footprint of energy from biomass is significantly larger than that of fossil fuels (refer back to Figure 4.6), it is clear that these mandates, unless carefully planned and implemented, have the capability to significantly constrain water resource availability. In fact, in 2011, 10 organisations including both the International Monetary Fund (IMF) and World Bank called on G20 countries to scrap biofuel mandates and subsidies altogether (Oxfam 2012).

The large water demand of biofuel production is almost entirely reflective of water consumed during crop growth. For example, it takes 14,000 litres of water on average to produce 1 litre of biodiesel from soybean or rapeseed (Gerbens-Leenes *et al.* 2008). The inherent energy content of green plants is also relatively low and so large quantities of crop are needed to make small quantities of fuel. This means that cropping of biofuels is unsuited to those localities with low RWR. Conversely, in locations were the RWR is large, biofuel cropping has the potential to be conducted in an entirely sustainable manner. Of course, other resource constraints affect this debate, not least the need for large areas of land and the potential for the displacement of crops for human consumption.

In any discussion regarding the environmental impact of biofuels, an important distinction must be made between first-generation and advanced variants (see Breakout Box 4.5). First-generation biofuels use the starch, sugar or oil fraction in the crop (typically corn, wheat or sugar cane) to produce the energy. These components of the crop are what provide the nutritional value in our foods, and so first-generation biofuels

Breakout Box 4.5 First-generation and advanced biofuels

Biofuels can be broadly divided into two categories – first-generation and advanced – on the basis of the source from which the fuel is derived. First-generation biofuels are produced directly from food crops, whereas advanced biofuels are not sourced from food crops unless the crop has already served its food purpose.

Advanced biofuels are also sometimes referred to as either second generation or third generation. Second-generation biofuels include grasses, seed crops, waste vegetable oil, agricultural and forestry residues (lignocellulosic feedstocks) and municipal solid waste, while third-generation biofuels represent those derived from algae. Advanced biofuels typically require different and/or additional processing in comparison to first-generation variants.

Advanced biofuel applications are growing in number and appear to have the potential to support a transition away from petroleum-based fuels, not least because the road transport system can shift to biofuels without insurmountable system changes (most schemes are typically at the pilot stage, however).

Microalgae, an advanced biofuel, have much higher oil yields than terrestrial biomass and so provide a much larger return on energy per hectare, while also demanding less water (Amaro *et al.* 2012). As of the mid 2010s, while production of biofuels from microalgae has been shown to be technically feasible, the high costs of the production process means that the economic viability of the process has yet to be proven (Amaro *et al.* 2012).

displace human food crops. In contrast, some advanced biofuels are produced from crop residues. This means that they have the potential to support, rather than compete with, the production of food. They also share the water demands as well as the demands for land and chemical fertilisers. This means that the water footprint of some advanced biofuels is far less than that of first-generation alternatives (see Figure 4.7).

Hydrogen Hydrogen represents another emerging technology towards which early optimism is being directed. The IEA (2006) go so far as to claim that hydrogen could

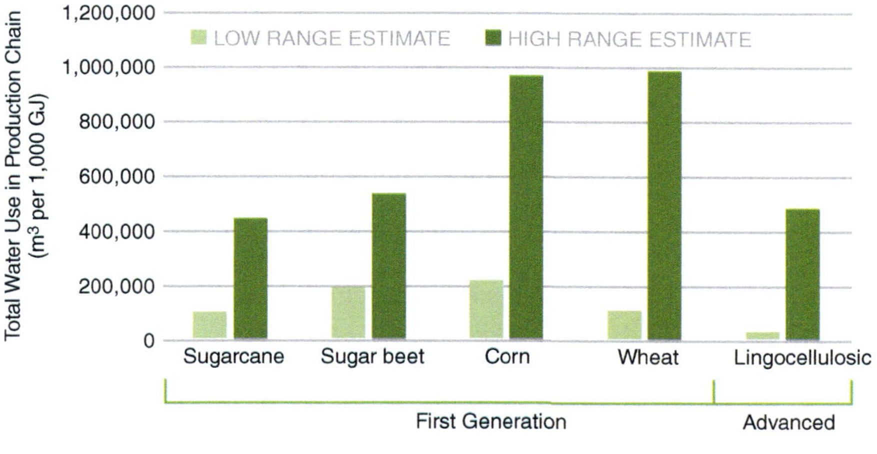

Figure 4.7 Water consumption for first-generation and advanced biofuels.
Source: Adapted from UNWWAP (2014).

become the dominant component of our future energy systems. Hydrogen is a versatile energy carrier that is capable of powering almost all end-use needs. It also exists in many organic compounds and can be separated from them in a number of ways, including from:

- hydrocarbons through heating;
- water via the process of electrolysis; and
- sunlight and water via biological organisms.

These techniques have the potential to produce essentially unlimited quantities of hydrogen that can be used to generate energy in one of two ways: to generate heat through combustion; or to produce electricity via chemical reactions in a fuel cell. In both cases, when pure hydrogen is used, the main by-product is water vapour. If the hydrogen itself can be produced from renewable means, the total system is therefore environmentally benign. Current applications of these technologies are limited and will take time to refine. Reducing costs, scaling up hydrogen production systems and achieving efficient storage in lightweight, transportable systems are key research priorities.

If and when storage systems allow, hydrogen fuel cells could provide a solution to the problem faced by intermittent renewable energy sources such as solar PV and wind. The proliferation of these energy forms is currently constrained by the start–stop nature of their weather-dependent energy production and the fact that storage technologies are currently expensive and energy-intensive to deploy at scale (Carbajales-Dale *et al.* 2014). One means of more effectively harnessing solar PV and wind could be to use the electricity they generate to, in turn, generate hydrogen. If stored efficiently, hydrogen fuel could then be used to service a broad range of energy demands.

The intermittency of solar PV and wind can also be addressed to some extent by the careful coupling of other renewable energy sources. Through improved data management and system operation, a portfolio of renewable energy sources might be able to overcome their singular weaknesses. Improved collation, analysis and application of data are therefore an important means by which energy and water security can be enhanced.

4.2.3.6 Future Energy Portfolios

The make-up of future energy portfolios will be decided by government responses to a number of competing influences, including the following:

- whether or not economic growth can be decoupled from increasing energy consumption;
- the extent to which individual countries and the international community reduce greenhouse gas emissions in order to limit global climate change. In this and many other respects, the choices made by a few countries will be key. For example, China used as much coal in 2010 as the rest of the world combined (IEA 2011);
- whether governments choose to pursue energy self-sufficiency or seek to use trade as a means to achieve security of energy supply for their citizens; and
- whether decision makers choose or are forced to consider the full range of natural resource inputs required to support energy production (such as, but not limited to, water) and the impacts of energy policy on the health of these resources.

In order to create a future energy system whose impacts on the environment are sustained within safe limits, it is clear that the water management requirements of energy production need to be taken into account. To do this effectively requires research into new technologies that are mutually beneficial to the management of multiple resources. Table 4.5 highlights a selection of past and recent innovations, describes their implications for water resource management, and explains the constraints to their more widespread application.

The IEA estimates that by 2035, global final energy consumption will have grown by more than 30%, with half this growth coming from electricity generation and around 90% expected to occur in emerging, typically Asian economies (IEA 2013a).

Table 4.5 Innovations in energy and water management.

Innovation	Advantages	Constraints
Waterless fracturing Replacement of water in the hydraulic fracturing process. Approaches being considered include the use of gels and gases.	• Less water required for fracturing. • Less water flowing back to the surface that otherwise requires management.	• Waterless fracturing techniques are currently more expensive than water-based alternatives, and this may limit their uptake to regions of severe water scarcity.
Carbon capture and storage CCS systems use a combination of technologies to capture carbon dioxide before it is emitted to the atmosphere. The captured carbon is typically injected into deep geological formations for long-term storage. In some cases, the carbon is reused in applications that include enhancing oil extraction.	• Integrating CCS into thermal power plants can reduce greenhouse gas emissions by between 80% and 85% (WEC 2010).	• CCS increases the water demands of a thermal power plant by up to 90% (WEC 2010). • CCS also reduces the efficiency of the plant, meaning that more fuel is required to produce a given amount of energy. • The first commercial-scale CCS power plants are only expected to come online in the mid-2010s, in part due to their high capital and operating costs.
Wastewater for cooling Using wastewater, treated as required, for cooling of thermal power plants. This process is already employed in around 50 power plants in the USA, including the largest nuclear power plant in the country (UNWWAP 2014).	• Reduced freshwater demand. • A major advantage of using wastewater is that it is an abundant resource in cities that are also the dominant source of energy consumption. • In developing countries that currently lack adequate sanitation, this represents an excellent opportunity to integrate wastewater treatment and energy infrastructure.	• Corroding substances present in wastewater typically have to be removed prior to use and this increases the cost of the cooling system.

(Continued)

Table 4.5 (Continued)

Innovation	Advantages	Constraints
Combined heat and power (CHP) plants The waste heat from the energy production process is used to warm homes and businesses.	• Dual functionality (heat and power) is more efficient than standalone power and heating systems. The combined efficiency can be as high as 90% compared to the 55% more typically achieved by conventional power plants (UNWWAP 2014). • CHP plants rely on existing technologies and are already in operation around the world. For example, 50% of Denmark's power is produced at CHP plants (Rodriguez *et al.* 2013).	• The benefits of CHP plants are optimised when they are located close to the demands for heat and power. • CHP plants require high initial capital investments and pay-back times may be lengthy. • CHP plants require complex operating procedures in order to balance the demands for heat and power.
Combined power and desalination plants Hybrid desalination plants that integrate desalination with thermal power generation. The waste heat from the power plant is used as the heat source for the desalination process.	• Simultaneous production of water and energy. • The integrated system is more efficient than that achieved when the two technologies operate in isolation.	• Demand for energy will vary seasonally, whereas demand for water is relatively constant. This makes efficient operation of the plant difficult to balance.

Figure 4.8 illustrates how the projected mix of sources used to generate electricity is predicted to change between 2011 and 2035. It shows a decrease in the relative importance of coal and an increase in the importance of renewables, particularly wind and solar PV. Geothermal energy remains significantly underutilised, although in some countries it can be an important component of the energy system. In Iceland for example, 25% of the country's electricity needs and 90% of its heating needs are met by geothermal energy (Bertani 2010). The UN highlights the benefits of this energy source, drawing attention to the fact that it is climate independent, produces minimal or near-zero greenhouse gas

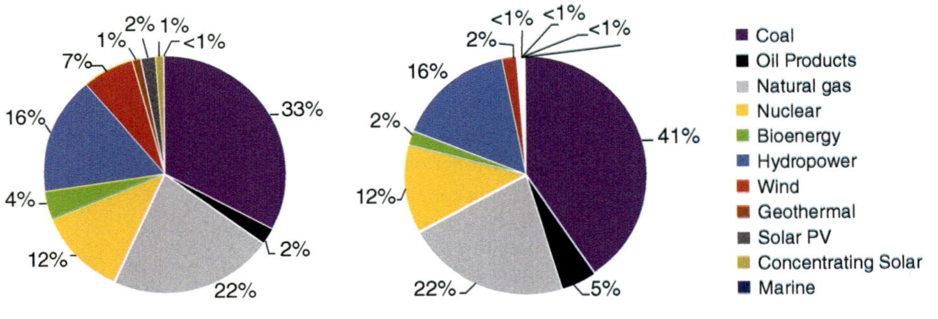

Figure 4.8 Electricity generation by fuel as percentage of total in 2011 (left) and 2035 (right).
Source: Adapted from IEA (2013a).

emissions, does not consume water and exhibits infinite availability over human timescales (UNWWAP 2014).

Table 4.6 considers some of the potential water resource implications of the IEA (2013a) energy projections.

Table 4.6 Energy predictions for 2035 and their implications for water resources.

IEA New Policies Scenario 2035 predictions as compared to 2011	Implications for water resources
Electricity	
• Global electricity demand increases by two-thirds, largely as a result of growth in non-OECD countries. • All energy sources see use for electricity generation grow in absolute terms. • Under a business-as-usual scenario, the number of people without access to electricity in 2030 may rise to 1 billion (UNWWAP 2014).	• Globally, water consumption by the electricity generation sector is expected to more than double in the next 40 years (WEC 2010). • Climatic trends towards increasing water temperatures and decreasing water availability will pose increased risks to electricity generation in many regions (Rodriguez *et al.* 2013). • Half of planned thermal power plants in southeast Asia are located in areas likely to face water shortages in the future (Rodriguez *et al.* 2013). • Only Latin America appears to have sufficient RWR to meet increased water demand from the energy sector (WEC 2010).
• Proportion of global electricity generation from fossil fuels falls from 68% to 57%. • Role of unconventional gas in electricity generation increases in importance.	• The extraction of unconventional gas both requires and produces water. On a local scale, the quantities may be significant. • Requirements for the treatment of produced water may increase local energy demands.
• Environmental concerns are likely to lead to more thermal power plants adopting closed-loop cooling systems (WEC 2010). • By 2035, only 1% of global fossil fuel power plants are expected to be equipped with CCS.	• As compared to once-through cooling systems, closed-loop systems withdraw 95% less water but consume between 10 and 15 times more (WEC 2010). • CCS can reduce greenhouse gas emissions by up to 85%; however, water consumption also rises by between 50% and 90% depending on the type of thermal power plant (WEC 2010). • Current CCS technologies reduce the efficiency of the power plant meaning that more fuel has to be burnt to generate the same amount of electricity (WEC 2010).
• Renewable energy sources account for nearly half the total increase in electricity generation in 2035 and around 50% of this increase comes from solar PV and wind. • Global wind and solar PV capacities are currently growing at 20% and 60% a year respectively (Carbajales-Dale *et al.* 2014).	• Solar PV and wind have negligible requirements for water (WEC 2010).
• Absolute electricity generation from bioenergy triples. The share of bioenergy from traditional biomass falls from 57% to 37%.	• Bioenergy has the potential to consume significant volumes of water and is therefore hydrologically suited to only a few regions where RWRs are sufficiently abundant.

(Continued)

Table 4.6 (Continued)

• Absolute electricity generation from hydropower increases by two-thirds; its share of total electricity generation remains at around 16%.	• Hydropower schemes may be jeopardised by changing surface water flows, particularly in glacier-fed regions (Rodriguez *et al.* 2013).
• Nuclear electricity generation doubles and maintains its 12% share of total electricity generation.	• Extraction and processing of uranium requires less water per unit of energy than other fossil fuel primary energy sources (WEC 2010). • The cooling requirements of nuclear power plants are greater than those for plants using fossil fuels (WEC 2010).
Transport sector	
• Energy consumption by the transport sector rises by a third. The sector's proportion of total final energy consumption remains at around 27%.	• Emerging technologies such as hydrogen fuel cells offer significantly reduced water demands but require many more years of research, development and demonstration before they can be proven viable.
• Production of conventional crude oil fails to keep up with global demand. Its share of total oil production falls from 80% to 65%. • The gap is likely to be filled by an increasing reliance on unconventional oils.	• Oil production from unconventional sources uses around three times more water than oil production from conventional sources (WEC 2010).
• By 2035, biofuels will meet 8% of road transport fuel demands, up from 3% in 2011. Consumption of biodiesel in road transport triples. • Advanced biofuels make up 20% of all biofuels in 2035, up from 1% in 2011.	• First-generation biofuels have the highest water demands of any energy source. • Lignocellulosic-based biofuels that rely on agricultural crop residues (one type of advanced biofuel) have much lower water demands than first-generation biofuels.

Source: Adapted from IEA (2013a), unless otherwise stated.

4.3 Urbanisation

4.3.1 The Rise of the City

Urbanisation is one of the overriding demographic trends of the last 100 years. In 1900, just one in ten of the global population lived in cities (Grimm *et al.* 2008). By 1950, this proportion had grown to one-third (Grubler *et al.* 2012), by the early 2010s it was more than a half and by 2025 it is expected to exceed 60% (UNISDR 2012).

Recently observed and future projected urban growth is overwhelmingly the result of the net movement of people from rural to urban areas, a pattern which was first witnessed in Europe and North America in the nineteenth and early twentieth centuries. In these regions, between 75% and 80% of all citizens now live in urban areas and the extent of further urban growth is expected to be limited (World Bank 2012). In some instances, such as in the case of Detroit, when a locally important employment sector declines, urban populations may even fall (Sattherthwaite *et al.* 2010).

In contrast, rapid urban growth is predicted for Asia and Africa. In these regions, urbanisation is currently below 40% of the national population (World Bank 2012);

however, the urban population is expected to double by 2030. Such a scenario would mean that 80% of the world's population lives in developing country cities (UNFPA 2007). Figure 4.9 shows the growing dominance of urban populations over the last 50 years. It highlights the slow rates of recent urban growth in Europe and North America and the comparatively rapid trends seen in Asia, the Middle East, North Africa, Latin America and the Caribbean.

The predicted dynamics of global urban growth are also expected to be accompanied by the rapid proliferation of medium-sized cities, those home to between 5 million and 10 million people. Between 2005 and 2015, the number of people living in cities of this size is expected to grow by 34% (UN 2006). Although increasing at a lesser rate, the number of megacities (agglomerations with populations in excess of 10 million people) will continue to rise. In 1970, New York and Tokyo were the only global megacities. By 2000, this number had grown to 18 and by 2025, it could be as high as 37, with 12 located in China and India alone (Lucci 2014).

Many of the world's rapidly growing cities are already experiencing challenges related to water. Urban vulnerabilities related to flooding, sea-level rise or an inability to provide widespread access to water are common. For example, in Kathmandu, Nepal only one-third of the population receives a domestic water supply. Furthermore, a lack of water for cooling means that power plants often function poorly, leading to power cuts that can last for up to 14 hours a day (UNWWAP 2014).

Historically, urbanisation accompanied economic industrialisation. In pre-industrialised eras, a high proportion of the population maintained a direct dependence on the land for basic human resources (e.g. water, food, land and energy). This encouraged communities to remain small and dispersed in order to limit competition. Put another way, in rural settings there was, and often remains today, a direct relationship between ecosystem services and human well-being (da Silva *et al.* 2012).

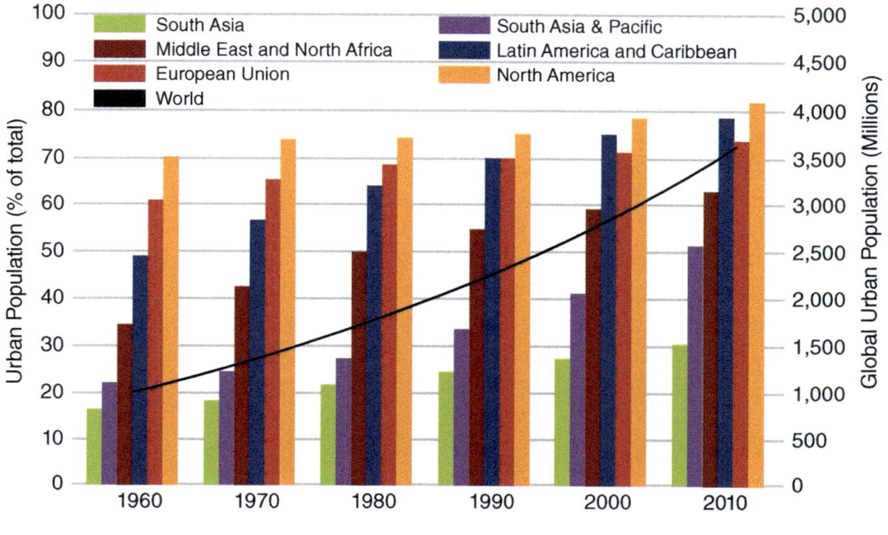

Figure 4.9 Urban population as a percentage of total population.
Source: Adapted from World Bank (2014).

As industrialisation took hold, advances in infrastructure made the resources that humans depend on more mobile, thereby allowing communities to congregate and grow. In turn, the density of people in these new population centres fuelled innovation and created opportunities for rapid economic advancement, largely predicated on the exploitation of fossil fuels. Seeing the benefits of urban living, more and more people were encouraged to move out of agriculture and into more profitable off-farm activities. A conveyor belt of rural to urban migration was therefore established, and a fundamental shift in the structure of economies was experienced as a result.

Because of their high population densities, urban planners can often achieve substantial efficiencies in the provision of many of the resources and services essential to public health and happiness. These include water supply and wastewater management as well as food distribution, energy supply, the provision of healthcare and education, and access to social and cultural pursuits. Managed effectively, with clear strategies and integrated policies, urbanisation can therefore allow governments to sustain high living standards with lower environmental footprints than would otherwise be required in rural settings.

Notwithstanding these advantages and opportunities, urbanisation also concentrates the strain that human populations place on the environment. Cities are epicentres of consumption. Covering just 2% of the global land area, they are responsible for 60% of all residential water use (Grimm *et al.* 2008) and between 55% and 80% of energy consumption (Grubler *et al.* 2012). Furthermore, the population density of urban centres makes them highly vulnerable to natural disasters and potential future changes in climate that are likely to promote increased flooding, heat stress and water scarcity (C40 Cities 2014). Cities can therefore be considered both a key driver and primary impact bearer of climate change.

A city's high demand for resources also means that it generates vast quantities of waste. Urban areas are responsible for around 80% of all carbon dioxide emissions (Han *et al.* 2012) and around 1.3 billion tonnes of municipal solid waste a year, equivalent to 1.2 kg per person per day (Hoornweg and Bhada-Tata 2012). As the local natural environment loses out to man-made infrastructure and impermeable land cover, its ability to effectively assimilate these wastes reduces, increasing the risk of further environmental degradation. Storage of water in urban aquifers, of heat retained in the urban environment, of toxic materials in building stock and of nutrients in urban wastes are all therefore important management issues for cities (Kennedy *et al.* 2007).

When infrastructure planning fails to account for potential urban futures, the opportunities for delivering service provision through economies of scale are more likely to be forgone and higher levels of resource access disparity more likely as a result. Short-sighted planning can also lock in inflexible infrastructure solutions, thereby preventing effective response to future change. Figure 4.10 shows the percentage of urban areas serviced by improved water sources and sanitation in select global regions (see Breakout Box 4.6 for definitions). The figure highlights the comparatively poor current performance of those Asian and African regions expected to witness the greatest future urban growth, particularly with regards to the provision of safe sanitation.

4.3.2 Peri-Urban Communities

Peri-urban areas (a loosely defined region on the edge of a city) represent a particular challenge for urban planners. While suburban environments in developed countries are typically regarded as privileged settings (and usually grow as a result of the expansion of

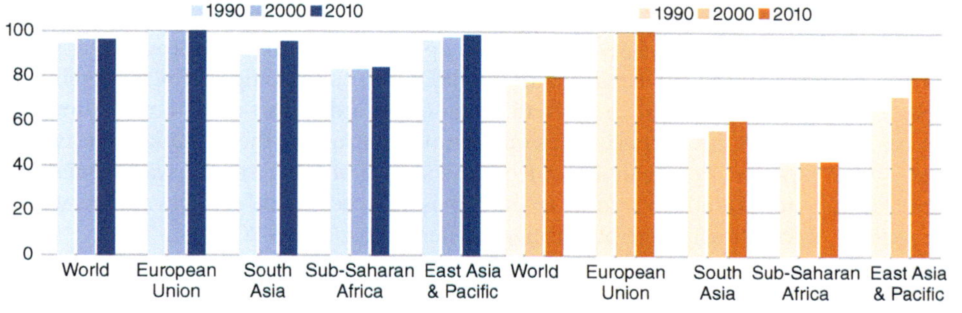

Figure 4.10 Percentage of urban population with access to improved water sources (left) and improved sanitation facilities (right).
Source: Adapted from World Bank (2014).

the middle class outwards from the crowded city centre), in developing and emerging economies, poorly planned urban sprawl can lead to the proliferation of slums. Most shantytowns and illegal settlements are located in peri-urban areas, and often these peripheral communities develop on vulnerable or low-quality land such as that used by the city to dispose of its waste.

In developing countries, peri-urban areas are often inadequately serviced by water and sanitation systems, energy infrastructure and food distribution networks. Their uncertain legal status (many homes are built haphazardly and without planning permission) also acts as a major disincentive to investments that might otherwise increase living standards. Without secure land tenure, any investment is exposed to significant risk, meaning that financial institutions may be less inclined to issue loans or be less capable of collecting fees (Marshall *et al.* 2009). This limits access to sources of safe water and, as a result, marginalised individuals or communities are often forced to rely on poor-quality water supplies.

In a study of cities in Egypt, Venezuela, India, Tanzania and Mexico, Allen *et al.* (2006) found that the water needs of the respective peri-urban populations were met through a broad array of disjointed and unconventional mechanisms that included informal operators, privately operated wells, gifts, rainwater harvesting and illegal connections. This lack of formal service provision means that developing country peri-urban communities are often forced to spend more than their urban counterparts to obtain safe water. In their study, Allen *et al.* (2006) found this cost disparity to be as high as 1,200%.

Working out how to effectively and efficiently support peri-urban communities is a key challenge for urban planners, and one that requires the careful piecing together of elements

Breakout Box 4.6 UN definitions of improved drinking water and sanitation

The UN defines an *improved drinking water source* as one that, by nature of its construction or through active intervention, is protected from outside contamination, in particular from contamination with faecal matter.

The UN defines an *improved sanitation facility* as one that hygienically separates human excreta from human contact.

of rural, regional and urban planning (Marshall *et al.* 2009). A number of municipalities (such as Mumbai in India) have invested in slum upgrade programs to improve water supply, road, drainage and waste management (UNISDR 2012). In doing so, the vulnerability of the associated communities to a broad range of hazards (man-made and natural) is reduced. Managed effectively, peri-urban environments also have the potential to deliver a range of vital functions to urban areas, from the supply of food, energy, water, building materials and other essential resources to the provision of ecological services such as wildlife corridors, microclimates and buffers against flooding (UNFPA 2007).

Many of the problems experienced in urban areas reflect a complexity of resource and capacity constraints, inadequate government policies and a failure to acknowledge and adequately plan for urban growth. In order to improve the sustainability of global water resource exploitation, decision makers and planners will need to re-focus their attention not just towards urban centres, but also towards the increasingly important relationship between the city and its immediate and more distant hinterlands.

4.3.3 Traditional Approaches to the Management of Urban Water Supply and Demand

The traditional approach to urban water and wastewater management has been to centralise treatment processes at a restricted number of facilities and to connect these facilities to sources of freshwater, to customers and to disposal points via an increasingly complex network of pipes. This centralised and linear approach to water service provision is well-suited to densely populated urban centres and has been highly effective at improving public health. However, as urban populations continue to grow, urban sprawl proliferates and the impacts from climate change intensify, the constraints of a linear approach to water and wastewater management are becoming increasingly apparent. These include:

- high-energy consumption due to the heavy reliance on extensive pumped pipeline networks to connect centralised treatment facilities;
- a natural tendency to only consider policies aligned to centralised water management, which can prevent and the emergence of potentially more efficient alternatives; and
- an inherent failure to acknowledge the value of the resources embedded in wastewater, for example nutrients, and thus a foregoing of their potential benefits.

These constraints are compounded by the increasing profligacy of urban water use. Figure 4.11 shows that in countries where the rural population is in the majority, domestic water use is typically less than 150 L/capita/day (data from UNDP 2006 and World Bank 2014). In contrast, most countries where more than 70% of the population live in urban areas tend to experience domestic water consumption rates that exceed 300 L/person/day. In the USA for example, urbanisation is around 80% and domestic water consumption exceeds 550 L/person/day. It should be noted that in some cities enhanced consumer awareness and the adoption of water saving measures can limit and eventually reduce domestic water consumption. In New York for example, water consumption fell by 5% between 2002 and 2006 (UNWWAP 2014).

Any reductions in domestic water consumption also result in associated reductions in energy demand, especially when water efficiency measures reduce the extent to which

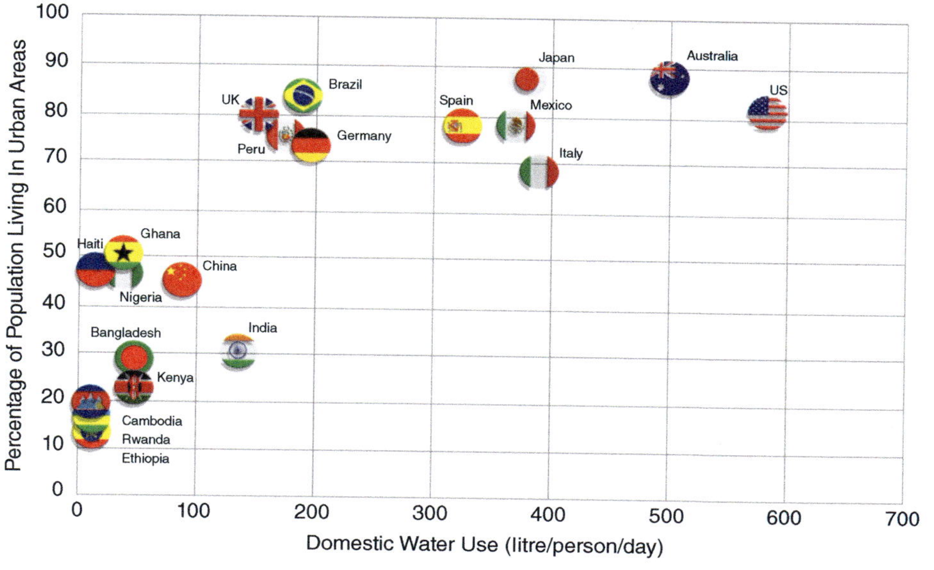

Figure 4.11 Percentage of urban population versus domestic water use.
Source: Adapted from World Bank (2014) and UNDP (2006).

water is heated. Hussey and Pittock (2012) calculated that a 15% reduction in Australian residential hot water use would completely offset the energy used in providing water to those households.

4.3.4 Alternative Approaches to Urban Water Supply

4.3.4.1 Cyclical Water Management Systems

Novotny (2012) postulates that the water sector's energy demands and volumes of waste can be significantly reduced by moving from centralised, linear management systems to semi-closed alternatives based on the objectives of reclaiming, reusing and recycling water and wastewater.

Cities are hotspots for the accumulation of nutrients, metals and other resources and should therefore be regarded as a highly valuable pool of materials. Furthermore, if linear systems could be transitioned to cascading or circular alternatives, it is possible that water demands could be met much more efficiently than is currently the case. For example, domestic wastewater returns (defined here as those not related to sanitation), represent a significantly underutilised resource in many cities. By volume, this resource can equate to 70% of domestic water consumption (Pidou *et al.* 2007). Its quality is also typically suitable for a range of applications including garden irrigation and car washing. As recognition of the value of domestic wastewater has grown, several countries are now taking steps to incentivise its exploitation. For example, many Japanese cities now mandate domestic wastewater recycling for buildings over a certain size.

The key characteristics of cyclical water management systems are closely aligned to the concept of 'urban metabolism', the idea that technical and socioeconomic processes in cities can be made more efficient by considering the inputs, transformations, stores and outputs of key materials. Water is regarded as by far the largest of these fluxes (Kennedy *et al.* 2007); by managing wastewater flows as resources in their own right, water reuse streams can be established. For example, cyclical supply chains in the Danish city of Kalundborg have enabled annual waste exchanges of some 2.9 million tonnes and an overall reduction in water consumption of 25% (Chertow 2004). For over four decades, Kalundborg's companies and industries have collaborated and co-located (Figure 4.12) to exploit synergies in their respective businesses that have realised improvements in efficiency far beyond those which could have been achieved had they continued to operate in isolation. The investments necessary to establish these cyclical systems have been paid back many times over (Domenech and Davies 2011).

Another defining characteristic of a cyclical approach to urban water management is an integration of the engineering and social sciences. Without community support and participation (guided by research from the social sciences), engineering enhancements to physical infrastructure are far less likely to succeed. For example, without water demand management, rates of water consumption will always be suboptimal. More

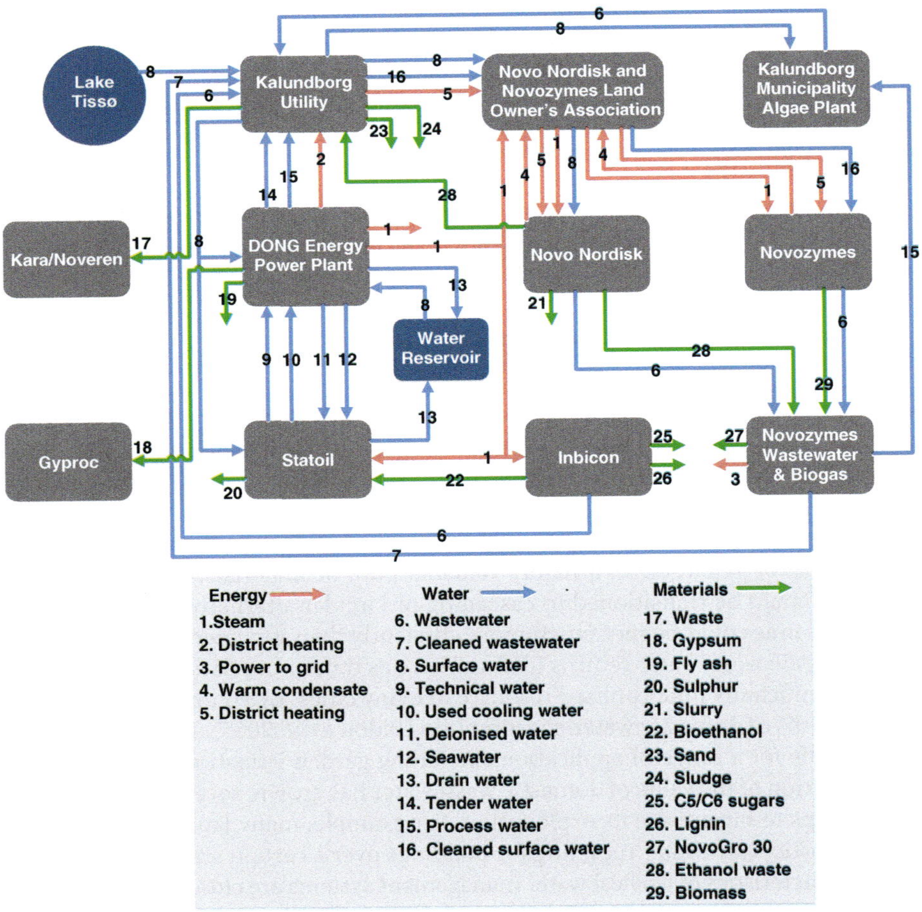

Figure 4.12 Industrial symbiosis in Kalundborg.
Source: Kalundborg Symbiosis (2015).

recently, researchers have begun to consider how the processes and interdependencies that characterise natural systems might also be able to be harnessed and applied to the built environment. By mimicking natural processes, researchers are increasingly realising that large efficiency gains in a broad range of processes can be achieved (see Section 4.3.8.2 for further discussion on the topic of biomimicry).

4.3.4.2 Hybrid Systems and Localised Networks

Another alternative to centralised water management is to use a hybrid approach, whereby localised networks within an overarching system supply small groups of customers with similar water demands (particularly in relation to water quality). The size of the local network could range from a large high-rise building to a shopping centre, commercial complex or even a district of a city. Because of the similar nature of their customer's demands, localised networks can exploit a locally abundant water source far more efficiently than would be the case using a purely centralised system. For example, the domestic wastewater generated from a residential high-rise, or the surface water runoff from a car park, could be captured and reused for irrigating recreational space. By developing local reuse cycles such as these, the energy requirements of the overarching system can be significantly reduced (UNWWAP 2014). With more water being reused, the volume of natural resources (water, soil and air) required to assimilate wastewater is also reduced.

The Pimpama-Coomera conurbation of the Gold Coast, Australia provides a good example of this concept. From 2009 to 2014, the development captured domestic wastewater and directed it to a nearby water recycling facility. Once processed, the recycled water (750,000 L/day) was piped back to residences for toilet washing, outdoor irrigation, car washing and firefighting (Stinchcombe and Brennan 2014). This approach to water recycling is also implemented in several other developments in Australia, where savings in potable water use have reached as high as 50% (Willis *et al.* 2010). Examples in other countries include the Solaire residential complex in New York, USA. This development comprises five residential water-use systems servicing six high-rise buildings. The recycled wastewater is used for toilet flushing and cooling, realising a 48% reduction in water consumption and a 56% reduction in wastewater discharge (NSU 2014).

In examples such as Pimpama-Coomera and Solaire, by using a localised network to supply a specific demand, water can be treated to a quality fit for its specific purpose. Of course, many urban areas are constrained by the fact that their water and wastewater infrastructure is already constructed and relied upon in its current form by a large number and broad variety of customers. Localised network approaches are therefore naturally suited to emerging economies or to newly developing peri-urban environments where this path dependency has not yet developed. For example, Masdar in the United Arab Emirates will supply two water streams (generated from solar-powered desalination) to its residents: one fit for drinking and the other fit for showering and dish washing (Novotny and Novotny 2011).

It should also be noted that in many of those countries that were first to experience industrialisation, water and wastewater management systems are now reaching the end of their design lives. Gradual retrofitting of new ideas therefore has the potential to achieve incremental enhancements that can still realise important water management benefits. In 2001, Copenhagen embarked on a scheme to replace its entire water main network. Although the process is expected to take 100 years, within just the first decade water leakage had fallen to 5% compared to the 20% or 25% experienced in most European cities (Bouton *et al.* 2013).

Notwithstanding their potential advantages, it should be acknowledged that localised networks are also characterised by a number of constraints, particularly where they are isolated from an overarching system. For example, because they typically rely on a small number of water sources, localised networks are vulnerable should one source fail. In contrast, exposed to the same failure, a broader system can compensate by drawing on an alternative water source in a separate region of the network. In fact, with increasing water demands and the growing pressures of climate change, the need to broaden the diversity of urban water portfolios, either by exploiting more distant sources or by harnessing new sources of water, is strong.

4.3.4.3 Inter-Basin Transfers

Many countries experience significant variations in water abundance within their boundaries that create inequalities in water availability and service provision. For example, although national supply capability in England is sufficient to meet demand, regional imbalances mean that while much of the west possesses a water surplus, a number of water resource zones in the southeast are classified as 'seriously water stressed' (Environment Agency and Natural Resources Wales 2013).

In order to balance out disparities such as these, large inter-basin transfers are often proposed as a means of linking water-abundant localities with their water-scarce counterparts. Most of these transfer schemes necessitate the construction of long-distance pipelines and, as a result, require significant expenditure, both for infrastructure and for their ongoing operation (high-energy demands from pumping). Potential environmental and social impacts (such as those associated with the introduction of invasive species and different water qualities) are often significant and together, these factors usually combine to render potential projects infeasible (with some notable exceptions, see Breakout Box 4.7). In some situations however, it is clear that existing water infrastructure

Breakout Box 4.7 The Chinese South-to-North Water Transfer Project

Northern China is home to more than 45% of the country's population and a rapidly growing urban populous, but has less than 20% of the nation's water resources (Jiang 2009). In response to the explosion in water demand that accompanies this urban population growth, the Chinese government turned to a hugely ambitious engineering solution, the South-to-North Water Transfer Project, by far the largest inter-basin water transfer scheme in the world.

Once fully operational in around 2050, the project's series of canals and diversions will direct up to 45 billion m^3 of water a year from basins of the Yangtze River in the south of the country to the comparatively water-scarce catchments of the Hai, Yellow and Huai rivers in the north (Jiang 2009). The project is expected to cost in excess of US$ 60 billion and will require the relocation of some 300,000 people. In so doing, however, the government hopes that the scheme will help to alleviate water scarcity for as many as 300 million people (Gupta and van der Zaag 2008).

In December 2014 the middle leg of the project was commissioned, a series of canals directing water some 1,400 km to service the urban and industrial water needs of Beijing.

could be better utilised to facilitate regional water sharing. For example, many industrialised countries possess extensive canal networks that provide well-established transport routes for a variety of goods (e.g. the Dutch canal network is more than 6,500 km long). Through targeted interventions, it is conceivable that systems such as these could be operated to broaden their functionality from one primarily of industrial transport to one that includes water supply. A significant advantage of such an approach lies in the fact that many canal networks were developed specifically for the purposes of linking major cities.

A key disadvantage of large water transfers is their potential to sustain or even accelerate water-intensive practices in the receiving location. For example, while the Roman Empire's networks of aqueducts is rightly recognised as one of the most advanced water management systems of the ancient world, some authors argue that the urbanisation that was sustained as a result likely pushed the empire closer to the limit of its water resources (Dermody *et al.* 2014). Examples such as this provide valuable lessons for those societies considering implementing the large and complex transfer schemes now permitted by our modern engineering capabilities.

4.3.4.4 New Sources of Water Supply

Our rapidly accelerating technological capabilities have allowed many water service providers to establish new and advanced sources of water supply.

Water Reuse Water reuse is an increasingly common means of reducing pressure on conventional water supplies; however, it is currently largely constrained to agriculture or indirect (and unplanned) applications (e.g. downstream re-abstraction of treated domestic wastewater discharged further up in the catchment). Stillwell *et al.* (2010) estimate that more than 26 million people in the USA use drinking water that contains between 5% and 100% treated wastewater effluent from upstream discharge. Notwithstanding this fact, cultural inertia and an inherent distrust of reusing waste products impede widespread uptake of planned indirect and particularly direct reuse schemes, despite the ability of treatment facilities to reliably produce water of the requisite quality. Windhoek, the capital of Namibia, is the best-known exception to this rule and has been successfully practicing direct water reuse since 1968. The city's wastewater treatment plant serves a population of 220,000, reclaiming municipal wastewater to potable quality standards. After treatment, reclaimed water is mixed with water from other sources so that around one-quarter of the city's drinking water is reclaimed wastewater (2030 WRG 2013).

As well as reducing pressure on traditional water supplies, water reuse initiatives also have the potential to deliver mutually beneficial outcomes to energy systems. For example, because wastewater is typically warmer than the water produced at water treatment plants (Novotny 2012), it can be used in a heat pump or exchanger to pre-heat water supplies and therefore reduce the volumes of energy used in the home. At a youth centre in Berlin, pre-heating using domestic wastewater reduces the energy demands associated with water heating by 20% (Schuetze *et al.* 2013).

Desalination Desalination is an increasingly important component of the water source portfolio of a number of cities, especially those in arid regions. Rapidly advancing technologies are making such systems increasingly more cost-effective; however, because of their high-energy demands, except in the most arid of regions, desalination is typically employed as a contingency measure, drawn on to supplement the water source portfolio during times of drought. For example, Thames Water's Beckton desalination plant in London, UK provides 150 million litres a day of redundancy capacity to the city, enough to supply one million residents (Thames Water 2014c). Often, schemes such as these are committed to by governments following crises of water supply such as major droughts or cyclones. In a number of cases, this near-sighted approach is caught out by the vagaries of short-term climate variability and the influence of political cycles, with the result that schemes may no longer be required once they are eventually commissioned.

In response to prolonged drought in the mid-2000s, the state and local governments of Queensland, Australia invested AUD$ 1.2 billion in the construction of a desalination plant to secure supplies to the regions' residents. However, in 2011, after less than two years of operation, the large energy demands and associated operating costs of the plant were considered an unacceptable financial burden by a new state government. With the drought also weakening, reliance on the plant was significantly reduced.

Managed Aquifer Recharge A further drought contingency measure now employed to safeguard the water supplies of a number of cities is managed aquifer recharge (MAR), the recharging of water into an underground aquifer for storage and subsequent withdrawal. The water recharged might be treated wastewater or may have been abstracted from rivers during periods of high flow. A wide variety of methods can be used to recharge the water; however, one of the most common involves the injection of water into the target aquifer via a well, a process known as aquifer storage and recovery (ASR).

Where suitable aquifers and injection capabilities exist, ASR provides a viable means of 'banking' volumes of water away for later use during times of scarcity. Furthermore, because the store lies underground, evaporative losses, unlike those for reservoirs, are negligible. Other impacts typical of reservoirs, such as community displacement, are also avoided. ASR can however be technically challenging and will almost always influence the natural groundwater environment. Because our knowledge of these underground systems is incomplete, ASR and other types of MAR have the potential to cause unintended impacts. In shallow aquifers for example, the increase in the water table caused by MAR may influence the hydrological characteristics of surface water systems. Where MAR is achieved by ASR, additional energy demands are also introduced for pumping. For example, groundwater pumping at a depth of 120 m is estimated to require energy at a rate of around 0.5 kWh for each cubic metre of water extracted. The potential for energy savings from ASR in comparison to establishing alternative water sources has however been demonstrated in a number of studies, for example in the San Francisco Bay area of the USA (see UNWWAP 2014).

In regions where the requisite geology exists, the number of MAR schemes is increasing. In Western Australia for example, the state government is investing in the development of a MAR scheme (via ASR) to provide a new, climate-independent water source for the city of Perth (Water Corporation 2014). The City of Salisbury, Adelaide also uses a ASR scheme alongside a series of constructed wetlands to store up to 14,000 million litres of water in the wet season for use during the following dry season (2030 WRG 2013). The municipality combines this scheme with a non-potable water distribution network.

4.3.5 Demand Management and the Role of Water Pricing

As alluded to throughout this chapter, the important role to be played by the social sciences in improving the sustainability of water management systems has often been overlooked. Engineers and social scientists must break out of their respective silos and learn to involve one another in collective research. While circular water management infrastructure and more diversified water source portfolios may realise step-change benefits in water and wastewater service provision, in order to ensure their long-term success social buy-in is essential. In this respect, a fundamental flaw of many existing approaches to water and wastewater management is the inability or failure of the body responsible to charge a price for water that is reflective of its full cost of supply, or relative scarcity. This introduces two severe impediments to the optimisation of the water management system:

1) the customer experiences no incentive to reduce their water consumption and gains no understanding of the consequence of their water use; and
2) revenues that could otherwise be used to fund infrastructure improvements and to ensure the financial sustainability of the system, a characteristic that is absolutely fundamental to maintaining long-term security of supply, are forgone.

Of course, a small volume of water use is essential and, as previously discussed, is a recognised human right. Support mechanisms must be established to ensure that this water is physically and financially accessible to all citizens. Above and beyond this volume however, water use (like the use of the vast majority of resources and commodities) is discretionary, and its cost should therefore account for all those factors involved in its provision.

Incremental block tariffs represent one potential pricing mechanism whereby different prices are charged for discrete volumetric blocks of water supply. Incremental block tariffs thereby allow differentiation between essential and discretionary water use and help to communicate a clear message to consumers. Furthermore, the fees and block sizes can be changed geographically, seasonally or in response to prevailing water availability. Seattle Public Utilities has charged its customers based on volume from as early as 1920 and now implements seasonal tariffs as well as charging higher rates to customers located further from centralised infrastructure (Stinchcombe and Brennan 2014). However, in many countries such as the UK, residents can still be charged a flat annual fee for water regardless of the amount they use. In the Republic of Ireland, between 1997 and 2014, domestic water use was entirely free of charge. In such systems, it is not surprising that consumers use more water than they really need to.

When exposed to accurate water prices, customers are far more likely to adopt measures that conserve water. Novotny (2012) considered the benefits of a range of domestic water-saving devices and found that they had the potential to achieve a 65% reduction in total water use, a significant financial saving for the customer and also a significant reduction on the supply burden imposed on the water service provider. Novotny (2012) found that the greatest water savings were achieved by initiatives outside the home. By switching to the growing native plants that don't require irrigation, a reduction in water use of more than 80% was achieved. Measures such as the installation of water-efficient shower heads and more efficient toilets reduced indoor water use by more than 40% (Novotny 2012).

As highlighted in Section 4.2, the links between water and energy management are particularly strong in the home. Table 4.7 describes a broad variety of measures that can help to reduce residential water heating.

While cutting back on outdoor irrigation is one of the most effective means of reducing urban water use, it is an indication of the complexity of urban interdependencies that this initiative may also promote unintended and detrimental consequences for other resources. Larson *et al.* (2013) suggest that curbing irrigation of public spaces may also remove an indirect means of urban cooling, thereby leading to increased energy demands for air conditioning. Grimm *et al.* (2008) estimated that up to 8% of electricity demand in the USA was used to compensate for the urban heat island effect; temperatures in urban areas can be 5°C higher than those in their surrounds (UNFPA 2007). For American cities with populations in excess of 100,000, Kennedy *et al.* (2007) suggest that peak electricity loads increase by about 1% for every degree increase in outside temperature.

This example serves to demonstrate how cities can often represent microcosms of the kinds of complexities, changes and tradeoffs experienced at far broader geographic scales (Grimm *et al.* 2008), for example those associated with global climate change. An important benefit of improved urban planning is therefore the application of lessons learned to other locations. Singapore, and the strategies of the country's national water agency PUB, provide an excellent example of progressive water and wastewater management (see Breakout Box 4.8). Singapore's experiences highlight how successful urban water management requires concurrent initiatives that address both supply and demand management. Institutional effectiveness is also crucial, and a unified and clear overarching strategy is vital to foster public understanding and support. While the political

Table 4.7 Energy efficiency measures related to residential water heating. Examples from USDoE (2004) unless otherwise stated.

Measures	Description
Repairing leaks in taps and showers and installing low flow heads and faucet aerators.	Installing a low-flow showerhead can reduce the hot water consumption of a shower by up to 30%.
Use more water-efficient dishwashers.	Contrary to popular belief, washing dishes by hand several times a day can actually use more hot water than modern dishwashers, particularly when the dishwasher is only operated with full loads. Dishwashers with specialised water heating systems provide the greatest energy savings.
Practice more water-efficient clothes washing.	Front-loading machines use less water and consequently less energy than top-loaders. Washing only full loads and using cold rinses will also reduce energy consumption.
Lower your thermostat setting.	Some manufacturers set water heaters at 60°C; however, around 50°C is suitable for most household needs. For every 5°C reduction in water temperature, energy consumption can be reduced by between 3% and 5%.
Installing more effective insulation.	Insulating hot water pipes and storage tanks as well as improved insulation around the home all help to reduce energy demands.

structure in Singapore is different to that in many other countries, the initiatives described in Breakout Box 4.8 provide useful examples that can be optimised for other locations and settings.

Breakout Box 4.8 Singapore's approach to water management

Reflecting Singapore's size (just 714 km^2), and therefore the limited area of land over which water can be captured and stored, the nation is water scarce despite receiving average annual rainfall of 2,400 mm (Tortajada and Joshi 2013). While the total national water demand in 2015 was around 2,000 million litres a day, it is predicted that this figure could double by 2060 (PUB 2016).

Singapore has faced water scarcity challenges since it gained independence in 1965. Singapore lacks native water resources and, as such, water has always been a top priority in government policy. It has required decades of continuous investment in technology and innovation for Singapore to turn its water vulnerability into a strength. Today, Singapore has its Four National Taps: local catchment water, imported water, recycled water (termed NEWater) and desalinated water. Measures implemented by Singapore's national water agency (PUB) to achieve this goal have included the following:

- the multi-agency effort to clean up the Singapore River (led by the Ministry of Environment) took a decade to complete, followed by a sustained focus on watershed protection. Singapore's river systems were grossly polluted when the city-state become independent. However, with strong political backing, during the 1980s the Singapore River was cleaned of pollutants and wastewater connections were installed to prevent overflows to watercourses (Joshi *et al.* 2012). Polluting industries were also relocated away from sensitive catchments;
- more reservoirs were also built. In the 1960s, Singapore only had three reservoirs; today, Singapore has 17 reservoirs with a water catchment covering two-thirds of the nation's land area. A significant proportion of Singapore's rainfall is now collected via an extensive network of drains, canals, rivers, stormwater collection ponds and reservoirs before it is treated for drinking water supply. This makes Singapore one of the few countries in the world to harvest urban stormwater on a large scale for its water supply (PUB 2016);
- Singapore implements universal collection and treatment of all domestic wastewater for treatment at water reclamation plants before discharge to the sea. Since 2003, Singapore has been reusing part of the treated used water by sending it to NEWater plants for further purification to produce ultra-clean NEWater. The vast majority of this water is used by industry and commerce with the remainder blended in the aforementioned reservoirs to undergo subsequent treatment and potable consumption (Tortajada 2006). The NEWater produced can meet up to 30% of Singapore's total demand, a figure that PUB estimates will rise to 55% by 2060 (PUB 2016);
- Singapore has also invested heavily in desalination. Today, it operates two desalination plants that can produce 455,000 m^3/day, enough to meet up to 25% of Singapore's current water needs (PUB 2016);
- unaccounted-for water in Singapore is around 5% of total supply, a very low figure compared to the 40–60% common for most Asian economies (Tortajada 2006). This success has been achieved through the maintenance of a reliable and efficient pipe

network, the accurate metering of water production at treatment plants and of customer consumption, strict legislation to deter illegal siphoning, and public vigilance in the reporting of leaks. Whereas in 1985, Singapore's water pipeline infrastructure experienced 95 leaks per 100 km of pipework, in 2014 this figure was reduced to an average of 6 leaks per 100 km (PUB 2016). 100% of Singapore's water supply system is metered and bill collection stands at 99% (Tortajada 2006);

- Singapore implements an incremental block tariff structure for its domestic customers, whereby an increase of around 20% is charged once household water consumption exceeds 40 m^3 a month (PUB 2016). The government also levies a water conservation tax and additional fee to reflect the cost of treating water and of maintaining the sewerage system (Tortajada 2006). Those that cannot afford to pay the fees receive a targeted subsidy. In response to these policies, average domestic water consumption per household fell by more than 10% between 1995 and 2004 (Tortajada 2006). Over the same period, the average household water bill doubled in absolute terms (Tortajada 2006), but has stayed relatively constant as a percentage of average household income (SingStat 2014);
- some parts of Singapore use a smart water grid to make the process of supplying water more efficient. The smart water-monitoring system uses a wireless sensing network, data-mining algorithms and real-time hydraulic modelling to inform PUB of what is happening in the water distribution network. Operational since April 2013, the smart water grid system allows PUB to perform hydraulic simulations and provides alerts on events such as network water pressure and water quality anomalies;
- Singapore's water management initiatives have been accompanied by the implementation of a coherent and evolving communication strategy. The NEWater Visitor Centre opened in 2003 and is the focal point for public education on NEWater and the Singapore Water Story (i.e. the Four National Taps). In order to encourage collaboration between academia and industry, a WaterHub has been established as a centre for exchanging research and best practice in water (Rygaard *et al.* 2011); and
- governance has been crucial to the success of Singapore's urban water management. The PUB is responsible for the entire water cycle (the 'Water Loop' shown in Figure 4.13), from the sourcing of water to its collection, treatment and distribution to consumers (Xi and Poh 2013). This single point of responsibility has supported the development of a holistic water management strategy and has also promoted transparency and accountability to Singapore's citizens. In turn, residents feel engaged in a national water management movement and are therefore more likely to respond positively to specific initiatives.

4.3.6 Using Water to Meet Urban Demands for Other Resources

Typically, as cities grow their ability to be supported by resources harnessed from their immediate surrounds progressively declines and resource supply chains increase in both their length and complexity. Initial reliance on the immediate hinterland is progressively replaced by the transport of resources from rural communities, often many hundreds of kilometres away. In turn, as global transport and communication networks have grown, many countries now increasingly rely on supplies of resources (and their embedded production inputs) from outside their national borders. Almost all resources

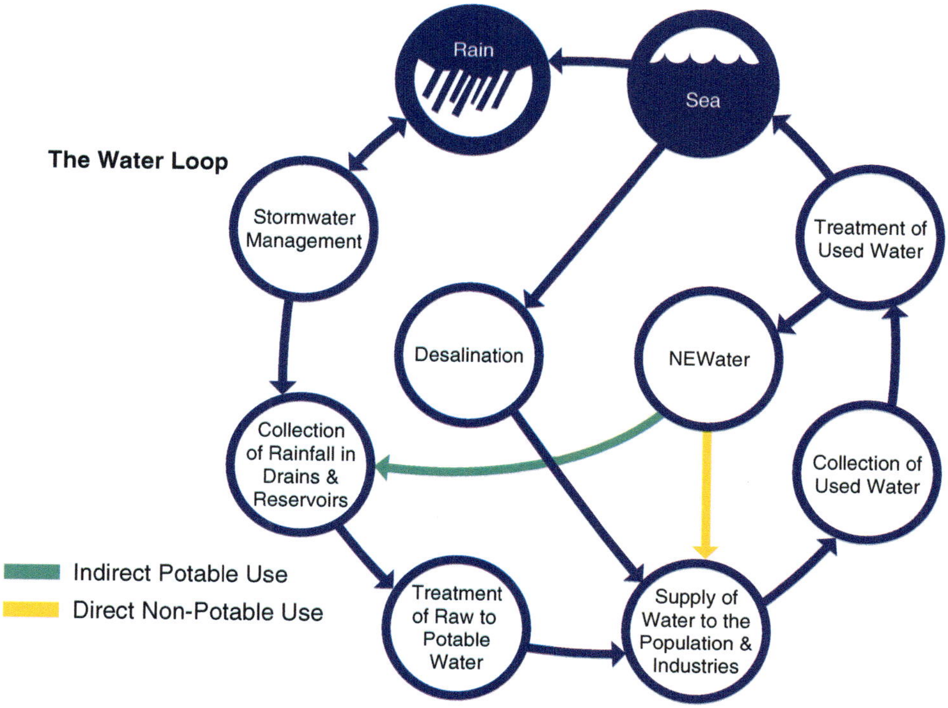

Figure 4.13 Singapore's urban water management: The Water Loop. *Source:* PUB (2016).

require inputs of water at some point in their production, so this globalisation of urban resource supply has ensured that a significant proportion of the overall water demand of a city is outsourced beyond its boundaries. For example, 80% of London's food is imported from outside the UK (World Bank 2010). Without these inflows of resources and their embedded virtual water, urban growth would have long since been constrained by the growing pains that we are now beginning to witness.

When products are imported from water-scarce locations where water management is inadequate, although the city and is residents may not experience any immediate impact, the host environment will. The water system in the source location may eventually fail and damage the product supply chain; however, with a modern society so full of globalised resource networks, the buyer of the product or resource can typically switch his or her supply chain with little or no impact. Eventually however, if key water sources and supply chains continue to be disrupted, the city may suffer serious repercussions. For example, over the course of 2007 and 2008, 14 African countries witnessed rioting that in part reflected anger at spikes in food prices (Berazneva and Lee 2011). The price rises were caused by the combined impacts of a range of events; however, poor harvests caused by drought were a contributing factor, particularly of wheat in Australia and Ukraine (Wiggins *et al.* 2010). Berazneva and Lee (2011) found that this burden of rising food prices was felt particularly severely in urban settings where populations were more reliant on market conditions and less able to resort to growing their own food. This example serves to illustrate how water availability in one location can influence urban food security and social stability many thousands of kilometres away. Accounting for virtual water flows is therefore a vital means of reducing the risks present within urban resource networks.

Reducing the length of supply chains and supporting local resource provision is one way of building resilience into urban resource management. In many wealthy cities, because very little land is devoted to agricultural production, the reliance on external food supply sources can be absolute (C40 Cities 2014). Despite this, demands for local produce linked to ideals of supporting local farmers, improving animal welfare and reducing the carbon footprint of food supply networks are growing, especially in developed country cities, thereby driving renewed focus towards urban and peri-urban agriculture. At the same time, planners are now realising that urban and peri-urban food supply networks provide an ideal opportunity to harness the abundant nutrients present in wastewater, helping to convert the linear water management system into a cyclical resource stream and reinstating an important link in global nutrient cycles (Daigger 2009). While the recovery of these nutrients and resources does require energy, Lofrano (2012) suggests that the embedded thermal and chemically bound energy of the recovered resources can exceed this expended fuel four times over.

A good example of the potential benefits of wastewater recovery is phosphorus, a non-fungible element and one that is crucial to agricultural production. Existing global rock phosphate reserves are estimated at between 50 and 100 years (Cordell *et al.* 2009); while much larger mineral resources do exist, the ease with which they may or may not be able to be extracted is currently unknown. An alternative and rich source of phosphorus is domestic sewage. Virtually none of the phosphorus present within food is used by the human body and so wastewater streams contain high concentrations of the resource that could be harnessed for human benefit. Urine contains 50% of the phosphorus of domestic sewage (and around 90% of the nitrogen), and its separation from the wastewater stream would also reduce scaling of pipework (Schuetze *et al.* 2013) and the potential for ecological damage in receiving environments (where residual phosphorus may otherwise promote algae growth and de-oxygenation). Yuan *et al.* (2012) suggest that, theoretically at least, up to 20% of the global demand for phosphate rock could be met by recovering phosphorus from domestic sewage.

Research and development in this field is increasingly focusing on the recovery of crystals high in phosphorus from both domestic sewage and farming effluent (Cordell *et al.* 2009). The first such nutrient-recovery reactor in Europe was commissioned in the UK in 2014. By promoting the settling of struvite crystals in a controlled manner, phosphorus is removed from wastewater and half a tonne a day of high-quality sustainable fertiliser generated as a result (CIWEM 2014). The role of wastewater in urban food production is discussed further in Section 5.3.1.2.

4.3.7 Flooding in Urban Environments

Flooding is the most common of natural disasters. Globally, the yearly number of flood events doubled between 1995 and 2010 (Jha *et al.* 2012), and in 2005 flooding caused an average of US$ 6 billion in total annual losses to 136 cities (Hallegatte *et al.* 2013). By 2050, and considering the impacts of demographic trends and expected climate change, the annual losses experienced by these cities could rise by an order of magnitude to US$ 63 billion (Hallegatte *et al.* 2013).

Cities are highly susceptible to flooding from a number of sources and for a number of reasons. Overtopping rivers, rising sea levels, coastal storms, the failure of man-made water distribution infrastructure and rising groundwater levels all pose acute risks for urban residents and businesses. Although rural flooding tends to affect large areas of land, urban floods are typically more costly and, because of the number of potential influences and range of receptors and vulnerabilities, difficult to manage (Jha *et al.* 2012).

4.3.7.1 Riverine and Coastal Flooding

Early urban populations coped with riverine and coastal flooding by locating important infrastructure on higher ground; however, as cities have grown, space constraints and the exploitation of increasingly marginal areas have left growing numbers of people and an increasing number of pieces of critical infrastructure badly exposed.

Governments have typically responded to these risks by constructing expensive flood defence schemes with the objective of routing water away from areas of concern as quickly as possible. Linear, man-made waterways are a common site in most cities. While broadly successful at reducing flood risk, such schemes invariably just move the problem downstream. Furthermore, they breed complacency in both the communities that establish themselves on floodplains behind defences, and in decision makers whose attention is drawn away from identifying and rectifying the ultimate cause of the problem. This means that if and when flood defences do fail, the impacts can be catastrophic. When the storm surge caused by Hurricane Katrina broke through flood levees and inundated 85% of the American city of New Orleans in 2005, 1,836 people were killed, more than 1,000,000 people were made homeless and damage equivalent to more than US$ 160 billion was caused (van Heerden 2007, World Bank 2013) (see Breakout Box 4.9).

More holistic approaches to flood risk management place emphasis on improved land-use planning and protecting natural ecosystems. In the UK for example, floodplains are zoned according to the types of infrastructure permitted within certain areas. Critical infrastructure must be located well outside typical flood event extents, whereas land adjacent to rivers is retained for recreational facilities and parkland that can serve as water storage areas in the event of a flood. The potential benefits of restoring natural ecosystems and of improving urban drainage are also significant. By slowing the passage of water across the landscape, upland watershed management can reduce both the magnitude and flashiness of downstream flood events. These characteristics have vitally important implications for effective flood response, especially in the face of future climate changes that point to an increased frequency of intense rainfall events. Agricultural

Breakout Box 4.9 New Orleans and its susceptibility to flooding

New Orleans lies at the mouth of the Mississippi River and the seven deltas it has created. Under natural conditions, these deltas would accrete and provide protection to the land. However, thanks to the 5,700 km of levees built to protect New Orleans, this natural process no longer occurs. Oil and gas extraction and other human activities have also combined to mean that the Louisiana coastline is now losing its wetlands at a rate of 100 acres a day (van Heerden 2007). Without these natural buffers, the levees surrounding New Orleans were much more susceptible to failure under the 2005 storm surge caused by Hurricane Katrina (Beatley and Newman 2013).

users could be encouraged to adopt measures that capture surface water runoff or to abstract more water when river or groundwater levels are high for storage and subsequent use when natural supplies are depleted. Augmenting the natural process of infiltration by temporarily capturing water in shallow basins can also help to reduce peak river flows, while acting to sustain base flows and soil moisture levels. By slowing the passage of water across the landscape, measures such as these also reduce soil erosion, pesticide runoff and, in turn, improve the quality of natural water sources.

Although logical and scientifically sound, because the benefits of watershed management are indirect, proving the link between the necessary investment and a reduction in flood risk to a particular community or business is challenging. This hampers investment and highlights the need for improved means of accounting for ecosystem services. As an indication of the importance of sound catchment management, the Colombian Government was able to estimate that the ecosystem services provided by the Magdalena River basin equated to 86% of its national GDP (UNWWAP 2014).

An increasing number of cities are now realising the advantages that can be gained by protecting their upstream catchments. For example, New York has invested in the acquisition and management of 70,000 acres of its upper watersheds in the Catskill Mountains (Beatley and Newman 2013) while Denver Water has invested US$ 16.5 million in upstream forest and watershed protection. It is hoped that this investment will reduce soil erosion and subsequent sedimentation of reservoirs and other water infrastructure, while also reducing the risk of wildfires (Denver Water 2014).

4.3.7.2 Stormwater Flooding

As cities have urbanised, the gradual loss of green space and proliferation of impermeable land cover has increased the frequency of flood events caused by heavy rainfall. The impacts of these stormwater events are often manifest as acute inundations, limited in their geographical extent but no less damaging as a result, especially when they disrupt water supply networks, transport systems, energy grids or other critical infrastructure.

The highly localised and varied nature of stormwater flood risk makes large-scale physical defences ineffective as a mitigation measure. Instead, localised initiatives are required to slow the rate at which water moves across the urban landscape. In a manner analogous to upland watershed management, measures such as permeable paving, swales and wetlands (see Table 4.8) all act to slow down water, providing incremental benefits that can add up to a highly effective approach at the city scale.

Combined sewer overflows (CSOs) are a particularly damaging consequence of the increasing volumes of surface water runoff experienced in some cities. While many modern cities implement separate stormwater and wastewater management networks, in older cities these two waters may be combined in a single pipe network. This means that during even light rainfall, stormwater runoff can overwhelm the capacity of the system and cause acute pollution incidents. In London, while separate systems for rainwater and foul sewage are now a requirement for all new developments, CSOs to the River Thames occur around 60 times a year, often in response to just a few millimetres of rainfall. The frequency of these events breaches the European Urban Wastewater Treatment Directive, causing adverse impacts to fish species and increasing health risks to recreational users of the river (Thomas and Crawford 2011). To help tackle the problem, Thames Water is investing £4 billion in the construction of the Thames Tideway

Table 4.8 Sustainable drainage systems.

Example	Description
Source control options	
Green roofs	• Covering rooftops with vegetation laid over a drainage layer with additional layers for waterproofing and insulation. • Green roofs are mandated in Toronto for roofs over a certain size (Beatley and Newman 2013).
Permeable paving	• Either the construction of hardstanding using permeable materials or the careful design of infiltration strips between impermeable blocks. • Appropriate permeable materials include gravel and geo-synthetic systems. • An underlying storage layer can be incorporated to promote infiltration to the ground or controlled release to surface water. • Chicago has achieved a 20% increase in permeable area and Philadelphia will replace at least one-third of all impervious surfaces with soil and plant systems that intercept stormwater and allow infiltration or evaporation (C40 Cities 2014).
Rainwater harvesting	• By disconnecting downpipes, inflows to the sewer system are reduced and a source of water for garden irrigation becomes available.
Permeable conveyance systems	
Filter strips	• Wide and gently sloping areas of vegetation that help to treat and slow runoff from adjacent impermeable areas such as roads.
Swales	• Broad, shallow and vegetated channels designed to convey, store and/or promote infiltration of runoff.
Passive treatment systems	
Detention basins	• Normally dry, but may incorporate a permanent pool at the inlet or outlet. • Designed to detain a certain volume of runoff as well as improving water quality and promoting the settling out of suspended sediment.
Constructed wetlands	• Ponds with shallow areas and wetland vegetation that act to improve pollutant removal and to reduce the suspended solid content of water. • Constructed wetlands also provide valuable habitat for flora and fauna.

Source: Adapted from Woods-Ballard *et al.* (2007) unless otherwise stated.

Tunnel to intercept CSOs before they reach the Thames. Washington DC, Portland, Oregon, Paris and the Rhine Ruhr region of Germany are also implementing tunnel solutions to CSO problems (Thames Water 2012). Supporting such schemes with sustainable drainage systems will be crucial to ensuring that the frequency of damaging CSOs does not rise again in future.

4.3.7.3 Groundwater Flooding

Groundwater flooding represents a growing trend in many developed country cities. While their rapid growth often encouraged the unsustainable exploitation of groundwater reserves and the rapid depletion of groundwater levels, as cities have expanded overexploited local water sources have been replaced with increasingly distant alternatives. Furthermore, leaks from the water distribution system, often more than one-quarter of total supply, add to groundwater recharge. As a result, urban groundwater tables have, in many cities, recovered to pre-urbanised levels and, in some, now threaten

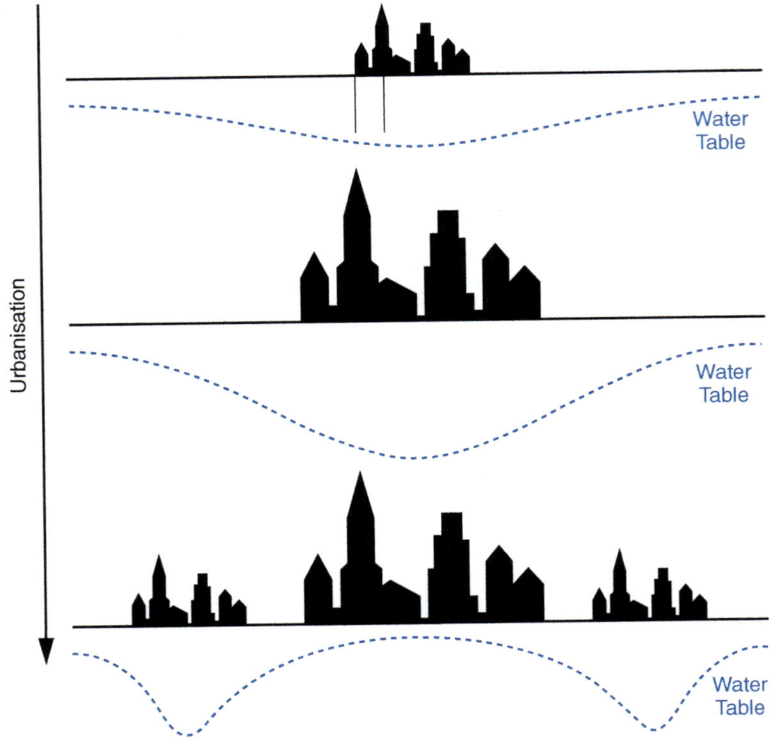

Figure 4.14 Urban exploitation of groundwater.
Source: Adapted from Kennedy *et al.* (2007).

important subsurface infrastructure such as metro systems and road tunnels. The groundwater table beneath central London is rising at between 1 m and 2.5 m a year, for example (Kennedy *et al.* 2007). Similar instances of groundwater rebound have been witnessed in Barcelona, Berlin, Birmingham, Budapest, Houston, Liverpool, Milan and Moscow (UNEP 2003). In some cases, rising groundwater tables can also intercept and mobilise the wastes and toxic substances discarded by the expanding city. In 1970s New York, oil released to the land more than a decade previously was mobilised by rising groundwater, causing it to damage local surface watercourses (UNEP 2003).

Kennedy *et al.* (2007) developed a model for the urban exploitation of groundwater systems which shows how the typical stages of urban growth can be linked to trends in groundwater level (Figure 4.14). They postulate that the physical integrity of cities depends to a large extent on achieving a sustainable equilibrium in the groundwater component of urban metabolism.

4.3.8 Opportunities and Challenges of Urban Water Management

The case studies and concepts presented in this chapter have highlighted the complexity of the relationships between a city and its management of water. In many instances, water acted as the initial catalyst for urbanisation, allowing populations to grow and trade to flourish. However, water was also taken for granted, depleted and degraded until its intrinsic importance to so many other urban interactions became apparent only through increasingly acute and detrimental impacts. In danger of being constrained by water availability, cities have been able to continue their expansion by exploiting advances in technology, transport networks and trading systems to maintain security of supply, both directly (through new water sources such as desalination, for example) and indirectly (via the embedded virtual water within products traded on regional and international markets). Acknowledgement of the limitations of traditional, linear urban water management and recognition of the need to move to more pragmatic and efficient approaches is now encouraging planners to adopt systems that integrate water and wastewater service provision with the management of other resources in order to achieve mutual benefits.

This cyclical approach to resource management is perhaps the key tenant of modern, sustainable urban environments. Figure 4.15 shows how the approach enables inputs of water, food and energy from outside the city's boundaries to be reduced. By considering wastes as resources rather than as management constraints, symbiotic relationships between different urban stakeholders emerge. In turn, by exploiting these relationships, water and other resources can be continually reused, and the wastes generated and ultimately discharged to the environment significantly reduced.

A systems approach to management also promotes layered resilience, an essential characteristic for urban environments exposed to the pressures of demographic trends and expected future climate change. By developing a portfolio of resource supply options, and by incorporating a large number of resource connections, city planners are afforded flexibility in how they respond to consumer demands. Spikes in demand or the failure of a particular water source or treatment facility can be compensated for by isolating the portion of the system in question and by increasing reliance on alternatives. Failures or operational issues can therefore occur safely in controlled portions of the resource management network, allowing the overall functionality of the city to be maintained and enabling the affected area to be more quickly repaired.

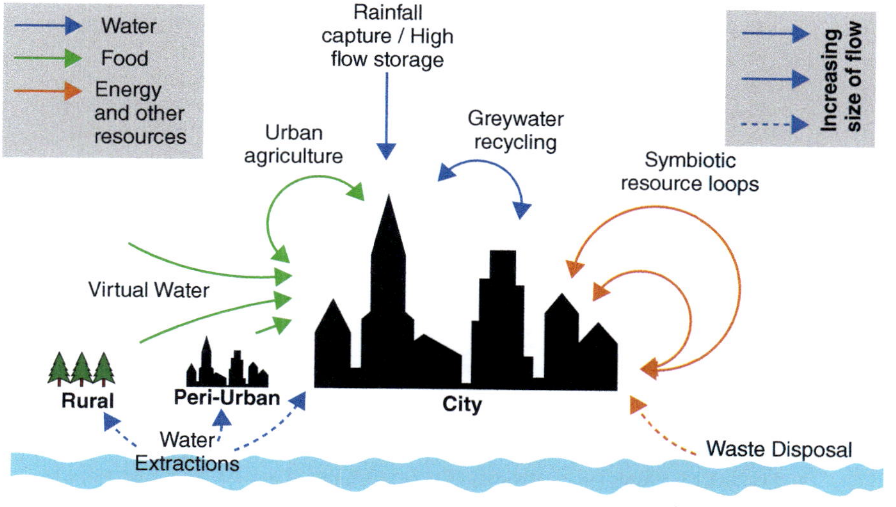

Figure 4.15 Urban resource flows.

A number of important enabling activities are required to support the development of a cyclical urban resource system, and the following subsections address a few of the key issues.

4.3.8.1 Improved Data Management

Cities generate billions of pieces of data describing all facets of urban life. In most instances however, this data is siloed, used for discrete tasks, un-transferrable (so cities find it hard to learn from one another) and poorly maintained (lack of meta-data for example). Its ultimate value as a means of establishing trends and synergies in urban activities is therefore lost. By contrast, smart data systems consider the relationships between data groups. They utilise recent rapid advances in remote monitoring, cloud computing, data storage and analytical processing to quickly identify trends in consumer demand, supply networks or external influencing factors in order to configure responsive infrastructure and thereby improve the overall efficiency of resource management.

Smart electrical grids are already being rolled out in some developed-country cities where they are helping to integrate increasingly distributed energy sources (such as home solar) and to track electricity usage (Moura *et al.* 2013). While the application of smart systems to the water sector has been slower to evolve, the potential benefits are significant (see Table 4.9).

Table 4.9 Smart water systems.

System element	Description
Sensing devices	• Water flow, pressure and contaminant sensing devices that collect and transmit data on a real-time basis are the foundation of a smart water grid. For example, sensors can be used to monitor pipe deterioration, ensuring that maintenance is undertaken in the most efficient manner possible. • Mutchek and Williams (2014) estimate that between 30% and 60% of water contamination events occur in the water distribution network, and that these events are only often detected by consumers once they have been exposed to the associated health risks. Biosensors and contaminant sensors within the pipe network could be used to alert authorities to off-specification water supplies, while pressure meters could be used to detect the leaks and pressure differentials that often lead to contamination.
Cloud services	• Cloud computing works by concentrating resources (such as hardware and software) in a limited number of locations and offering remote access to those resources. • The cloud allows users to pay for only the service they require rather than the physical resource; the cloud approach therefore provides significant cost savings. It also promotes more efficient sharing of sensed data (Perera *et al.* 2014).
Smart pumps and valves	• By adjusting power levels based on monitored conditions, smart pumps reduce the energy intensity of the pumping process. Leak detection sensors also save energy and biosensors can locate biofilms that slow water flow. • Variable speed pumps can ramp up or down depending on sensed flow conditions, and can also be equipped to identify clogs in the system and respond by reversing flow to break up blockages. Mutchek and Williams (2014) claim that this advance has the potential to save up to 70% of the lifecycle cost of a pump.

Table 4.9 (Continued)

System element	Description
Smart irrigation controllers	• Smart irrigation controllers can receive and/or collect weather data or sense soil moisture levels and other parameters to optimise water scheduling, and prompt valves and pumps to implement the watering. • Research by the University of Melbourne shows that smart approaches to irrigation could achieve improvements in economic water productivity of between 25% and 75% (PMSEIC 2010).
End-use sensors	• End-use sensors allow suppliers and users of water to monitor usage in real-time. This allows consumers to become more responsive to demand control measures and also allows the supplier to better understand peaks in demand. • Smart end-use metering is also essential for effective incremental volumetric water pricing.

Source: Adapted from Mutchek and Williams (2014) unless otherwise stated.

4.3.8.2 Learning from Nature

Nature does not waste resources. It seeks to use the minimum amount of energy for a given task and, as such, delivers its services with incredible efficiency. In contrast, energy wastage can be seen throughout human activities and processes. Coal-fired power stations operate at efficiency levels of not much greater than 35% for example, and irrigation water pumps have an efficiency of around 70%. Furthermore, human approaches to controlling nature have often acted against its energy-efficient and energy-conserving tendencies. While river engineers and planners have traditionally sought to construct linear channels and to route water away from the city as quickly as possible, natural watercourses evolve in an entirely opposite manner, avoiding straight lines and instead forming meanders that act to slow water passage across the landscape. The benefits of studying and learning from natural processes are evidenced throughout human history; however, it is only now that concerted research is being directed towards this field of biomimicry. As such, significant improvements in many essential urban activities are being identified. For example, the activated sludge treatment process developed in the early 1990s takes its lead from the role of bacteria in breaking down contaminants. New systems are now being studied that expand on this natural treatment process to use worms, beetles and other microorganisms to convert waste into structured humus which then acts as a filter to turn raw sewage into irrigation water (much like the decomposition of leaf litter in the rainforest) (Biomimicry 3.8 Institute 2014a). Continued research and development of the biomimicry discipline will enable further enhancements in the efficiency of urban resource management. Example innovations are as follows:

- a mixing impeller modelled on frozen whirlpools is now being used in more than 200 cities to prevent water stored in reservoirs from becoming stagnant. The process is much more efficient than traditional mixers and, as a result, energy consumption is reduced by 90% and disinfectant residues in the water distribution network are reduced by up to 80% (Harman 2013); and
- shark skin is covered with tiny scale-like denticles that cause turbulence in the surrounding layer of water, thereby reducing drag. A paint that mimics this characteristic

has been developed that can be used to reduce the drag experienced by the fuselage of airplanes and the hulls of ships. The developers claim that if every airplane in the world were to be painted with sharkskin paint, 4.48 million tons of fuel could be saved each year. Drag on ships could also be reduced by up to 5%, potentially saving a large container ship around 2,000 tons of fuel a year (Harman 2013; Biomimicry 3.8 Institute 2014b).

4.3.8.3 Integrating the Management of Urban Resources

Managing different urban resources in unison helps to improve the sustainability of their exploitation, while at the same time improving quality of life. For example, open space can be made multifunctional so that it captures excess surface water runoff during heavy rainfall, while also acting as recreational land (such as a cycle path or sports pitch) during dry weather. Examples of water-sensitive urban design such as this can be further enhanced to simultaneously optimise energy demand and use. For example, the concept of green roofs can be expanded to incorporate designs that also act to cool buildings. As such, energy consumption to cool homes (which in turn requires water at the power plant) is reduced. Novotny and Novotny (2011) postulate that water-centric cities of the future will consider a range of integrated issues, including:

- water conservation;
- distributed stormwater management, including sustainable drainage systems (SUDS);
- distributed wastewater treatment that generates water for reuse;
- incorporating landscape components into urban design in order to attenuate diffuse pollution (buffer strips for example);
- heat, energy and nutrient recovery (including biogas recovery and hydrogen generation from biogas and wastewater); and
- use of renewable energy sources.

4.3.8.4 Leadership and Social Action

As with almost all the initiatives identified in this book, leadership and social action are prerequisites for achieving sustainable urban resource management. Strong political will and long-term holistic planning are necessary to grasp and best manage the complexities of the urban water management system. The benefits of leadership such as this are only surpassed in their importance by those of social agency and community action. Urban dwellers are increasingly engaged in large social networks that allow individual ideas to rapidly expand into coordinated calls for change. While cities may have traditionally been the seats of political power, they can now also be considered focal points for social movement. Where effective communication (on issues such as potable reuse and water pricing, for example) that is supported by smart data systems can create an urban population well-informed of the importance of effective water management, the likelihood of sustainable initiatives being successfully implemented is significantly increased.

References

2030 Water Resources Group. 2013. Managing Water Use in Scarce Environments: A Catalogue of Case Studies. 2030 Water Resources Group, Washington DC, USA.

Accenture. 2012. Water and Shale Gas Development – Leveraging the US Experience in New Shale Developments. Available at: https://www.accenture.com/t20150527T211219__w__/it-it/_acnmedia/Accenture/Conversion-Assets/DotCom/Documents/Local/it-it/PDF_2/Accenture-Water-And-Shale-Gas-Development.pdf (accessed February 2017).

Allen, A., Davila, J.D. and Hofmann, P. 2006. The peri-urban water poor: citizens or consumers? *Environment and Urbanisation*, 18, 333–351.

Amaro, H.M., Macedo, A.C. and Malcata, F.X. 2012. Microalgae: An alternative as sustainable source of biofuels? *Energy*, 1, 1–9.

Beatley, T. and Newman, P. 2013. Biophylic cities are sustainable, resilient cities. *Sustainability*, 5, 3328–3345.

Berazneva, J and Lee, D.R. 2011. Explaining the African Food Riots of 2007-2008: An Empirical Analysis. Cornell University, New York, USA.

Bertani, R. 2010. Geothermal energy: an overview of resources and potential. In Proceedings of the International Conference on National Development of Geothermal Energy Use, Slovakia. Available at: https://www.geothermal-energy.org/pdf/IGAstandard/ISS/2009Slovakia/I.1.Bertani.pdf (accessed February 2017).

Bildstein, C. 2007. Power price rise alarms industry. *The Advertiser*. 14 May, Adelaide, South Australia, Australia.

Biomimicry 3.8. 2014a. Biolytix® Water Filter. Available at: www.asknature.org/product/f9d2ab73e15de8d6e44bc15cac4549a3 (accessed January 2017).

Biomimicry 3.8. 2014b. Shark Paint. Available at: www.asknature.org/product/09b2f1ecaf4ba1641a99fbb9a9df6f59 (accessed January 2017).

Bouton, S., Cis, D., Mendonca, L., Pohl, H., Remes, J., Ritchie, H. and Woetxel, J. 2013. How to Make a City Great. McKinsey & Company, New York.

BP. 2014. BP Energy Outlook 2035. BP, London, UK.

Byers, 2012. Modelling and Transitions of the UK Water-Energy Nexus. Available at: www.water.ox.ac.uk/120416conference/posters/byers.pdf (accessed January 2017).

C40 Cities. 2014. Climate Action in Megacities. C40, London.

Caffour, I. 2008. Energy Efficient Water and Wastewater Treatment. Environmental Knowledge Transfer Network, Cranfield University, UK.

Carbajales-Dale, M., Barnhart, C.J. and Benson, S.M. 2014. 'Can we afford storage? A dynamic net energy analysis of renewable electricity generation supported by energy storage', Energy and Environmental Science, Royal Society of Chemistry.

Chertow M.R. 2004. Industrial symbiosis. In Cleveland, C.J. (ed.) Encyclopaedia of Energy. Elsevier, San Diego, USA.

CIWEM. 2014. Food for thought. *The Environment*, 19, 38–39.

Cooley, H. and Donnelly, K. 2013. Water-Energy Synergies: Coordinating Efficiency Programs in California. Pacific Institute, California.

Copeland, C. 2014. Energy-Water Nexus: The Water Sector's Energy Use. Congressional Research Service, USA.

Cordell, D., Drangert, J-O. and White, S. 2009. The story of phosphorus: global food security and food for thought. *Global Environmental Change*, 1, 1–14.

CSIRO. 2012. Coal Seam Gas Developments – Predicting Impacts. CSIRO, Canberra, Australia.

da Silva, J., Kernaghan, S. and Luque, A. 2012. A systems approach to meeting the challenges of urban climate change. *International Journal of Urban Sustainable Development*, 1, 1–21.

Daigger, G. 2009. Evolving urban water and residuals management paradigms: water reclamation and reuse, decentralisation and resource recovery. *Water Environment Research*, 81, 809–823.

Denver Water. 2014. From Forests to Faucets: U.S. Forest Service and Denver Water Watershed Management Partnership. Available at: www.denverwater.org/supplyplanning/watersupply/partnershipUSFS/ (accessed February 2017).

Dermody, B.J., van Beek, R.P.H., Meeks, E. *et al.* 2014. A virtual water network of the Roman world. *Hydrology and Earth System Sciences*, 18, 5025–5040.

Domenech, T. and Davies, M. 2011. Structure and morphology of industrial symbiosis networks: The case of Kalundborg. *Procedia – Social and Behavioural Sciences*, 10, 79–89.

Dominguez-Faus, R., Powers, S. Burken, J.G. and Alvarez, P.J. 2009. The water footprint of biofuels: a drink or drive issue? *Environmental Science and Technology*, 43, 3005–3010.

Elimelech, M. and Phillip, W.A. 2011. The future of seawater desalination: energy, technology and the environment. *Science*, 333, 712–717.

Energy Sector Management Assistance Program. 2012. A Primer on Energy Efficiency for Municipal Water and Wastewater Utilities. The International Bank for Reconstruction and Development/The World Bank Group, Washington DC, USA.

Environment Agency and Natural Resources Wales. 2013. Water stressed areas - final classification. Environment Agency, Bristol, UK.

ESTAT (Statistical Office of the European Union). 2013. Urban waste water treatment. Available at: www.eea.europa.eu/data-and-maps/indicators/urban-waste-water-treatment/urban-waste-water-treatment-assessment-3 (accessed January 2017).

Gerbens-Leenes, P.W., Hoekstra. A. Y. and Van der Meer, Th.H. 2008. The water footprint of bio-energy: global water use for bio-ethanol, bio-diesel, heat and electricity. Available at: http://www.utwente.nl/ctw/wem/organisatie/medewerkers/hoekstra/reports/report34.pdf (accessed January 2017).

Granit, J. 2011. Will growth in energy demand compete with scarce water resources? SIWI, Stockholm, Sweden.

Grimm, N.B., Faeth, S.H., Golubiewski, N.E., Redman, C., Wu, J., Bai, X. and Briggs, J.M. 2008. Global change and the ecology of cities. *Science*, 319, 756–760.

Grubert, E. and Kitasei, S. 2010. How energy choices affect fresh water supplies: a comparison of US coal and natural gas. WorldWatch Institute, Washington DC, USA.

Grubler, A., Bai, X., Buettner, T., Dhakal, S., Fisk, D. J., Ichinose, T., Keirstead, J.E., Sammer, G., Sattherthwaite, D., Schulz, N.B., Shah, N., Steinberger, J. and Weisz, H. 2012. Urban energy systems. In Global Energy Assessment – Towards a Sustainable Future. Cambridge University Press, Cambridge, UK

Gupta, J. and van der Zaag, P. 2008. Interbasin water transfers and integrated water resources management: where engineering, science and politics interlock. *Physics and Chemistry of the Earth*, 33, 28–40.

Hallegatte, S., Green, C., Nicholls, R.J. and Corfee-Morlot, J. 2013. Future flood losses in major coastal cities. *Nature Climate Change*, 3, 802–806.

Han, J., Fontanos, P., Fukushi, K., Herath, S., Heeren, N., Naso, V., Cecchi, C., Edwards, P. and Takeuchi, K. 2012. Innovation for sustainability: towards a sustainable urban future in industrialized cities. *Sustainability Science*, 7, 91–100.

Hancock, N.T., Xu, P., Roby, M.J., Gomez, J.D. and Cath, T.Y. 2013. Towards direct potable reuse with forward osmosis: technical assessment of long-term process performance at the pilot scale. *Journal of Membrane Science*, 445, 34–46.

Harman, J. 2013. The Shark's Paintbrush: Biomimicry and How Nature is Inspiring Innovation. Nicholas Brealey Publishing, London, UK.

Hoornweg, D. and Bhada-Tata, P. 2012. What a Waste: a global review of solid waste management. Urban Development Series Knowledge Papers, The World Bank, Washington DC, USA.

Hussey, K. and Pittock, J. 2012. The energy-water nexus: managing the links between energy and water for a sustainable future. *Ecology and Society*, 17(1), 31.

International Desalination Association. 2011. The State of Desalination 2011. IDA, USA.

International Energy Agency. 2006. Hydrogen Production and Storage: R&D Priorities and Gaps. IEA, Paris, France.

International Energy Agency. 2011. IEA Coal Information. IEA, Paris, France.

International Energy Agency. 2012. Technology Roadmap: Hydropower, IEA, Paris, France.

International Energy Agency. 2013a. World Energy Outlook 2013. IEA, Paris, France.

International Energy Agency. 2013b. Key World Energy Statistics 2013. IEA, Paris, France.

International Renewable Energy Agency. 2012. Renewable Energy Technologies: Cost Analysis Series: Hydropower. IRENA, Bonn, Germany.

IRIN. 2013. Powerful storm and power shortages fill Gaza with waste water. Available at: www.irinnews.org/report/99358/powerful-storm-and-power-shortages-fill-gaza-with-waste-water (accessed January 2017).

Jha, A.K., Bloch, R. and Lamond, J. 2012. Cities and Flooding: a guide to integrated urban flood risk management for the 21st century. The World Bank, Washington DC, USA.

Jiang, Y. 2009. China's water scarcity. *Journal of Environmental Management*, 90, 3185–3196.

Joshi, Y.K., Tortajada, C. and Biswas, A.K. 2012. Cleaning of the Singapore River and the Kallang Basin in Singapore: human and environmental dimensions. *Ambio*, 41, 777–781.

Kalbar, P.P., Karmakar, S. and Asolekar, S.R. 2013. Assessment of wastewater treatment technologies: life cycle approach. *Water and Environment Journal*, 27, 261–268.

Kalundborg Symbiosis. 2015. Diagram. Available at: http://www.symbiosis.dk/en/diagram (accessed January 2017).

Kennedy, C., Cuddihy, J. and Engel-Yan, J. 2007. The changing metabolism of cities. *Journal of Industrial Ecology*, 11, 43–59.

Kennedy, S.J., Priestley, S., Cook, S., Inman, M., Gregory, A. and Hall, M. 2008. Energy use in the provision and consumption of urban water in Australia and New Zealand, CSIRO, Canberra, Australia.

Larson, K.L., Polsky, C., Gober, P., Chang, H. and Shandas, V. 2013. Vulnerability of water systems to the effects of climate change and urbanisation: a comparison of Phoenix, Arizona and Portland, Oregon (USA). *Environmental Management*, 52, 179–195.

Lenzen, M., Bhaduri, A., Moran, D., Kenmoto, K., Bekchanov, G. A. and Foran, B. 2012. The role of scarcity in global virtual water flows. ZEF – Discussion Papers on Development Policy No. 169. Centre for Development Research, University of Bonn.

Lofrano, G. 2012. Green Technologies for Wastewater Treatment: Energy Recovery and Emerging Compounds Removal. Springer, London, UK.

Lucci, P. 2014. An urban dimension in a new set of development goals. Overseas Development Institute Working Paper, ODI, London, UK.

Marshall, F., Waldman, L., MacGregor, H., Mehta, L. and Randhawa, P. 2009. On the edge of sustainability: perspectives on peri-urban dynamics. STEPS Working Paper 35, STEPS Centre, Brighton, UK.

Matichich, M., Byers, B., Pitzler, D. and Ahmed, S. 2012. The Changing Value of Water to the US Economy: Implications from Five Industrial Sectors. CH2M HILL, Denver, USA.

Moura, P.S., Lopez, G.L., Moreno, J.I. and De Almeida, A.T. 2013. The role of Smart Grids to foster energy efficiency. *Energy Efficiency*, 6, 621–639.

Mutchek, M. and Williams, E. 2014. Moving towards sustainable and resilient smart water grids. *Challenges*, 5, 123–137.

National Water Commission. 2008. Emerging trends in desalination: a review. Australian Government, Canberra, Australia.

Natural Systems Utilities. 2014. Battery Park City NYC – Water Reuse. Available at: www.naturalsystemsutilities.com/battery-park-city-water-reuse-system/ (accessed January 2017).

Novotny, N. 2012. Water and energy link in the cities of the future – achieving net zero carbon and pollution emissions footprint. In Lazarova, V., Choo, K.H. and Cornel, P. (eds) Water/Energy Interactions of Water Reuse. IWA Publishing, London, UK.

Novotny, V. and Novotny, E.V. 2011. Water centric cities of the future – towards macro scale assessment of sustainability. In Howe, C. and Mitchell, C. (eds) Water Sensitive Cities. IWA Publishing, London, UK.

Oxfam. 2012. The Hunger Grains. Oxfam GB, Oxford, UK.

Perera, C., Zaslavsky, Z., Christen, P. and Georgakopoulos, D. 2014. Sensing as a service model for smart cities supported by internet of things. *Transactions on Emerging Telecommunications Technologies*, 1, 1–12.

Perrone, D. and Hornberger, G.M. 2014. Water, food, and energy security: scrambling for resources or solutions? *WIREs Water*, 1, 49–68.

Pidou, M., Memon, F.A., Stephenson, T., Jefferson, B. and Jeffrey, P. 2007. Greywater recycling: treatment options and applications. *Proceedings of the Institution of Civil Engineers: Engineering Sustainability*, 160, 119–131.

Prime Minister's Science, Engineering and Innovation Council. 2010. Challenges at Energy-Water-Carbon Intersections. PMSEIC, Canberra, Australia.

PUB (Singapore's National Water Agency). 2016. The Singapore Water Story. Available at: https://www.pub.gov.sg/watersupply/singaporewaterstory. (accessed February 2017).

Reardon, S. 2012. Water Shortages hit US Power Supply. Available at: http://www.newscientist.com/article/dn22178-water-shortages-hit-us-power-supply.html#.U5MwimxwZl8 (accessed January 2017).

Rodriguez, D.J., Delgado, A., DeLaquil, P. and Sohns, A. 2013. Water Papers: Thirsty Energy. World Bank, Washington DC, USA.

Rygaard, M., Binning, P.J. and Albrechtsen, H-J. 2011. Increasing urban water self-sufficiency: new era, new challenges. *Journal of Environmental Management*, 92, 185–194.

Sanders, K.T. and Webber, M.E. 2012. Evaluating the energy consumed for water use in the USA. *Environmental Research Letters*, 7, 1–11.

Sattherthwaite, D., McGranahan, G. and Tacoli, C. 2010. Urbanization and its implications for food and farming. *Philosophical Transactions of the Royal Society B*, 365, 2809–2820.

Schuetze, T., Lee, J.-W. and Lee, T-G. 2013. Sustainable urban (re-) development with building integrated energy, water and waste systems. *Sustainability*, 5, 1114–1127.

SingStat. 2014. SingStat Table Builder. Available at: www.tablebuilder.singstat.gov.sg/publicfacing/mainMenu.action (accessed January 2017).

Stanford Woods Institute for the Environment (SWIE). 2013. Water and Energy Nexus: A Literature Review, Stanford University, Stanford, California, USA.

Stillwell, A.S., King, C.W., Webber, M.E., Duncan, I.J. and Hardberger, A. 2010. The energy-water nexus in Texas. *Ecology and Society*, 16.

Stinchcombe, K. and Brennan, L. 2014. Blue City: The Water Sustainable City of the Near Future. Blue Economy Initiative, Canadian Water Network, Canada.

Thames Water. 2012. Why does London need the Thames Tideway Tunnel? Thames Water, London, UK.

Thames Water. 2014a. Facts and Figures. Available at: https://corporate.thameswater.co.uk/media/facts-and-figures (accessed February 2017).

Thames Water. 2014b. Reducing Leakage. Available at: https://www.thameswater.co.uk/help-and-advice/bursts-and-leaks/tackling-leakage (accessed February 2017).

Thames Water. 2014c. Thames Gateway Water Treatment Works. Available at: https://www.thameswater.co.uk/help-and-advice/water-quality/where-our-water-comes-from/thames-gateway-water-treatment-works/how-the-treatment-works-operates (accessed February 2017).

Thomas, G.B. and Crawford, D. 2011. London Tideway Tunnels: tackling London's Victorian legacy of combined sewer overflows. *Water Science Technology*, 63, 80–87.

Tortajada, C. 2006. Water management in Singapore. *Water Resources Development*, 22, 227–240.

Tortajada, C. and Joshi, Y.K. 2013. Water demand management in Singapore: involving the public. *Water Resources Management*, 27, 2729.

UNFPA. 2007. State of the World Population 2007: Unleashing the Potential of Urban Growth. UNFPA, New York, USA.

United Nations. 2006. World Urbanization Prospects: The 2005 Revision. UN, New York, USA.

United Nations Development Programme. 2006. Human Development Report 2006 - Beyond Scarcity: Power, Poverty and the Global Water Crisis. UN, New York, USA.

United Nations Environment Program. 2003. Groundwater and its Susceptibility to Degradation. UNEP, Nairobi, Kenya.

United Nations International Strategy for Disaster Risk Reduction. 2012. Making Cities Resilient Report 2012. UNISDR, New York, USA.

United Nations World Water Assessment Programme. 2014. The United Nations World Water Development Report 2014: Water and Energy. UNESCO, Paris.

United States Bureau of Land Management. 2012. Oil Shale and Tar Sands Final Programmatic Environmental Impact Statement. USBLM, Washington DC, USA.

United States Department of Energy. 2004. Energy-Efficient Water Heating. USDoE, Washington DC, USA.

United States Department of Energy. 2009. Estimating Freshwater Needs to Meet Future Thermoelectric Generation Requirements. USDoE, Washington DC, USA.

United States Energy Information Administration. 2014. New York: State Profile and Energy Estimates. Available at: www.eia.gov/state/?sid=ny (accessed January 2017).

United States Environmental Protection Agency. 2013. Energy Efficiency in Water and Wastewater Facilities. USEPA, Washington DC, USA.

Van Heerden, I.L. 2007. The failure of the New Orleans levee system following Hurricane Katrina and the way forward. Public Administration Review, Special Issue.

van Vliet, M.T.H., Yearsley, J.R., Ludwig, F., Vogele, S., Lettenmaier, D.P. and Kabat, P. 2012. Vulnerability of US and European electricity supply to climate change. *Nature Climate Change*, 2, 676–681.

Water Corporation. 2014. Groundwater replenishment. Available at: http://watercorporation.com.au/water-supply-and-services/solutions-to-perths-water-supply/groundwater-replenishment (accessed January 2017).

Wessex Water. 2011. Catchment management: managing water – managing land. Wessex Water, Bath, UK.

Wien Energie. 2014. Energy Policy Guidelines. Wien Energie, Vienna, Austria.

Wiggins, S., Keats, S. and Compton, J. 2010. What caused the food price spike of 2007/08? Lessons for world cereals markets. Overseas Development Institute, London, UK.

Willis, R.M., Stewart, R.A. and Emmonds, S.C. 2010. Pimpama-Coomera dual reticulation end use study: pre-commission baseline, context and post commission end use prediction. Available at: http://www98.griffith.edu.au/dspace/bitstream/handle/10072/34322/64327_1.pdf?sequence=1 (accessed February 2017).

Woods-Ballard, B., Kellagher, R., Martin, P., Jefferies, C., Bray, R. and Shaffer, P. 2007. The SUDS Manual. CIRIA, London, UK.

World Bank. 2010. Cities and Climate Change: an urgent agenda. World Bank, Washington DC, USA.

World Bank. 2012. Renewable Energy Desalination: an emerging solution to close the water gap in the Middle East and North Africa. World Bank, Washington DC, USA.

World Bank. 2013. Improving the Assessment of Disaster Risks to Strengthen Financial Resilience. World Bank, Washington DC, USA.

World Bank. 2014. World DataBank. Available at: databank.worldbank.org/data/home.aspx# (accessed January 2017).

World Energy Council. 2010. Water for Energy. World Energy Council, London, UK.

World Health Organisation/UNICEF. 2008. A Snapshot of Sanitation in Africa. WHO/UNICEF, USA.

Wu, M., Mintz, M., Wang, M. and Arora, S. 2008. Consumptive Water Use in the Production of Bioethanol and Petroleum Gasoline. US Department of Energy, Washington DC, USA.

Xi, X. and Poh, K.L. 2013. Using system dynamics for sustainable water resources management in Singapore. *Procedia Computer Science*, 16, 157–166.

Yuan, Z., Pratt, S. and Batstone, D.J. 2012. Phosphorus recovery from wastewater through microbial processes. *Current Opinion in Biotechnology*, 23, 878–883.

5

Eat

The UN's FAO defines food security as having been secured when 'all people at all times have physical, social and economic access to sufficient safe and nutritious food to meet their dietary needs and food preferences for an active and health life'. All signatories to the UN's Universal Declaration of Human Rights have an obligation to ensure their citizens achieve an adequate standard of living, including adequate food. To achieve this goal, countries employ a variety of measures, including subsidies and the transfer of resources, that target either the producers or consumers of food and which vary considerably from country to country.

Achieving food security, indeed producing any food product, is impossible without water. Knowledge of just how much water is involved is, however, largely absent from the public conscience. While consumers understand that crops must be sustained by rainfall or irrigation, the majority are unaware of the large volumes of water required or the impacts which that water use has on societies, economies and the environment, often in communities and countries many thousands of kilometres away.

This lack of awareness prevails against a backdrop of both hunger and overconsumption. Amazingly, of the global population of 7.2 billion, 60% suffer a poor diet. Over 800 million people are undernourished, 2 billion suffer micronutrient deficiencies and a further 1.4 billion are over-consuming at dangerous levels (FAO 2013a).

To improve global food security in a manner that also sustains global water resources, the following two objectives should be a high priority for decision makers:

1) **Improve the water efficiency of food production, particularly rain-fed agriculture**. Despite 80% of agriculture being rain-fed, water efficiency initiatives have typically focused on irrigated systems where absolute yields are higher. However, thanks to the increasingly globalised nature of food trade, the water resources that support agricultural production are now accessible from almost anywhere on the planet. This highlights the need for the careful stewardship of all water resources, regardless of local water abundance or scarcity. The current focus of water efficiency initiatives towards irrigated agricultural systems should therefore be rebalanced to give rain-fed agriculture greater attention. Efforts are also required to ensure that, for a given country or region, trade policies and agreements do not inadvertently incentivise the unsustainable use of water.
2) **Ensure that water is one of the primary influences on food choice**. The characteristics of the global food system reflect choices made by both consumers and governments. Consumers in developed and emerging economies increasingly demand

Water Resources: A New Water Architecture, First Edition.
Alexander Lane, Michael Norton and Sandra Ryan.

protein-rich diets that require large volumes of water to produce. They also demand year-round product availability and so the global food system has responded by establishing complex international trading networks. Whilst satisfying consumer desires, these trading networks often fail to account for their impacts on water resources. This situation needs to change and consumer decision making, through its effect on demand, has the potential to act as a highly effective mechanism. In contrast to the developed world, consumers in developing countries often lack the opportunity to make their own food choices. When they eventually receive this right, they must be empowered to make sustainable decisions.

Rather than directly stimulating demand, government decisions establish the framework within which resources are managed. Unfortunately, departmentalised responsibilities mean that food and water strategies are typically developed in isolation, often leading to the implementation of food policies that inadvertently promote the degradation of water resources, either domestically or abroad. Cross-department policy making is crucial if mutually beneficial outcomes are to be secured.

To stimulate the collective initiative needed to achieve these objectives, this chapter highlights the vital role played by water in the global food system. It draws attention to those trends that will have the greatest influence on our ability to feed future populations and focuses in particular on the increasingly globalised ways in which food and water systems interact.

5.1 The Hidden Water in Food

5.1.1 How Much Water is Hidden in Food?

The global food system uses more water than any other industry. To maintain basic health, we each require 2–4 litres of drinking water per day in addition to around 20 litres for safe sanitation (SIWI 2012). Direct per capita water use in most European countries is between 200 and 300 litres a day (UNDP 2006). Contrast that to the 2,000 to 5,000 litres required to produce a person's food for that one day (SIWI 2012), and the scale of the demands for water from food production become apparent. In fact, of the estimated 7,452 Gm^3 of water required to sustain humanity for one year (our global water footprint), 85% is related to food (Hoekstra and Chapagain 2006).

This relationship between water and food is replicated when national water footprints are considered. Table 5.1 shows how agriculture is responsible for 56% of Australia's water footprint, 59% of the water footprint of the USA, 86% of that of China, and 90% of the water footprint of Brazil. In fact, in all countries, agriculture is responsible for the largest share of the national water footprint (Hoekstra and Mekonnen 2011).

Although all nations have a proportionally large agricultural water footprint, different components of the global food system are more water intensive than others. Figure 5.1 shows the reliance of the global population on different food groups alongside the respective water footprint of each group.

Considered on a global scale, cereals are by far the most important source of dietary energy, delivering almost 50% of total calorific intake (FAO 2012). The water footprints of products in this food group are relatively small; for example, it takes 1,644 m^3 of water to produce 1 tonne of cereals, equivalent to 0.51 litres for every kilocalorie (kcal)

Table 5.1 The contribution of agriculture to national water footprints.

		Agricultural water footprint	
Region	**Water footprint (Gm^3/yr)**	**Gm^3/yr**	**%**
Global total	7,452	6,391	86
		Agricultural water footprint	
Country	**Water footprint ($m^3/cap/yr$)**	**$m^3/cap/yr$**	**%**
USA	2,488	1,459	59
Australia	1,393	777	56
UK	1,245	810	65
China	702	605	86
Brazil	1,381	1,242	90
Jordan	1,303	1,209	93
Egypt	1,097	919	84
India	980	921	94
Thailand	2,223	2,131	96
Global average	**1,243**	**1,067**	**86**

Source: adapted from Hoekstra and Chapagain (2006).

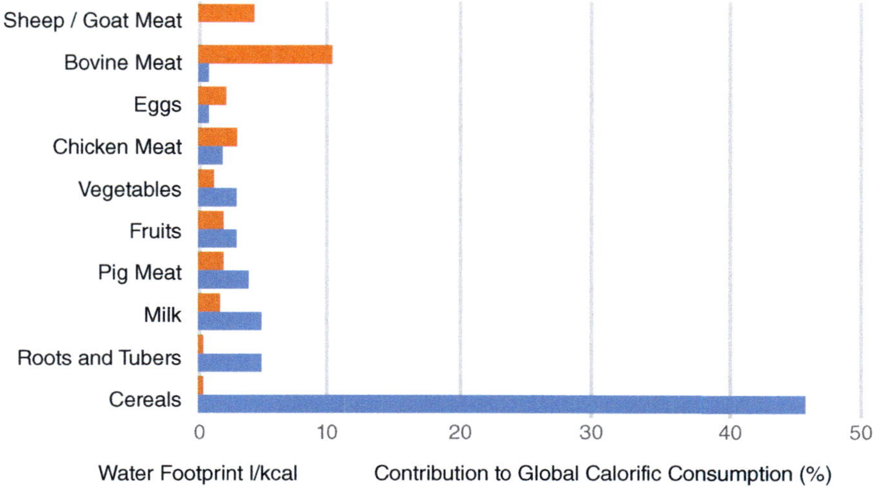

Figure 5.1 Water footprint of food groups. Water footprint data from Mekonnen and Hoekstra (2010). Calorific consumption data from FAOSTAT.

(Mekonnen and Hoekstra 2012). Contrast this water footprint to that of dairy products and meat. It takes 1.8 m^3 of water to produce 1 kcal of milk and as much as 10.2 m^3 of water to produce 1 kcal of beef.

The differences in these figures are explained by the fact that while vegetable products provide a direct food source, animal products require water not only for livestock drinking but also, crucially, to grow the livestock's feed. Mekonnen and Hoekstra (2012) found that 98% of water use along the supply chain of animal products is attributable to growing feed.

Somewhat fortunately, the current global population relies on meat and dairy products to satisfy only 13% of total calorific intake. However, meat and dairy consumption is growing, and growing at an increasingly rapid rate. Historically, diets based on animal products have been portrayed as those enjoyed by persons of high stature within society. If the populations of developing economies seek to emulate this lifestyle, humanity will have a far greater challenge in achieving future food and water security. Thirty-five percent of total global crop production is already dedicated to producing feed for animals (IGEL 2013).

As this discussion illustrates, analytical advances are increasingly enabling researchers to quantify the water footprint of food products and, in turn, the water footprint of different components of the global food system. Understanding the impacts of this water use is however a far more complex task, one that requires a thorough evaluation of where that food is produced, the methods used in its production and the 'types' of water consumed in the process.

5.1.2 The Impact of Water Use in the Global Food System

At this point it is necessary to return to the concepts of green, blue and grey water that are introduced in Chapter 3. The vast majority of agricultural systems are rain-fed, that is, they rely on green water to sustain their crops and livestock. Globally, this figure is around 80% but in some regions it is much higher; in Sub-Saharan Africa for example, 95% of agriculture is rain-fed (Wani *et al.* 2009). Throughout most of human history, green water flows have been sufficient to sustain our populations. Where green water is available in sufficient volumes, supporting the production of water-intensive foods such as rice, (see Figure 5.2 for water footprint) can be entirely sustainable and can occur without detriment to other users of water. This is because green water cannot easily be allocated to other uses. Its so-called 'opportunity cost' is low (see Breakout Box 5.1).

In comparison to green water, the opportunity cost of blue water use is high. Producing water-intensive food products where green water is scarce necessitates the artificial diversion of sources of blue water such as rivers and underground aquifers for the purposes of irrigation. With this infrastructure in place, blue water becomes a highly mobile resource, readily allocated and reallocated between municipal, industrial, recreational or environmental needs.

In addition to using significant volumes of blue water, irrigated agricultural systems are often characterised by the widespread application of fertilisers and pesticides. Where these substances enter the environment, usually as a result of runoff caused by rainfall or by poor irrigation and land management practices, water is required to dilute the pollutants so that water quality standards are maintained. This means that fertiliser- and pesticide-intensive agricultural systems are often characterised by large grey water footprints, particularly when the application of these substances is poorly planned.

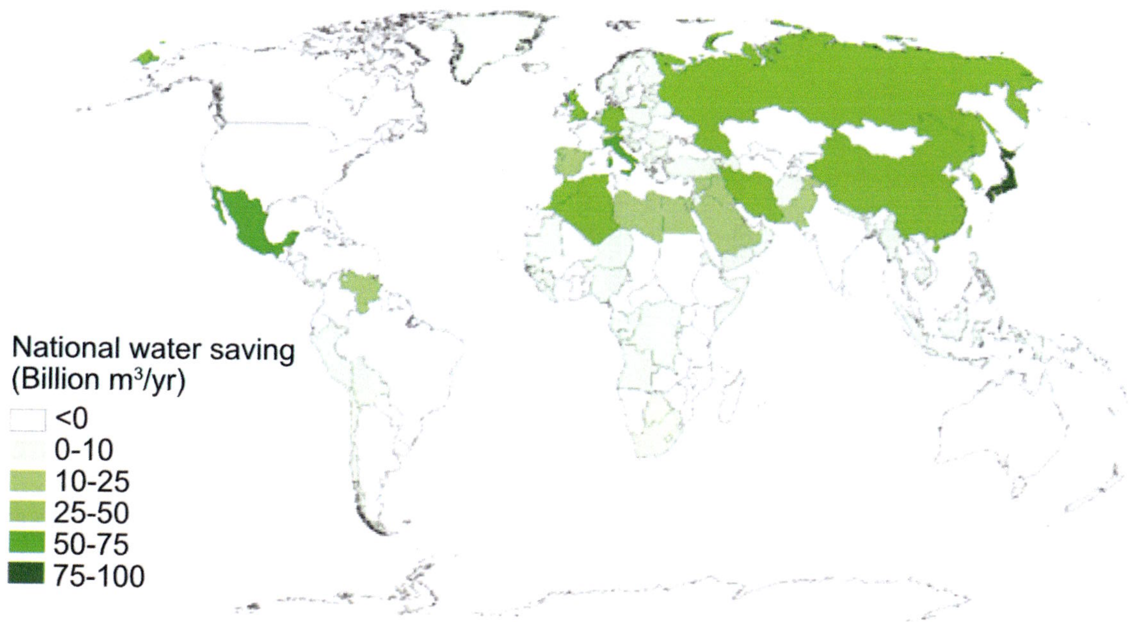

Figure 5.2 National water gains related to international trade in agricultural products.
Source: Chapagain *et al.* (2006); data collected over period 1997–2001.

Breakout Box 5.1 Opportunity cost

The opportunity cost of a decision is the value of the next best alternative forgone. For example, if a farm can produce one million tonnes of wheat or two million tonnes of corn, the opportunity cost of producing one million tonnes of wheat is two million tonnes of corn. The concept of opportunity cost is very important for the management of resources. Used appropriately, the concept can help to ensure that the most sustainable option is chosen.

Although rain-fed farming dominates global agriculture in terms of absolute acreage, the 20% of global farmland under irrigation is of vital importance for food production. Because irrigation increases the yields of most crops by between 100% and 400%, irrigated cropping is responsible for 40% of global food production (SIWI 2012) and half of agriculture's economic value (Langford *et al.* 2012).

It is highly unlikely that the rapid increases in global population and standard of living witnessed over the last 100 years could have been achieved without irrigated agriculture. In the twentieth century, large public investments in agricultural research led to rapid increases in crop yields in many industrialised and emerging economies. Between 1967 and 2007, while the global area of land under cultivation grew by 8%, global yields increased by 115% (Overseas Development Institute *et al.* 2012). While modern plant breeding techniques, improved agronomy and the development of inorganic fertilisers and modern pesticides all helped to fuel these advances (IFPRI 2002), irrigation was also a crucial supporting factor, reducing the constraining effect of otherwise relying on the vagaries of rainfall for water. This period of agricultural advancement, often referred to as the 'Green Revolution', enabled most industrial countries to eliminate the threat of starvation for their citizens and to achieve sustained food surpluses. Furthermore, farmers' incomes increased, in turn stimulating rural economies and promoting better nutrition (IFPRI 2002).

The socioeconomic advances witnessed as a result of the Green Revolution highlight the vitally important role that a strong agricultural sector can play in improving livelihoods. Furthermore, the increased yields achieved by irrigation provide strong justification for the continued diversion of blue water for this purpose. Similarly, the diversion of blue water for domestic uses and for industrial consumption also have clear social and economic advantages that are easy to comprehend and relatively straightforward to quantify.

The concept of allocating water (or any resource for that matter) to the demand which provides greatest economic benefit is a sound one; in practice however, the concept only works effectively if the benefits of all competing demands can be appropriately quantified. Unfortunately, this is rarely the case. Not all uses of blue water can be as easily valued as those of agriculture and domestic use. Quantifying the benefit of allocating blue water to environmental needs is particularly challenging, and a failure to accurately achieve this often leads to instances of environmental degradation and the depletion of biodiversity. Breakout Box 5.2 considers how these issues have been managed in the Murray-Darling Basin of Australia.

Rockstrom *et al.* (2009) calculate that between 20% and 50% of all river flows must be retained within the river system if riverine environments are to be sustained. However, because the benefits to society of naturally functioning ecosystems are largely intangible,

Breakout Box 5.2 The Murray-Darling River

The Murray-Darling is the world's 16th longest river and its vast basin drains much of the Australian states of Victoria and New South Wales, as well as a large portion of southern Queensland (a combined area that is more than four times larger than the UK). Australia's early leaders were quick to identify the economic value of the Murray-Darling and a program of dam construction and flow regulation was initiated. These projects supported a vast expansion in irrigated agriculture that now provides one-third of Australia's food supply, a large export industry and AUS$ 5 billion a year to the Australian economy (Beddington *et al.* 2012). By the late 1960s however, the continued diversion of blue water was leading to widespread environmental impact. High nutrient and salinity levels were damaging ecology and, due to channel siltation, the flow of the Murray-Darling often failed to reach the sea in drought years.

Recognising the need to arrest this deterioration, the Australian Government commenced a series of restorative initiatives. In November 2012, the Murray-Darling Basin Plan became law and the Australian Government began a process of buying back water allocations to meet environmental requirements. The management of the Murray-Darling basin and the allocation of its available water resources remains a highly divisive issue however. Furthermore, population growth and increasing competition for water from the mining and petroleum industries may well increase these pressures in the future.

environmental requirements for water are often the first to be compromised. Overuse of blue water has the potential to cause salinisation, water-logging and soil degradation, processes that ultimately result in the pollution of more blue water, in turn promoting a spiral of decline in water quality. Globally, about 30% of irrigated land is already severely or moderately affected by salinisation (FAO 2002).

Section 5.1.1 highlights how the water footprints of animal food products far exceed those of plant products. However, the location and type of the animal production system can have significant implications on both the total water footprint and, importantly for its subsequent impacts, its green, blue and grey constituent parts.

While the proliferation of intensive animal production systems has attracted criticism on the grounds of animal welfare, the water-related impacts are also worthy of note. Mekonnen and Hoekstra (2012) evaluated the total water use of various animal production systems and found that intensive systems generally had smaller water footprints than grazing or mixed systems, principally reflecting the improved feeding efficiencies achieved in the former. The authors found that in intensive systems, less feed was used to produce a given amount of food and hence total water savings were achieved as a result. The authors also found however that dependence on blue and grey water increased as production systems moved from grazing to intensive systems. Steinfeld *et al.* (2006) argue that the US livestock sector is responsible for around one-third of the nitrogen and phosphorus loads in its freshwater environments, a situation that often causes eutrophication and dead zones in coastal areas.

The differences in water intensity between different production systems mean that the water footprint of a given product can vary considerably from country to country. Hoekstra and Mekonnen (2012) found that the average water footprint (blue, green and grey combined) of beef produced and consumed in the US was more than 50% larger than that in the UK; the authors also found that in Niger, the water footprint of cereals was six times the global average. The production of a certain crop or food product will therefore be more suited to particular countries depending on the types of water required (green, blue and grey) and their respective availabilities. Notwithstanding this observation, and despite many countries being located in water-scarce environments, the majority of these nations still manage to deliver a level of food security to their citizens. To do so, they rely heavily on the global food system and, specifically, the import of goods from countries where they can be produced using less water than would otherwise be the case if that food were produced at home. When analysed at a global level, the benefit of this trade in 'virtual water' is clear. Chapagain *et al.* (2006) calculated that if all imported agricultural products were to have been produced in their respective countries of consumption, they would require somewhere in the order of 1,605 Gm^3 of water a year. In reality, the global trade in food means that humanity actually produces these products using around 1,250 Gm^3 of water, saving the planet 355 Gm^3 of water every year (Chapagain *et al.* 2006).

By studying virtual water flows at a finer level of detail, it is possible to identify those countries that, through trade in food, act as virtual water exporters or importers. For example, Egypt saves 3.6 Gm^3 of water a year by buying water-intensive food products from its trading partners. Contrast this to Thailand, which exports 28 Gm^3 of water a year, mostly through its flourishing trade in rice (Aldaya *et al.* 2010). Considering proportions as opposed to volumes, some countries exhibit an extreme dependence on water sources outside their national boundaries. In Malta for example, Hoekstra and Mekonnen (2012) calculate this dependency to be 92%. It is also greater than 80% in Kuwait, Jordan and Israel (Hoekstra and Mekonnen 2012). By analysing the global trade in food products, Chapagain *et al.* (2006) have compiled maps depicting the global flow of virtual water between countries. Figures 5.2 and 5.3 show those nations that import and export most virtual water through food.

At this point, it should be remembered that for certain countries, exporting large volumes of virtual water can be entirely sustainable. If green water resources are plentiful, then establishing a trade surplus in virtual water through food exports makes sense. The low opportunity cost of green water use means that, from an economic perspective, the choice of a government to export food products grown using green water is unlikely to be contested. Problems do arise however when countries use blue water to sustain water-intensive export industries. Ensuring that food trade policies give appropriate regard to water will therefore be crucial for improving the food and water security of the global population.

5.2 An Increasingly Important Problem

5.2.1 Population Growth

The global population has doubled in the space of 45 years, reaching 7.3 billion in the middle of 2015. It could be 11.2 billion by the end of the century (UNDESA 2015). In addition to the obvious influence of absolute increases in population, our ability to provide food for future populations in a manner that also protects and maintains water resources will be influenced by several other elements of demography and social change.

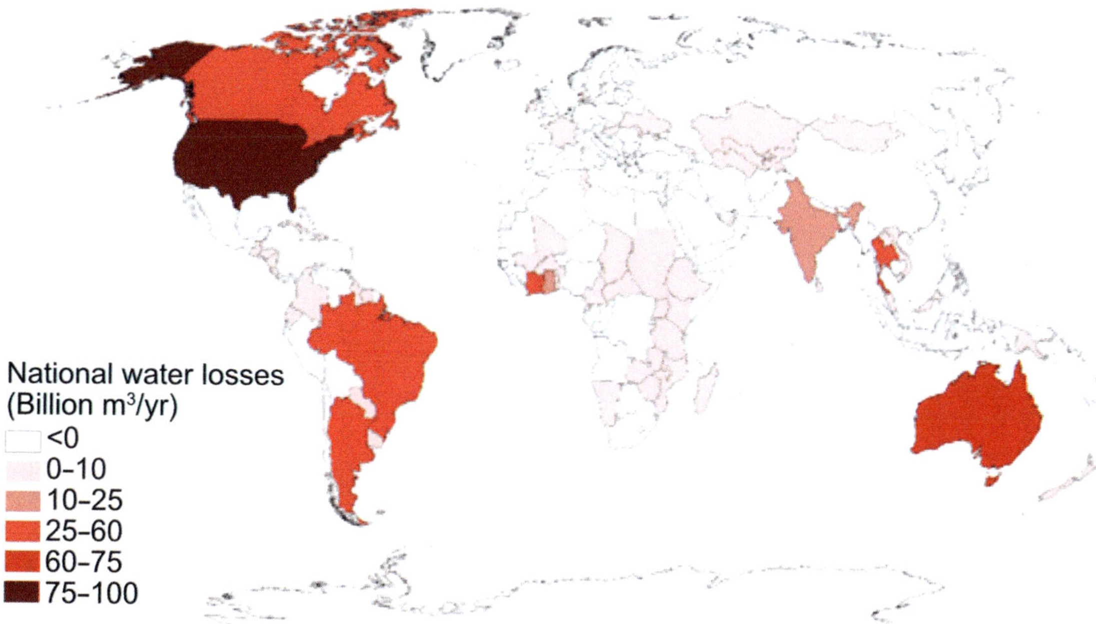

Figure 5.3 National water losses related to international trade in agricultural products.
Source: Chapagain *et al.* (2006); data collected over period 1997–2001.

Diet is a crucial factor and is explored thoroughly in Section 5.2.2. Urbanisation is another important influence. While systems of urban food provision can benefit from the ability to adopt centralised supply networks, they lack the available land to support local food production. Problems associated with the disposal of urban waste and wastewater, and the maintenance of air and water quality, are often experienced (FAO 2008). Typically, crops and livestock must be grown either overseas or in other domestic regions, and then transported over often long distances to the city. Intra-national as well as international virtual water flows therefore result, adding a further layer of complexity to the virtual water system. By way of example, the external water footprint of the city of London is almost four times the size of the water footprint that resides within its geographical boundaries (Feng *et al.* 2011). In Australia, the regional movement of food is vitally important to both food security and rural economies; 90% of jobs in food production and around half of all jobs in food processing and manufacturing are located in non-urban areas (Australian Government DAFF 2001).

China provides an excellent case study on the management of intra-national water flows, both physical and virtual. In the early 1990s, government investment in support of manufacturing and tertiary industries in the south of the country initiated a labour shift away from agricultural industries in those regions. This change, coupled with rising living standards, meant that food demands could no longer be supported by local farming. To compensate, trade links were established that moved food (and its associated virtual water) from the north to the south of the country. In 1999, this intra-national virtual water flow equated to 52 billion m^3 (Ma *et al.* 2006); however, paradoxically, it is the north of the country that is by far the most water scarce. The north supports 65% of the country's total arable land but possesses only 18% of the country's water (Piao *et al.* 2010), a situation that has encouraged the depletion of groundwater to sustain production (Ma *et al.* 2006). To address this issue, the Chinese government could have sought greater reliance on the import of water-intensive goods from outside its borders. Keen to achieve national self-sufficiency however, the government has instead embarked on the most significant water transfer project in human history, the South-to-North Water Transfer Project described in Breakout Box 4.7. When considered with an appreciation of the potential benefits of the international trade in virtual water, the South-to-North project appears illogical and a highly inefficient means of reducing water scarcity. Furthermore, considering the uncertainties of future climate change, it may well prove difficult for the project to achieve its aims. China is however also investing in a variety of other water-related initiatives. For example, the country's biotechnology capacity is the largest outside North America and its Seed Project aims to respond to the effects of climate change by developing more stress-resistant crops (Piao *et al.* 2010). China is also investing in more efficient farm infrastructure with the aim of reducing water use in irrigated agriculture to around 60% of 2010 levels (Liu and Yang 2012).

5.2.2 Changing Diet

The water footprint of different food products varies markedly; it therefore follows that diet imparts a large influence on the size of an individual's water footprint. Figure 5.4 shows the average water footprint of different food products and shows that meat and basic dairy goods are relatively water intensive, as are many of the luxury items enjoyed by those in developed countries (e.g. chocolate). Vanham *et al.* (2013) found that a diet

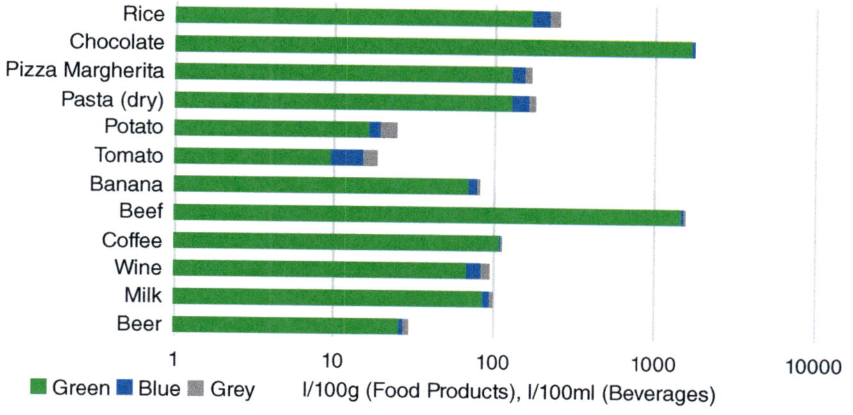

Figure 5.4 Water footprint of food products and beverages. Adapted from Water Footprint Network (2015).

high in animal proteins had a water footprint nearly 40% larger than that of a vegetarian diet. Furthermore, a 'healthy' diet (consisting of reduced amounts of sugar, meat and animal fats) was found to have a water footprint more than 20% lower than that of an average diet.

Figure 5.4 also shows how the size of the green, blue and grey water footprints varies between products. Items with relatively large blue and grey components indicate a more resource-intensive production process, requiring human intervention to divert water supplies, to produce fertilisers and pesticides, and to implement the farm management systems that may ultimately pollute surface watercourses and groundwater supplies. Interestingly, Figure 5.4 shows that it is often those products with a low overall water footprint, tomatoes for example (19 L/100 g), that have proportionally high blue water (30%) and grey water (20%) components. Beef in comparison, while having a significantly larger total water footprint (1,542 L/100 g), is only reliant on blue and grey water for 4% and 3% of this total, respectively. Once again, this example serves to highlight the need to explore the water footprint metric thoroughly before drawing conclusions as to the potential impacts of food production on water resources.

It should also be acknowledged that livestock raised on non-arable land without irrigation can have very little impact on water resources. When considered at the global scale, the majority of beef cattle are raised this way (Ridoutt and Huang 2012); industrial livestock systems however, which do have substantial water footprints, are expected to increase in number in order to support the dietary demands of increasingly middle-class populations (UNFAO 2006). It is predicted that the number of middle-class persons on the planet could more than double by 2030 (Kharas 2010) and that, as a result, per capita calorific intake from meat could rise by 40% by 2050 (IME 2013).

The numbers presented in Figure 5.4 suggest that an effective means of reducing the size of an individual's water footprint would be to reduce their consumption of animal products. Boersema and Blowers (2011) argue that the widespread adoption of vegetarian diets might even enable the global water footprint of food to be brought within the limits that can be safely and sustainably supported by the planet.

Initiating a transition in food consumption away from animal products would require a hugely significant shift in the social mindset. Meat is a highly symbolic food, and one intimately tied to success and prosperity. Figure 5.5 presents the cultural food hierarchy developed by Twigg (1984) alongside the respective water footprint of each product (or equivalent item). The pyramid shows how a diet based on red meat and animal products signifies success and prosperity, and how one based on cereals and root vegetables is symptomatic of a lower social class. With the exception of fish, the juxtaposition between this cultural hierarchy and the water footprints of the respective products is stark. The most aspirational food has a water footprint 20 times the size of the least. From a water resource perspective this cultural food hierarchy should be reversed, with grains and vegetables promoted to the top of the pyramid and meat relegated to a lowly position at its base.

The social constructs underpinning aspirational food groups also have deep roots in history and religion. Certain foods are sacrosanct in many countries, with governments even going so far as to subsidise their sale. The Egyptian government subsidises bread for example, and many of its population regard this subsidy as a basic right (IFPRI 2013). More than any other food item however, it is meat that is most frequently demanded by populations seeking to emulate their social peers. One need look no further than a restaurant menu to see how many dishes are named after the meat they contain. The dominant role that meat has over the sociology of food and eating is strong and will be very difficult to change.

Notwithstanding the challenge of influencing dietary choice, it should be remembered that diets can be changed and that this change can occur over short timeframes. Take organic food for example. Between 2000 and 2010, the global market in organic

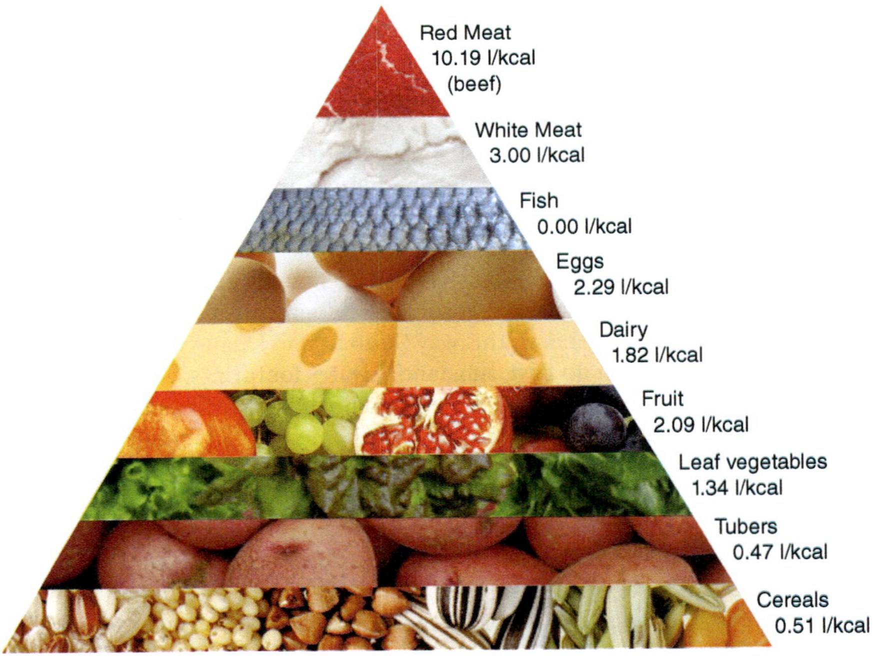

Figure 5.5 Cultural food pyramid and water footprints.
Cultural food hierarchy after Twigg (1984); water footprint data from Mekonnen and Hoekstra (2012).

products grew three-fold (Willer and Kilcher 2012) and organic products are now ubiquitous on supermarket shelves. The water footprint of an organic diet (particularly its grey water component) can often be less than that of a non-organic diet (Ercin *et al.* 2012).

Other dietary trends have also emerged over relatively short timeframes, founded on popular culture and celebrity endorsement. The 'functional food' market in the USA (including foods claimed to provide energy enhancement and weight management benefits) is estimated to be growing at a rate as high as 20% a year (PWC 2009). 'Slow food' and 'fair trade' are other examples of successful food movements. Health, social marginalisation, animal welfare and a rejection of intensive agricultural systems are all influencing factors behind the growth of these markets; however, disappointingly, water is absent as a driver, despite the 'sustainable' credentials claimed by so many recent dietary trends. This oversight means that many of those diets perceived to be 'sustainable' may in fact be relatively water intensive. Take the concept of 'localism', for example. The idea of reducing food miles and supporting local producers is well-founded and seems entirely sensible. However, assume you live in a water-scarce region and consider if supporting your local producer meant buying beef fed on forage grown on irrigated farmland. Under these conditions, it may in fact be far more water-efficient to purchase beef from a rain-soaked country many thousands of kilometres away. Of course, one must not forget the other elements of sustainability influencing this example: carbon emissions and animal and social welfare, for instance. However, the case does serve to illustrate the need to give much greater consideration to the role of water in our decisions around food.

5.2.3 Food Waste

Global food production stands at 4 billion tonnes a year, more than enough to feed the 7.2 billion people that currently inhabit the planet and probably sufficient to support one or two billion more (Ausubel *et al.* 2012). Current instances of hunger and malnutrition therefore reflect a lack of physical or economic access to food. The inability of people to reach food distribution points or the failure of market systems to supply populations are most frequently cited as the primary causes of widespread hunger (e.g. Vermeulen *et al.* 2012). Acknowledging the need to address this marginalisation is obvious, and yet certain elements of the current global food system appear to contradict this need.

Take food losses and food waste, for example. One-third of all food intended for human consumption is lost or wasted before it is eaten (FAO 2013b). One-third of all food is 1.3 billion tonnes, equivalent to four times the amount of food consumed each year in the USA. It is astounding that such a scenario can play out when over one-third of the global population goes hungry. Not only is the food itself wasted, but so are all the inputs that went into its production. The blue water footprint of annual lost and wasted food equates to the entire flow of the River Volga (the longest and largest, by discharge, in Europe) for a full year (FAO 2013b), or nearly three times the volume of water in Lake Geneva. The lost land equates to an area twice the size of Australia, and the wasted pesticides exceed the annual amount applied in Africa and Europe combined (Kummu *et al.* 2012). Imagine the lost and wasted food as a country, and its carbon emissions would be exceeded only by those of China and the USA (FAO 2013b).

While the proportion of food that goes unconsumed remains relatively constant around the world, the causes are very different. Figure 5.6 shows the percentage of total food waste incurred at each stage of the supply chain in developed and developing countries.

In developing countries, food losses are the issue and there is relatively little waste. Post-harvest losses reflecting inadequate storage facilities, lack of suitable cold chains and an inability to access markets (or knowledge of what products the market is currently demanding) all combine to mean that one-third of produce rots before it reaches a consumer. This situation is especially saddening when you consider how simple some of the technologies required to overcome this situation are. Lipinski *et al.* (2013) highlight an example from Afghanistan where the German government funded the distribution of small metal silos to households. By storing food in these containers, household food losses fell from around 20% to as little as 1% (Lipinski *et al.* 2013).

Food losses are less of an issue in developed countries. The majority of farming systems are highly advanced and use specialised technologies to ensure that crops are harvested at the optimum time. Cold chains are well established and transport networks are comprehensive. The problem in the developed world is waste at the point of harvest, at the point of sale, and in the home.

Modern consumers in developed countries not only demand year-round produce from all over the world regardless of season, they also require their produce to conform to strict aesthetic standards. Carrots must be perfectly straight, tomatoes blemish free, and oranges vivid in colour. These demands mean that a significant proportion of produce is never harvested. In the UK, this figure can be as high as 30% (IME 2013).

Of that food that does reach the supermarket and, in turn, the home, very large volumes are simply thrown away, either as a result of the careless attitude of those that can afford to behave wastefully, or due to misunderstandings over food safety. As much as 115 kg of food is wasted each year per person in Europe and North America compared

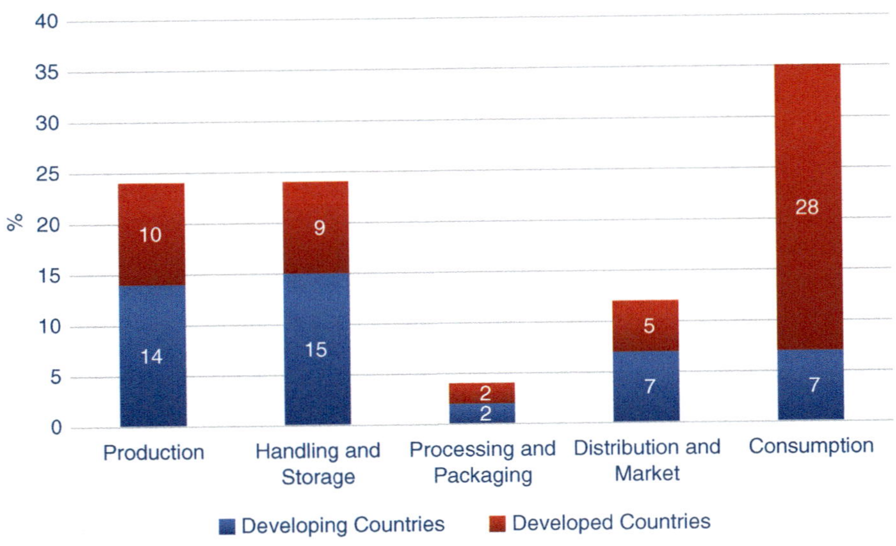

Figure 5.6 Percentage of food lost or wasted.
Source: Adapted from Lipinski *et al.* (2013).

to just 6 kg in Africa and 11 kg in southeast Asia (IGEL 2013). By another measure, the average American family of four wastes 6,000 kcal of food a day, equivalent to approximately US$ 1,600 a year (Lipinski *et al.* 2013).

Food labelling is an important issue in the food waste debate and the confusion between 'use by', 'best before' and 'display until' dates can leave many consumers uncertain as to whether or not their food remains safe to eat (see Section 5.3.2 for further discussion). It could be argued that this issue is as much the symptom of a disconnection to the origins of the food we eat. The senses of touch and smell should be sufficient to enable us to identify whether food is unsafe to eat; however consumers, especially in the developed world, now almost exclusively rely on the supermarket as their food safety guide. This reliance perhaps also explains the ignorance of most people towards the water footprint of food.

Several other elements of the modern supermarket system also promote food waste. For example, promotions linked to volume, particularly buy-one-get-one-free deals, encourage bulk buying that ultimately leads to waste in the home.

5.2.4 Food as a Globalised Commodity

Notwithstanding the need to improve the level of food and water security afforded to a significant portion of the global population, that the situation is not far worse reflects the considerable success of the international trade in food. By moving water-intensive goods from where they can be grown with high water productivity to where they would otherwise be grown at low productivity, international trade saves 28% of the international virtual water flows related to the trade of agricultural products, or 6% of the global water use in agriculture (Chapagain *et al.* 2006). Despite this benefit, the vast majority of trade in food occurs for reasons unrelated to water and, as a consequence, unintended impacts on water resources and the broader environment can result. The logical argument of producing food where it can be sustained by green water remains one that is significantly underutilised.

The European Union (EU) Common Agricultural Policy (CAP) is one of the best examples of how agricultural policy can affect trade. The scheme had its roots in 1950s western Europe whose societies and agricultural industries had been severely damaged by war and where, as a consequence, food security could not be guaranteed. The aim of the CAP was to boost agricultural productivity (essentially a move towards food self-sufficiency) and it did this by offering subsidies guaranteeing high prices to farmers (European Commission 2011). The CAP was extremely successful in achieving its aims; agricultural productivity increased rapidly, wheat yields doubled between 1962 and 1990 (European Commission Directorate-General for Agriculture and Rural Development 2013) and water productivity also improved. In fact, the water productivity of UK rain-fed agriculture tripled between 1950 and 2000 (Allan 2011).

However, the CAP also required huge financial expenditure. Seventy percent of the total EU budget went to service the CAP in 1984 and this figure remained at around 40% in 2013 (European Commission Directorate-General for Agriculture and Rural Development 2013). In addition, the use of subsidies tends to incentivise mono-cropping, the modification of hydrological processes and the increased use of farm inputs (Beddington *et al.* 2012). The EU also had to contend with almost permanent surpluses of farm commodities. To avoid a collapse in commodity prices, some of these surpluses were stored in vast stockpiles while others were exported (with the help of further subsidies) to developing countries in a process referred to as 'dumping'. It was cheaper for the affected developing countries to buy CAP-subsidised EU produce than to invest in indigenous agriculture. This meant that while the productivity of European farming increased rapidly, growth in the productivity of agriculture in the developing world stagnated. Between 1961 and 1991, African agricultural productivity grew at an annualised rate of just 1.3% (UNFAO 2001). Although the EU has now initiated systematic reforms to the CAP to address the issues of overproduction, detrimental environmental impacts and dumping, its legacy remains. Agricultural productivity in many areas of Africa has only recently returned to levels achieved in the 1970s (NIC 2012).

Notwithstanding the influence of agricultural subsidies on global trade, they are not the only reason why markets fail to adequately consider water. Take Australia for example, a country where subsidies represent just 3% of total farm income (as compared to an average for OECD countries of 19%; OECD 2013). Australia is the driest inhabited continent on Earth and yet its agricultural industry produces three times the food it needs to sustain its population. The rest of this food is exported and, as can be seen in Table 5.2, Australia is second only to the USA in terms of the volume of its agricultural virtual water exports (Chapagain and Orr 2008). Farming and agricultural activity occupy two-thirds of the Australian land surface, yet only 6% of the country is suitable for crop production. This constraint is overcome through irrigation, a process that consumes two-thirds of Australia's water (Australian Government DAFF 2011). The country's reliance on irrigation means that the blue water footprint of its exports are significant; one-third of the water footprint of its wheat exports are sustained by blue water, for example (Aldaya *et al.* 2010). Australia is therefore exporting a large portion of its blue water through food; and because the associated opportunity costs are high, this scenario promotes conflict between the agricultural industry and other waters users (refer back to Breakout Box 5.2 and the Murray-Darling basin, for example).

An excellent example of the manipulation of the virtual water trade in food can be seen in the history of Saudi Arabia. Water in the kingdom is extremely scarce and RWR amounts to just 2.4 Gm^3 a year (compare this to the 147 Gm^3 of RWR a year in the UK; Chapagain and Hoekstra 2004). Despite this scarcity, and thanks in no small part to its oil wealth, in the 1970s the Saudi government embarked on a series of agricultural

Table 5.2 Top agricultural virtual water exporters.

Rank	Net Virtual Water Exporting Country	Net Green Virtual Water Exporting Country	Net Blue Virtual Water Exporting Country
1	USA	USA	USA
2	Australia	Argentina	India
3	Argentina	Brazil	Pakistan
4	Canada	India	Australia
5	Thailand	Australia	Thailand
6	India	Canada	Argentina

Source: Adapted from Chapagain and Orr 2008 and Konar et al. (2012).

initiatives supported by subsidies for water, electricity and gasoline, all intended to help the country achieve food self-sufficiency. These subsidies incentivised farmers to pump groundwater from deep underground fossil aquifers and, in turn, through irrigation, allowed then to rapidly increase agricultural productivity.

In terms of achieving its primary aim, the agricultural investment program was hugely successful. Saudi Arabia became self-sufficient in wheat in 1984 and began exports of the crop shortly afterwards. However, with a growing population, the huge financial burden of sustaining subsidies and worries over ever-declining groundwater levels, it became clear that a reliance on domestic wheat production was unsustainable. Initially, attempts were made to address dwindling groundwater supplies through further investment, principally in desalination. The kingdom is the world's largest producer of desalinated water, using it to satisfy 70% of urban water demand but at huge financial expense. Increasingly, Saudi Arabia is realising that the aim of food self-sufficiency in such a dry country is unachievable and the government is gradually returning to agricultural imports to meet its food requirements.

This example highlights how for some countries – the vast majority of countries in fact – self-sufficiency in food is an unsustainable, potentially damaging and ultimately unachievable aim. It also indicates how, as is so often the case, realisation of the critical role played by water in maintaining societal functions typically only occurs once a monetary value (in this case associated with the cost of subsidies and desalination) can be ascribed to it.

Another useful example of the complexities associated with the trade of virtual water in food comes from the UK. The UK is the world's sixth largest virtual water importer (Chapagain and Orr 2008) and this enables it to maintain food and water security despite increasing levels of domestic water scarcity, particularly in the southeast of the country. The majority of the UK's external agricultural water footprint originates in Brazil (Chapagain and Orr 2008). This makes hydrological sense given that Brazil's RWR, at 8,233 Gm^3 a year, is significantly larger than the UK's (Chapagain and Hoekstra 2004). Other elements of the UK's agricultural water footprint are less sustainable, however. For example, 76% of the country's tomatoes are grown in Spain. This means that UK consumers use 13.3 million m^3 of Spanish water each year, 82% of which is drawn from blue water resources (Chapagain and Orr 2009). They also pollute an additional 1.3 million m^3 of Spanish water resources as a result of leached pesticides and fertilisers during the production process (Chapagain and Orr 2009). This water-intensive and environmentally damaging system persists because the importing country does not immediately feel the water-related impacts of its actions. It does however import a significant amount of risk. Unsustainable use of blue water cannot be maintained in perpetuity. Eventually the system will fail (e.g. when environmental degradation forces government action) and the importing country's food and water security will be damaged as a result. Making governments aware of this risk is important if damaging instances of virtual water trade are to be avoided.

5.2.5 Climate Change

The relationships between water and food will be increasingly influenced by the trends and events that accompany global changes in climate. These changes will vary both between and within geographic regions, and this variability makes the potential impacts on globalised systems such as those of food and water difficult to predict. The Fifth Assessment Report of the IPCC evaluated the impact of four climate change scenarios, which included atmospheric CO_2 equivalents (as a combined value for carbon dioxide, methane and nitrous oxide) reaching between 475 ppm and 1,313 ppm by 2100 (Barker *et al.* 2013). Considered in isolation, an increase in atmospheric CO_2 will result in increased crop yields. Modelling of Australian wheat yields found that a rise in CO_2 concentration from 350 ppm to 750 ppm resulted in a yield increase of between 23% and 34% (Crimp *et al.* 2008). However, the extent of this positive response depends on the photosynthetic pathway of the plant in question; the benefit is much less pronounced in C4 plants (such as maize, sorghum and sugarcane) as compared with C3 plants (such as rice, soybean and wheat) (Olesen and Bindi 2002). See Breakout Box 5.3 for definitions.

Climate change will also alter water and temperature regimes, and such changes have the potential to override any positive CO_2 fertilisation effect on crop yields. For example, there is medium confidence that while moderate warming will benefit crop yields in the mid to high latitudes, in low-latitude regions even a slight warming will result in a decrease in yields (Easterling *et al.* 2007). By 2100, there is also high confidence that while annual mean precipitation is likely to increase in the high latitudes and in the mid-latitude wet regions, mean precipitation will decrease in many mid-latitude and subtropical dry regions (IPCC 2013). Extreme precipitation events are also very likely to become more intense over mid-latitude land masses and wet tropical regions (IPCC 2013), both important agricultural production areas. Many climate systems will effectively migrate to other latitudes, and it is therefore crucial that best practice knowledge is transferred between regions.

The Fifth Assessment Report projections suggest that climatic variability is also likely to increase making it harder for farmers to plan their cropping regimes, particularly in rain-fed systems. The global patterns of drought and flooding that accompany the El Niño Southern Oscillation are already claimed to be behind up to 35% of the variation seen in global wheat yields (Ferris 1999). In Ethiopia, hydrological variability is estimated to cost the economy around one-third of its potential performance (Verhoeven 2013).

Breakout Box 5.3 Photosynthetic pathways of plants

The acronyms C3, C4 and CAM describe differences in the photosynthetic pathway of plants. C3 plants take up carbon dioxide to form molecules with three carbon atoms; C4 plants form molecules with 4 carbon atoms; and crassulacean acid metabolism (CAM) plants use the C3 and C4 pathways at different times of the day.

Because of biochemical and morphological differences between the three photosynthetic pathways, C3, C4 and CAM plants are suited to different environments. C4 plants such as maize and sorghum are most productive in warm climates with summer rains. They therefore dominate tropical and subtropical grasslands and savannahs, whereas C3 plants such as rice and wheat dominate cooler, temperate grasslands (Forseth 2010). C4 plants are also 30–35% more efficient in photosynthesis, especially when the concentration of CO_2 in the atmosphere is high (Nguyen 2002).

For many countries, the likely impacts of climate change point to a reduced ability to rely on rain-fed cropping. Evidence for this trend is apparent in the Middle East (World Bank 2009), Southeast Asia (Fischer *et al.* 2006) and in Mediterranean Europe (Vanham *et al.* 2013). In turn, this may force an intensification of agriculture in other areas, northern Europe for example (Olesen and Bindi 2002), and an overall increase in irrigation requirements. Fischer *et al.* (2006) suggest that this increase could be as high as 20%.

Agricultural systems in developing countries are likely to bear the brunt of the impacts from climate change. Their higher proportion of rural communities often rely for their livelihoods on the low-input farming of marginal land, land that is highly susceptible to poor productivity when exposed to climate variability. Furthermore, farmers in these regions often lack access to the support structures (such as functioning markets) that would otherwise promote adaptive capacity.

Large virtual water importers in the developed world will also feel the impacts of climate change. These countries harbour the risks of failed crops in the nations from which they import and, should climate change alter the location of agricultural bioregions as is likely to be the case, a significant reshuffling of the virtual water trading system will result. Countries such as the UK, which currently use just 1% of its blue water for irrigation (UK Environment Agency 2009), may have to increase this allocation, intensifying conflicts between water users in turn.

It should also be remembered that agriculture is a major cause of greenhouse gas emissions and therefore a significant contributor to climate change. Fertiliser production, ruminant digestion, rice cultivation, land use and land clearance all combine to mean that agriculture accounts for 13.5% of global greenhouse gas emissions (IPCC 2007). In fact, greenhouse gas emissions from livestock farming are 40% higher than the emissions from all cars, lorries, ships and aircraft combined (UNFAO 2006). The intensive farming systems of developed countries also typically contribute more emissions than the low-input systems commonly found in the developing world (Olesen and Bindi 2002). The challenge is therefore to improve the productivity of farming systems without increasing (and hopefully reducing) the associated carbon footprints.

5.3 How to Respond to the Water/Food Conundrum

5.3.1 Improving the Efficiency of Water Use in the Global Food System

In a 2011 report on the future of food and farming, the UK Foresight Programme (Foresight 2011) predicted that while global competition for all agricultural inputs is likely to increase, a growing pressure on water supplies is likely to be experienced first. Improving the water efficiency of food production should therefore be a key aim, and one that history shows can be achieved. As compared to 1977, the USA now produces 13% more beef from 30% fewer cattle using 12% less water, 33% less land and by generating 16% fewer carbon emissions (IGEL 2013).

5.3.1.1 Rain-Fed Agriculture

To date, strategies for improving crop yields have typically focused on irrigated systems where total yields are highest. The majority of global crop production is however rain-fed, so this research imbalance must be corrected. In 2011, the FAO analysed the role

Table 5.3 Yield gap in rain-fed agriculture.

Region	Yield Gap in 2005 (%)
North America	33
South America	52
West and central Europe	36
East Europe and Russia	63
West Asia	49
East Asia	11
Southeast Asia	32
Australia and New Zealand	40
North Africa	60
Sub-Saharan Africa	76

Source: Adapted from FAO (2011).

that water and soil play in the productivity of rain-fed agricultural systems. These factors are intrinsically linked; soils may have poor water-holding capacity or may too readily promote runoff. Soils that are deficient in nitrogen also have lower water-use efficiencies. For different global regions, Table 5.3 presents the yield gap in rain-fed cropping between current productivity and that which is achievable assuming soil and water conditions can be optimised. As the table shows, improvements in water and soil conditions could allow productivity to double in many areas. In Sub-Saharan Africa, soil and water constraints mean that agricultural productivity is currently less than one-quarter of what it could be. The reasons for this situation include limited water availability during critical growing periods, poor water-holding capacity of the soil, low uptake capacity of drought-damaged roots, poor soil nutrient availability, and an inability to manage pests and disease.

While highlighting the huge scope for improving rain-fed agricultural productivity in many global regions, Table 5.3 also suggests that the mechanisms to achieve these improvements already exist and that they can therefore be transferred from other areas. In East Asia for example, agricultural systems are well advanced and this allows them to achieve the majority of their yield potential. Significant gains in global productivity can be expected through improved communication and the sharing of existing knowledge and expertise; a step change in technological capability may not necessarily be required. Table 5.4 identifies a range of management techniques that could be employed to improve the productivity of rain-fed agricultural systems.

The concepts of agro-forestry and integrated crop-livestock systems introduced in Table 5.4 form part of a boarder approach to food production termed 'agro-ecology'. Agro-ecology aims to initiate a shift away from conventional agricultural techniques that focus on a single goal to agro-ecosystems that deliver a much broader range of benefits. Intensive agricultural systems use land for a single purpose, and often do so to the detriment of a host of other ecosystem services that the land might otherwise provide (e.g. water storage and carbon sequestration; Fitter 2012). In contrast, in an agro-ecological

Table 5.4 Improving the productivity of rain-fed agriculture.

Initiative	Description
Zero tillage	Avoiding disturbance of soil between crop cycles helps to improve moisture and organic matter content while also reducing erosion. In turn, this can remove the need for a fallow period between cropping cycles. The technique does however necessitate alternative approaches to sowing and weed control.
Rainwater harvesting	Harvesting rainfall for later use can help to minimise variations in water availability and thereby improve the reliability of agricultural production. Rainwater harvesting can increase yields by two to three times as compared with conventional dryland farming (FAO 2002).
Reusing crop residues	Use of crop residues as mulch helps to increase rainfall infiltration and limit runoff, in turn increasing soil moisture content and reducing the likelihood of surface water pollution. Crop residues can also be used for animal feed.
Zai systems	Zai is a traditional water harvesting system that consists of a series of man-made pits filled with organic matter that capture runoff and store moisture. Zai pits can then be planted with annual crops such as sorghum. The technique has particular benefit for helping to rehabilitate poor-quality land.
Agro-forestry	Growing trees within cropping systems, particularly those that fix nitrogen, helps to improve the efficiency of water, nutrient and carbon cycles while also protecting biodiversity. Agro-forestry has been shown to increase maize yields by us much as 280% (FAO 2011).
Integrated crop-livestock systems	This approach aims to take advantage of the synergistic relationships between crop and livestock systems. In silvopastural livestock systems, grasses are grown for grazing while shrubs and trees provide edible leaves and shoots for cattle. Silvopastural systems promote healthy soils with better water retention and less potential for runoff of pesticides and fertilisers. The mixture of grazing produces more food for animals per unit area of land and has been shown to increase muscle growth and milk production in ruminants (Broom *et al.* 2013).
Crop selection and rotation	There are broad differences in the water-use efficiency of different crops; for example, wheat is more water efficient than grain legumes or canola (Sadras and McDonald 2012). Crop selection must therefore take account of prevailing and potential future water regimes. Cropping cycles also influence soil moisture availability. Including legumes in the rotation cycle is an effective way of improving water-use efficiency as it increases available nitrogen and reduces the incidence of disease (Sadras and McDonald 2012). Time of sowing can also influence water efficiency depending on the crop in question.
Bio-fertilisers	Mycorrhizae are symbiotic relationships between the roots of plant species and fungi. The fungi take up nutrients and water and transport these to the plant root. In return, the fungi receive sugars from the plant (Ruane *et al.* 2008). Mycorrhizae fungi can be added as a bio-fertiliser to crops and are particularly beneficial for phosphorus acquisition (Fitter 2012).
Soil carbon sequestration	Enhancing soil organic carbon improves yield, increases the rate of water infiltration and reduces susceptibility to runoff and erosion (Lal 2011). The approach also increases the ability of the soil to sequester CO_2 in a self-enhancing process. Mechanisms to increase soil organic carbon content include zero tillage, mulching and soil amendment through application of biochar (a type of charcoal) (Lal 2011).

system, the intention is that productivity is increased by adopting a broader ecosystems approach and by promoting the cyclical management of resources. This can include, for example, the reuse of agricultural waste products such as crop residues for animal feed, the promotion of natural habitats and the holistic control of pests using natural techniques. The aim of agro-ecological systems is to account for the interaction of all agricultural resources in order to identify mutually beneficial approaches that optimise soil, water, air, economic, social and biodiversity outcomes.

Improved rain-fed agricultural productivity would directly benefit the world's poorest people. This demographic typically live in rural areas and directly rely on rain-fed agriculture to sustain their livelihoods. Buendra *et al.* (2011) found that if the global productivity of rain-fed agriculture could be improved to 80% of its potential, there would need to be only a 7% increase in the global area under cropping to meet global food demand in 2050.

It is important to note that it is not only in water-scarce areas where the efficiency of rain-fed agriculture could be improved. Vanham *et al.* (2013) found that the lowest agricultural water productivities can often be found in countries and regions were water is most abundant. This abundance breeds complacency and profligacy and, while harvests remain successful, the globalised nature of the international food system means that indirectly, others can suffer. The trade in virtual water contained in food means that freshwater resources are accessible from almost anywhere on the planet and therefore that inefficient water use, wherever it occurs, reduces the ability to deliver food and water security to the global population.

5.3.1.2 Irrigated Agriculture

Notwithstanding the detrimental environmental impacts that can result from poor irrigation management (Section 5.1.2), crops yields in irrigated systems can be more than double those achieved by rain-fed cropping. Irrigation performance and the potential for environmental impact do however vary significantly depending on the type of irrigation adopted (Table 5.5).

Globally, surface irrigation systems (flood, border and furrow) are by far the most common, particularly on small farms, and are likely to remain the dominant approach for many decades (FAO 2002). Despite the fact that they are wasteful of water and a major cause of waterlogging and salinisation, these systems are cheap to install and do not require the operation and maintenance of sophisticated hydraulic equipment. The relationship between the rate of water application and crop demand is however only very crudely controlled. Water is frequently applied to non-growth areas and may run-off the land, leaching any pesticides and fertilisers to surface watercourses or shallow groundwater. Evaporative losses are also high.

In contrast, drip irrigation systems require comparatively high initial capital outlay (too large for many farmers in developing countries) and continued maintenance but do provide a precise means of controlling water application that, if combined with monitoring of soil moisture deficits, can be operated in a manner that is highly responsive to crop water demand. Furthermore, drip irrigation can also be used for the dual application of water and nutrients in a process termed 'fertigation', thereby reducing the likelihood that the relationship between water and nutrient availability will constrain crop yield. In comparison to flood systems, drip irrigation has been claimed to increase water efficiency by one-third (IME 2013) and, on farms in Kenya, increased yields by more than 80% (UN

Table 5.5 Alternative irrigation systems.

Irrigation system	Description
Flood irrigation	Water entry to the area to be irrigated is uncontrolled with no measures to direct or manipulate water flow. Flood irrigation is simple to operate; however, water efficiency for most crop types is very low.
Border irrigation	Borders are constructed to ensure that water enters the area to be irrigated as a controlled sheet of water. Volumes of water loss are therefore less than those for flood irrigation.
Furrow irrigation	Simple channels are constructed using basic equipment into which water is directed.
Sprinkler irrigation	Water is applied by sprinklers that mimic precipitation. Water is distributed to sprinkler heads via pressurised pipe networks. Sprinkler irrigation systems can be fixed or portable and their operation can be optimised to achieve desired application rates and wetting patterns.
Drip irrigation	Water is precisely applied to each plant typically via on-ground perforated pipes. This ensures low rates of water loss due to evaporation, seepage or over-watering.
Subsurface irrigation	Water is directed to the root zone of the crop via perforated pipes. As for drip irrigation, rates of water loss can therefore be very low.

2005). Water efficiencies for subsurface irrigation systems are even higher. Precise irrigation techniques such as these also help to minimise the otherwise detrimental effects that surface irrigation systems can have on the environment (e.g. erosion of soils and runoff of pesticides).

The benefits of precise irrigation techniques can be further optimised if they are used to supplement rainfall. Annual crops have periods during their growing cycles when deficits of water can be highly detrimental to yields. By supplementing rainfall with precise irrigation during the periods when it is most required, both overall water efficiency and yield can be optimised. Examples in Syria have shown that by supplementing 300 mm of rainfall with 150mm of irrigation, wheat yields can be doubled (UNEP and IWMI 2012). Gordon *et al.* (2010) found that the highest gains in agricultural water productivity could be achieved by combining supplemental irrigation with improved tillage and nutrient management.

Despite the water efficiency benefits of precise irrigation, wasteful irrigation systems and management practices persist, often sustained by the low price paid by the majority of farmers for their water. Prices for irrigation water are almost always lower than the true economic value of the water, and frequently lower than the cost of supply. Farmers in Spain pay just 2% of the supply cost, for example (Rabobank 2008). In such a scenario, there is very little incentive for the farmer to invest in precise irrigation. Under-priced water means under-investment, poor system maintenance and depletion of assets. In turn, responsibility for the operation and maintenance of irrigation systems is often left to governments, a role they frequently struggle to achieve successfully (IWMI 2006a). Full cost-recovery pricing is essential if efficient irrigation practice is to be stimulated. Where water is attributed an appropriate price, irrigators are incentivised to conserve this precious resource.

With an accurate pricing system in place, markets can also be established to allow water to be traded to the use to which it provides greatest benefit. Australia's Murray-Darling basin has had a functioning water trading system since the 1990s and, despite lingering restrictions on interregional trading and some concerns over the level of proposed environmental buy-backs by the government, irrigators are increasingly accepting of and reliant on the system. During the 2006–2009 drought, many farmers sold their entitlements to growers of longer-lived species and shifted to dry-land cropping (ACCC 2013). They then bought back these entitlements when water availability increased, lessening the economic impact of the drought by almost 40% (NWC 2012).

Wastewater Reuse: One means of securing water supplies for irrigation is to use recycled wastewater, an approach that aligns with the concept of cyclical management raised in relation to agro-ecology and one that has been associated with land application and crop production for centuries. The fertiliser value of wastewater effluent can have significant benefits and the FAO (2002) suggests that there is potential for all of the nitrogen and much of the phosphorus and potassium normally required for crop production to be supplied in this way. Micronutrients and organic matter would also be valuable and, because most of the nutrients are absorbed by the crop, they are otherwise prevented from causing potential damage to the environment (e.g. through eutrophication).

Wastewater reuse in agriculture has particular promise in improving the food security of rapidly expanding urban populations. Wastewater is already used to support 20 million peri-urban farmers in Africa (IWMI 2006b) and in Hyderabad, India the FAO (2008) found that wastewater-irrigated peri-urban fodder and vegetable production contributed significantly to improving the livelihoods of poor urban and peri-urban communities. In Israel, almost three-quarters of urban wastewater is reused and most of this goes to support peri-urban agriculture (Rygaard *et al.* 2011). The diversification of the food system that results from the incorporation of local agricultural networks is increasingly recognised as a valuable food security resilience mechanism, and an important community development benefit. In order to support the concept, many cities are now incentivising urban and peri-urban agricultural schemes. For example, the cities of Chicago and San Francisco, USA have modified their planning laws to make the development of urban agricultural systems more straightforward (Beatley and Newman 2013). In the Brazilian city of Belo Horizonte, the local government preferentially procures peri-urban food crops in order to stimulate the industry; in Kathmandu, Nepal, rooftop cropping schemes have helped many families become self-sufficient in vegetables and herbs (Dubbeling 2013).

While a lack of space acts as the primary physical constraint on urban agriculture in developed countries, in the developing world, a variety of other factors must also be overcome. As raised in Section 4.3.2, the uncertain land tenure arrangements typical of the peri-urban environment often prevent investment in the infrastructure upon which food systems rely. Healthy soils are also often at a premium, so the protection of any existing agricultural land base surrounding cities is crucial if peri-urban agricultural initiatives are to succeed (Dubbeling 2013). Furthermore, while wastewater streams provide a safe source of fertiliser in many developed cities, safe effluent products in developing country cities are rare (Bahri 2009). Wastewater reuse in agriculture can represent a major hazard in these cities, not only jeopardising the health of the farmer but also those who purchase from urban markets. Marshall *et al.* (2009) highlight case studies from India where high concentrations of heavy metals were found in the water and soil used in peri-urban agricultural systems. More research is therefore needed to determine the risk of pollutants entering the human food chain (Ruane *et al.* 2008). For example, even after tertiary wastewater treatment, risks associated with enteric virus, toxic contamination and pollution of the environment remain (Jimenez *et al.* 2010; Ganoulis 2012).

Notwithstanding these concerns, it is important to remember that where appropriate procedures are carried out under controlled conditions, the use of wastewater in agriculture can be entirely safe. Guidelines published by the World Health Organisation (WHO) provide advice to farmers and policy makers and advocate a series of risk management strategies to protect human health (WHO 2006). Post-harvest measures are also an important component of health risk reduction, ensuring the presence of multiple barriers to prevent environmental or public health impacts. Ilic *et al.* (2010) advocate the need for multiple control points along the production chain, with an emphasis on local safety targets and education.

High Flow Storage: Uncertainties over future climate add another layer of complexity to agricultural water management. In many regions of the world, current projections point to a reduction in the ability to rely on rain-fed cropping. Any increased compensatory demand on blue water resources through irrigation will therefore raise levels of competition with existing users. Falkenmark (2013) identifies many agriculturally significant river basins that are 'closed' or 'closing' (a status reached when blue water allocations begin to impinge on environmental flows). These rivers are found in developed and developing countries and include the Colorado, the Murray-Darling, the Nile, the Indus, and the Yellow rivers (Falkenmark 2013). In these basins and many others, the scope for additional irrigation is very limited and it will therefore be crucial to promote agricultural practices that are adaptive to changing climates. In this respect, high flow storage presents a very logical initiative.

The construction of large on-line dams on major rivers is highly contentious. These structures simultaneously alter downstream flow patterns and permanently flood large upstream areas causing environmental degradation, land-use change and, in some cases, forcing community relocation. Large dams can however provide significant and multiple benefits via, for example, reliable water supply, hydropower generation and downstream flood control. In 2000, the World Commission on Dams (WCD) estimated that up to 40% of global irrigated land relies on dams for its water which in turn generate around 19% of global electricity. In some regions, there does remain scope to support irrigated agriculture through further dam construction; however, the most promising dam sites have already been developed on many major rivers (WCD 2000). In contrast, smaller off-line dams for the purposes of high flow storage are much less contentious.

Climate projections indicate high confidence in an increased frequency of extreme rainfall events, and it therefore makes sense to capture high river flows for use at a later date. During high flows, the diversion of blue water can occur without detriment to other users (including the environment); furthermore, by slowing down the passage of water across the landscape, the risk of high flows causing damage to downstream infrastructure is significantly reduced. For the farmer, a store of water allows him or her to plan cropping with confidence, providing a level of reassurance that helps to justify continued investment in farm improvements.

5.3.1.3 Research and Development

The significant improvements in yields achieved by many countries (mainly western countries) during the Green Revolution relied heavily on sustained investment in agricultural research and development. In other regions, similar phases of growth are now occurring. In China for example, investment in agricultural research has been rising at an annual rate of 10% since 2001. The projects borne out of this funding have enhanced the resilience of soils to drought, saved up to 2.5 billion m^3 of irrigation water and improved yields, while also reducing associated emissions of greenhouse gases (Beddington *et al.* 2012). Over the last 50 years, each harvested hectare of land in China has become 4.5 times more productive (Ausubel *et al.* 2012) with rice, maize and wheat yields growing by 90%, 150% and 240%, respectively (Piao *et al.* 2010).

In most global regions however, particularly low-income countries, investment in agricultural research has fallen and a knowledge divide with the West has grown as a result. By its nature, technological research requires significant upfront capital expenditure and is characterised by long payback periods. Because of this, the future potential profits from any innovation have more chance of being eroded by factors over which the investor can have only limited control (e.g. market prices). These risks are typically greater in developing countries and this often makes investment unattractive.

Table 5.6 highlights some of the emerging technical innovations that could help to improve agricultural productivity and water efficiency in the developing world. It also includes examples that, while not directly related to agricultural productivity, do have important influences on food security.

Table 5.6 Technological innovations.

Technological innovation	Description
Crop related:	
Selective breeding	Breeding of crops with beneficial traits such as drought resistance, high yield, pest resistance and heat tolerance. For example, a high yield and disease-resistant variety of wheat developed by research institutions and adopted at scale across Mexico has transformed the country into a secure wheat exporter (World Economic Forum 2010).
Genetic modification	Artificially removing or adding a specific gene to crops in order to promote more desirable characteristics.
Nanotechnologies	The manipulation of matter on an atomic and molecular scale. Nanotechnologies can be used in a wide variety of applications from agrichemicals to increase their efficiency, nanoporous materials to store water in the soil and nanotechnologies to increase the muscle mass of animals.
Hydroponics	The process of growing plants using nutrient solutions without soil, hydroponics has the potential to use only 10% of the water required in field cultivation systems (Bradley and Marulanda 2000). Hydroponic systems do however require significant upfront investment. Furthermore, potential applications outside temperate climatic regions are currently limited (Bradley and Marulanda 2000).

Table 5.6 (Continued)

Technological innovation	Description
Food/nutrition related:	
Fortification	Artificial addition of micronutrients to food products. For example, the addition of iron and folic acid to wheat flour in Egypt has been highly successful at addressing malnutrition among children and marginalised groups (World Economic Forum 2010).
Bio-fortification	Breeding crops to increase their nutritional value. This can be achieved by conventional selective breeding or genetic engineering. Bio-fortification differs from fortification in that it focuses on the fortification of the crop rather than the addition of nutrients to food products.
Laboratory-grown foods	Use of stem cells to grow animal tissue in the laboratory. Very few natural resource inputs are required to achieve this, but it is an unproven technology.
Climate and market forecasts:	
Mobile banking and market data	Improves rural liquidity and knowledge of market demands and prices.
Drought and flood forecasting	Seasonal climate forecasts provide the opportunity for farmers to choose whether to adopt new technologies and intensify production or to opt for alternative strategies (Vermeulen *et al.* 2012). Seasonal forecasts need to be provided at a scale appropriate to the farmer's requirements. Communication channels also need to be improved.

The artificial mimicking of selective breeding is an increasingly common phenomenon and while there are important ethical issues that require reasoned debate, genetically modified (GM) foods should not be dismissed outright. In 2011, approximately 160 million hectares of land worldwide were cultivated with GM crops. Figure 5.7 shows the countries with most GM cropland; the total is shared relatively equally between developed and developing nations (Falke-Zepeda *et al.* 2013).

The process of genetic modification has typically been used to address problems associated with weeds and pests (Fitter 2012); however, the potential benefits in terms of improved yields are signifit. Uga *et al.* (2013) identified a genetic modification to rice that acts to direct roots down instead of horizontally, thereby resulting in yields under severe drought conditions that are more than three times those of normal rice varieties.

In many countries, GM foods have already captured a significant portion of the market; however, consumers remain intensely sceptical of the industry. Risks do exist and these should be the focus of continued research. For example, GM species with advantageous traits could become invasive in unmanaged ecosystems or may transfer to crops with wild relatives (Fitter 2012). Suppliers must develop effective communication strategies that present fact and that in turn stimulate reasoned debate.

For farm-based technologies that possess direct links to food productivity and water efficiency, the importance of data collection and interpretation is paramount for effective decision making. Particularly for precision irrigation and deficit agriculture, system optimisation is highly dependent on the availability of data describing soil moisture levels and crop response. Data collection represents an area of agricultural research in

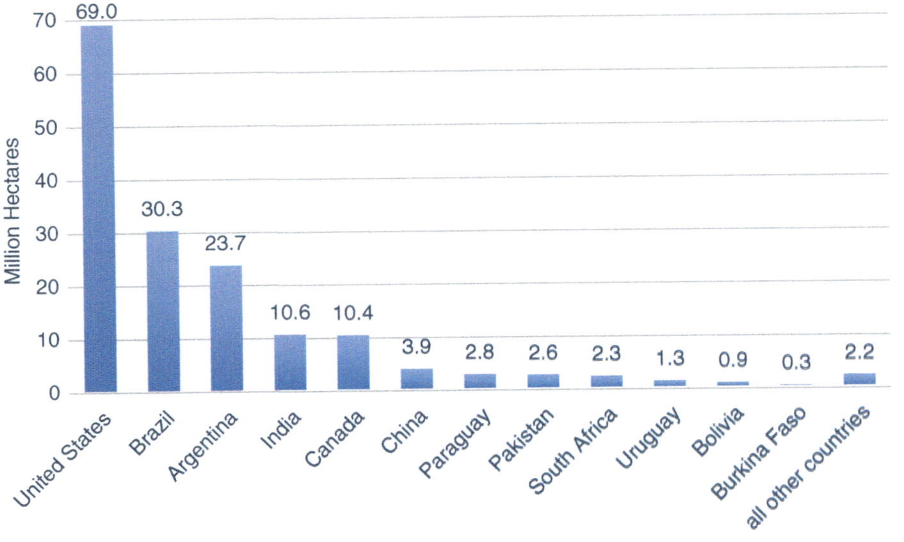

Figure 5.7 Area of land cultivated with genetically modified crops.
Adapted from Falke-Zepeda *et al.* (2013).

which collaboration has the potential to realise significant benefits; where landowners can begin to monitor soil moisture levels and pool their data, widespread awareness of irrigation best practice can grow quickly. The potential for community-wide benefits also makes data collection and dissemination an attractive funding initiative for corporations looking to improve and advertise their corporate social responsibility (Breakout Box 5.4). Finally, where agricultural data can be collated over a wide geographic area, the relationships between local, regional and global food systems and other human and environmental phenomena, particularly climate change, can be more accurately tracked.

Breakout Box 5.4 i-crop and AgMIP

The i-crop tool measures the inputs and outputs of farming activity based on data derived from a range of instruments including soil moisture probes in the field and local weather stations. Developed by PepsiCo and Cambridge University, the tool allows users to access information online and enables improved decision making on, for example, when to apply water to crops (PepsiCo 2010). In 2013, PepsiCo's trials of i-crop achieved a 5% increase in crop yields for 49% less irrigation water per tonne (2 Degrees 2015).

The Agricultural Model Inter-comparison and Improvement Project (AgMIP) is a major international effort linking the climate, crop and economic modelling communities with information technology to produce improved crop and economic models and, in turn, the next generation of climate impact projections for the agricultural sector (Rosenzweig *et al.* 2013). By establishing strong links between climate projections and economic and crop models, potential adaptation strategies can quickly be identified. Knowledge of markets also improves, helping to increase the overall efficiency of the crop production process (Rosenzweig *et al.* 2013).

5.3.2 The Importance of Consumer Education

Consumer demand acts as a primary influence on the global food system; it therefore follows that consumer education should be a key focus for those assigned responsibility for ensuring food and water security.

Most consumers are unaware of the virtual water content of their food, let alone its internal and external footprints or its green, blue and grey water components. As a result, consumers in the developed world happily purchase (and thereby continue to incentivise the production of) significant quantities of food produced via unsustainable irrigation of crops in water-scarce countries. Furthermore, we then throw away one-third of these purchases and thereby waste the water, energy, land and human capital that went into producing them. At the current time, consumer food choices are made with little or no regard to water and one cannot expect to establish mutually beneficial food and water policies, and production and supply chain systems, without addressing this oversight.

Fortunately, there are many ways in which consumers can be made aware of the relationships between water and food. Media outlets and public communication are obvious starting points. The rise of the celebrity chef provides a useful avenue through which to reach a wide audience, and recent examples have shown how social media can quickly turn a plea for change into a broad social movement. Take the issue of European fish discards, for example. The practice of throwing back fish at sea (often either dead, dying or badly injured) in order to keep catches within landing quotas means that discard rates in EU fisheries often reached as high as 98% (European Commission 2011). Discards have been an unfortunate trait of European commercial fishing practices for more than 40 years; however, in 2011, with the attention of a British celebrity chef and a high-profile media campaign, the numbers of those calling for change quickly snowballed. By May 2013, the European Government had enacted new laws to legislate against the practice and fish discards for most species are now limited to 5% of the catch (European Parliament 2013).

Notwithstanding their influence on the sustainability of commercial fishing, fish discards represent just one example of the detrimental environmental effects that have the potential to arise from the existing global food system. The vital role played by virtual water trade represents a far more wide-ranging issue; it therefore seems logical that, with the right backing, social movements calling for improved management of water and food could gain traction. Three areas of public communication show significant promise:

1) **The origins of food**. By explaining where food products come from, consumers can more readily understand the inputs required in their production. They can also begin to re-learn the concepts of seasonality and food safety that were engrained in previous generations but that have since been replaced with perceptions of year-round availability and rigid shelf life. Reverting back to making food choices on the basis of seasonality (and therefore, to a large extent, green water availability) and relying on the human senses to determine whether food is or isn't safe to eat will go a long way to ensuring that the public can make food choices that are more water efficient.
2) **The water content of food**. While this chapter has shown that when used in isolation the total volumetric water footprint can be a somewhat crude tool for interpreting the impact of water use, it can nevertheless help to make consumers more aware of the importance of water in food production. In turn, the public may be more willing to explore the relationships and associated impacts in greater detail. In any case,

making food choices on the basis of volumetric water footprint, despite not accounting for the 'colour' of the water used or water availability in the country of origin, does constitute an extremely important and beneficial first step.

3) **Food waste.** Wasted food not only represents a significant volume of wasted water but it also represents wasted money, wasted greenhouse gas emissions, wasted fertilisers, wasted land and wasted human effort. Making the consumer aware of these facts should provide sufficient incentive for households to reduce their volumes of domestic waste, not least in response to the immediate financial saving that will accrue. Simple punitive mechanisms could also add extra influence (see Section 5.3.3 for more discussion on this issue).

Labelling has the potential to act as a highly effective means of increasing consumer awareness. In principle, developing an indicator of water sustainability would appear to represent a logical and repeatable means of allowing customers to make reasoned food choices. In practice however, a number of difficulties arise. For example, the multiple factors that combine to determine the intensity of water use in food production are not easily condensed into a simple and reliable indicator. One could argue the need for indicators of domestic and international water footprint, and for measures of green water, blue water and grey water use. And how should water use be quantified? By volume? By the calorific value of the associated food product? A second important constraint relates to the need to secure international cooperation for any labelling scheme to have meaningful value. Without this international backing, variations in labelling requirements might diminish rather than improve consumer understanding. Finally, any water intensity label would have to compete with the increasing number of other food-safety- and health-related warnings that populate food packaging. Given these limitations, priority should be focused on the aforementioned media and public engagement initiatives. These schemes have a greater likelihood of achieving successful outcomes in a shorter timeframe and may naturally lead to labelling schemes, particularly if public interest grows.

One aspect of product labelling that can and should be the focus of immediate attention is the currently confused nature of food safety labelling. In Europe for example, a consumer purchasing a fresh food product is confronted with 'use by', 'best before' and 'display by' labels. Only the first relates to the safety of food, indicating when highly perishable goods may present a risk of food poisoning. A 'best before' label indicates the date before which a product can reasonably be expected to retain its optimal condition (Defra 2011). It therefore relates to the quality of food only, not its safety. The 'best before' label is however a legal requirement in the EU, a piece of legislation that thereby encourages consumers to throw away safe food.

In contrast to 'use by' and 'best before' labels, 'display by' dates have no legal mandate and are used by supermarkets solely as a means of helping staff to control stock. The 'display by' label therefore further dilutes the impact of the vitally important 'use by' date. Confronted with this variety of conflicting information, any consumer will no doubt follow the most conservative of the labelling instructions, increasing the volume of wasted but otherwise safe food. Lipinski *et al.* (2013) estimate that 20% of food wasted in the UK reflects label misunderstanding. Date labelling of food should be revised to include one date only, that which indicates when the food product is no longer safe for human consumption.

Figure 5.8 Most common reasons for food waste, based on survey of 1862 UK residents aged 16 years and over between November and December 2006.
Adapted from WRAP (2007).

Multi-buy offers, widely employed and advertised by supermarkets, also promote waste, encouraging the consumer to buy more than they need and leading to waste in the home. A survey of UK shoppers found that 30% of respondents blamed volume-based supermarket promotions for household waste (see Figure 5.8). Buy-one-get-one-free deals and other promotions linked to product volume should be banned, or else carefully restricted.

While reducing food waste has the potential to provide a range of benefits (e.g. to water use, to the use of other resources and to personal finance), human nature dictates that decision making can often be more strongly influenced by the threat of negative outcomes. Careful punitive measures may therefore be more effective in controlling food waste. These could include, for example, calculating domestic and hospitality refuse collection fees on the basis of weight. Alternatively, voluntary initiatives could include the reduction of restaurant portion sizes and the recirculation of supermarket food that has passed its 'display by' date to those in need within the community (Breakout Box 5.5).

Breakout Box 5.5 Waste not want not

Food Cycle is a UK charity that enlists volunteers to collect and redistribute surplus food within the community. Since its inception in 2009, FoodCycle has cooked over 73,000 meals, reclaimed over 74,000 kg of surplus food that would otherwise have been wasted and reached over 3,800 beneficiaries through the work of over 3,000 volunteers (Food Cycle 2013).

OzHarvest is a similar organisation based in Australia. Founded in 2004, it has expanded its operations to all of Australia's major urban centres. It has over 600 volunteers, more than 2,000 food donors and has delivered more than 20 million rescued meals. As the profile of OzHarvest and what it stands for have grown, major companies are beginning to take note. The charity signed an agreement with one of Australia's two major supermarkets in 2015.

Even very simple changes can have significant effects. For example, by removing trays from its canteens, students at Grand Valley State University Michigan were limited to only the portions they could carry in any one visit. In turn, food waste fell by 25% and the University saved US$ 79,000 a year (Lipinski *et al.* 2013).

5.3.3 Improve Governance of Water Use for Food Production

Governance, and its associated regulatory instruments, provides the overarching framework within which resources are managed. Governance therefore exerts a highly influential force and, where policies are poorly thought out, unexpected and often detrimental impacts can result. This chapter has shown how, for the resources of food and water, these detrimental impacts are most frequently felt by the environment. The inability of society to accurately value environmental services means that when competition for resources increases, environmental needs are often the first to be compromised.

While all countries trade water-intensive food products, few explicitly consider the process as a means of managing water (Hoekstra and Mekonnen 2012). Concepts such as virtual water and the water footprint have only entered the academic discourse within recent decades, and remain largely absent from contemporary politics. This means that many of the most important elements of the global food system have been planned without regard to water. As a result, international trade agreements often inadvertently incentivise the production of water-intensive goods in water-scarce countries. Furthermore, many countries have, and often continue to strive for, self-sufficiency in food, largely as a result of either perceived or real geopolitical threats. The example of Saudi Arabia elaborated in Section 5.2.4 highlights the detrimental impacts of what is ultimately an unachievable aim, at least in the long term.

Food security rather than food self-sufficiency should be the goal of governments and, in this respect, the international trade in food provides an excellent solution. So long as market distortions are removed, trade provides a rare mechanism to increase the volumes of food produced while at the same time decreasing pressure on water resources. Exchanges of virtual water already save the planet 355 Gm^3 of water a year (Chapagain *et al.* 2006); however, this mechanism remains an opportunity that is significantly underutilised for two primary reasons:

1) the failure of many governments to develop national food policies that appropriately consider their impacts on water resources; and
2) the difficulties associated with establishing markets that accurately account for water.

The latter of these two factors is very complex, reflecting historical arrangements and geopolitical factors that will, considering the necessarily large number of nations and organisations involved, be hard to overcome, at least in the short to medium term. National food policies can be more readily strengthened however, and focus should be directed towards encouraging greater cross-departmental communication. The siloed nature of most government departments naturally leads to discrete planning and, in turn, disjointed policy that fails to secure mutually beneficial outcomes. Efforts should be made to acknowledge the relationships between water and food, as well as other resources such as energy, in order to arrive at policy and regulation that is joined-up and inclusive.

In many countries, agricultural policies with the aim of supporting the domestic farming industry routinely promote the overuse of water; this can lead to an oversupply of produce that may be wasted or go on to distort global markets. Policy makers should develop a clear understanding of relationships such as these before deciding whether to allocate domestic water resources to the production of particular food products or to pursue water savings through trade in virtual water. Moreover, they must make these decisions with due consideration of the future challenges and uncertainties, such as population growth, changing diets and climate change, that are likely to affect the global food system.

While agricultural and food policy must account for water, it must also be viewed as a much broader tool that is able to address land, energy and environmental relationships. Where governments can establish overarching resource strategies, an enabling framework is created that naturally guides the management of resource systems down avenues that achieve a broad array of mutually beneficial outcomes.

Figure 5.9 presents some of the mechanisms that could be used to incentivise the more sustainable management of the global food system as it relates to water.

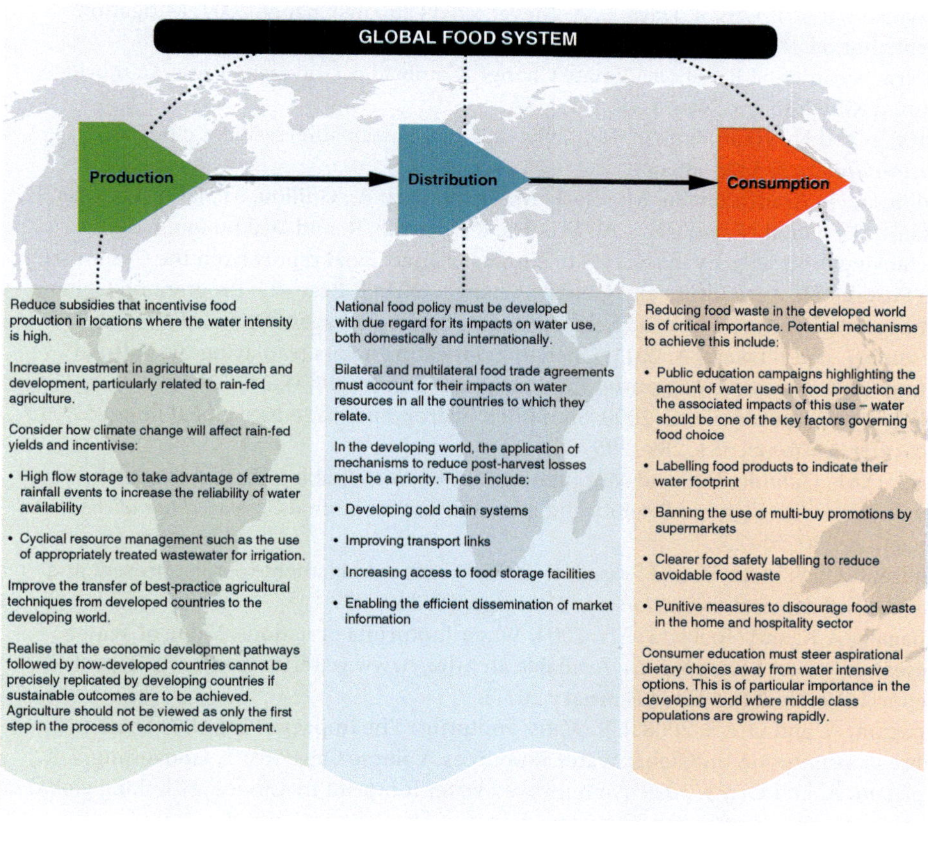

Figure 5.9 Linking the governance of food and water.

References

2 Degrees. 2015. Water – the essence of sustainable business? Available at: https://www.2degreesnetwork.com/groups/2degrees-community/resources/water-essence-sustainable-business/ (accessed January 2017).

Aldaya, M.M., Allan, J.A. and Hoekstra, A.Y. 2010. Strategic importance of green water in international crop trade. *Ecological Economics*, 69, 887–894.

Allan, T. 2011. Virtual Water: Tackling the Threat to Our Planet's Most Precious Resource. I.B. Tauris & Co Ltd, London, UK.

Australian Competition and Consumer Commission. 2013. ACCC Water Monitoring Report 2011–2012. ACCC, Australia.

Australian Government Department of Agriculture, Fisheries and Forestry (DAFF). 2011. Issues paper to inform development of a national food plan. DAFF, Canberra.

Ausubel, J.H. Wernick, I.K. and Waggoner, P.E. 2012. Peak farmland and the prospect for land sparing. *Population and Development Review*, 38, 217–238.

Bahri, A. 2009. Managing the other side of the water cycle: making wastewater an asset. TEC Background Papers No. 13, Global Water Partnership, Sweden.

Barker T., Bashmakov, I., Bernstein, L. et al. 2007. Technical summary. In B. Metz, O. R. Davidson, P. R. Bosch, R. Dave, L. A. Meyer (eds) Climate Change 2007: Mitigation. Contribution of Working Group III to the Fourth Assessment Report of the Intergovernmental Panel on Climate Change. Cambridge University Press, Cambridge, United Kingdom and New York, NY, USA.

Beatley, T. and Newman, P. 2013. Biophylic cities are sustainable, resilient cities. *Sustainability*, 5, 3328–3345.

Beddington, J., Asaduzzaman, M., Clark, M., Fernandez, A., Guillou, M., Jahn, M., Erda, L., Mamo, T., Van, B.N., Nobre, C.A., Scholes, R., Sharma, R. and Wakhungu, J. 2012. Achieving food security in the face of climate change: Final report from the Commission on Sustainable Agriculture and Climate Change. CGIAR Research Program on Climate Change, Agriculture and Food Security (CCAFS). Copenhagen, Denmark.

Boersema, J. and Blowers, A. 2011. Changing our eating habits by playing the cultural trump card. *Journal of Integrative Environmental Sciences*, 8(4), 243–252.

Bradley, P. and Marulanda, C. 2000. Simplified hydroponics to reduce global hunger. *ISHS Acta Horticulturae*, 554, 289–295.

Broom, D.M., Galindo, F.A. and Murgueitio, E. 2013. Sustainable, efficient livestock production with high biodiversity and good welfare for animals. *Proceedings of the Royal Society B*, 280, 20–25.

Buendra, K.C., Schultz, B. and Prasad, K. 2011. Water management to meet present and future food demand. *Irrigation and Drainage*, 60, 348–359.

Chapagain, A.K. and Hoekstra, A.Y. 2004. Water footprints of nations. Value of Water Research Report Series No. 16. Available at: http://www.waterfootprint.org/Reports/Report16Vol1.pdf (accessed February 2017).

Chapagain, A. and Orr, S. 2008. UK Water Footprint: The Impact of the UK's Food and Fibre Consumption on Global Water Resources. Volume One. WWF, Godalming, UK.

Chapagain, A. and Orr, S. 2009. An improved water footprint methodology linking global consumption to local water resources: A case of Spanish tomatoes. *Journal of Environmental Management*, 90, 1219–1228.

Chapagain, A., Hoekstra, A.Y. and Savenije, H.H.G. 2006. Water saving through the international trade of agricultural products. *Hydrological Earth System Science*, 10, 455–468.

Crimp, S., Howden, M., Power, B., Wang, E. and De Voil, P. 2008. Global Climate Change Impacts on Australia's Wheat Crops. Report prepared for the Garnaut Climate Change Review, CSIRO, Canberra.

Dubbeling, M. 2013. Cityfood: Linking cities on urban agriculture and urban food systems. RUAF Foundation and ICLEI, The Netherlands.

Easterling, W.E., Aggarwal, P.K., Batima, P. et al. (eds) 2007. Climate Change 2007: Impacts, Adaptation and Vulnerability. Contribution of Working Group II to the fourth assessment report of the Intergovernmental Panel on Climate Change. Cambridge, UK: Cambridge University Press.

Ercin, A.E., Aldaya, M.M. and Hoekstra, A.Y. 2012. The water footprint of soy milk and soy burger and equivalent animal products. *Ecological Indicators*, 18, 392–402.

European Commission. 2011. Studies in the Field of the Common Fisheries Policy and Maritime Affairs. Lot 4: Impact Assessment Studies Related to the CFP. European Commission.

European Commission Directorate-General for Agriculture and Rural Development. 2013. CAP expenditure in the total EU expenditure. Available at: ec.europa.eu/agriculture/cap-post-2013/graphs/graph1_en.pdf (accessed January 2017).

European Parliament. 2013. 2011/0195(COD) – 28/11/2013 Committee recommendation tabled for plenary, 2nd reading. Available at: www.europarl.europa.eu/oeil/popups/summary.do?id=1325811&t=e&l=en (accessed January 2017).

Falkenmark, M. 2013. Growing water scarcity in agriculture: future challenge to global water security. *Philosophical Transactions of the Royal Society A*, 371, doi: 10.1098/rsta.2012.0410.

Falke-Zepeda, J., Gruere, G. and Sithole-Niang, I. 2013. Genetically Modified Crops in Africa: Economic and Policy Lessons from Countries South of the Sahara. IFPRI, Washington.

Feng, K., Hubacek, H., Minx, J., Siu, Y.L., Chapagain, A., Yu, Y., Guan, D. and Barrett, J. 2011. Spatially explicit analysis of water footprints in the UK. *Water*, 3, 47–63.

Ferris, JNJ. 1999. An analysis of the impact of ENSO (El Nino/Southern Oscillation) on global crop yields. Available: http://ageconsearch.umn.edu/bitstream/21517/1/sp99fe01.pdf (accessed February 2017).

Fischer G., Tubiello F., Van Velhuizen H. and Wiberg, D. 2006. Climate change impacts on irrigation water requirements: effects of mitigation, 1990–2989. *Technological Forecasting and Social Change*, 74, 1083–1107.

Fitter, A. 2012. Why plant science matters. *New Phytologist*, 193(1), 1–2.

Food and Agriculture Organisation of the United Nations. 2002. Crops and Drops: making the best use of water for agriculture. FAO, Rome.

Food and Agriculture Organisation of the United Nations. 2008. Urban Agriculture for Sustainable Poverty Alleviation and Food Security. FAO, Rome.

Food and Agriculture Organisation of the United Nations. 2011. The State of the World's Land and Water Resources for Food and Agriculture. FAO, Rome.

Food and Agriculture Organisation of the United Nations. 2012. Statistical Yearbook. FAO, Rome.

Food and Agriculture Organisation of the United Nations. 2013a. The State of Food and Agriculture: food systems for better nutrition. FAO, Rome.

Food and Agriculture Organisation of the United Nations. 2013b. Food Wastage Footprint: Impacts on natural resources. FAO, Rome.

Food Cycle. 2013. Impact. Available at: www.foodcycle.org.uk/what-we-do/impact/ (accessed January 2017).

Foresight 2011. The Future of Food and Farming. Foresight, London.

Forseth, I. N. 2010. The ecology of photosynthetic pathways. *Nature Education Knowledge*, 3(10), 4.

Ganoulis, J. 2012. Risk analysis of wastewater reuse in agriculture. *International Journal of Recycling of Organic Waste in Agriculture*, 1, 3.

Gordon, L.J., Finlayson, C.M. and Falkenmark, M. 2010. Managing water in agriculture for food production and other ecosystem services. *Agricultural Water Management*, 97, 512–519.

Hoekstra, A.Y. and Chapagain, A.K. 2006. Water footprints of nations: water use by people as a function of their consumption pattern. *Water Resources Management*, 21, 35–48.

Hoekstra, A.Y. and Mekonnen, M.M. 2011. National Water Footprint Accounts: The green, blue and grey water footprint of production and consumption. UNESCO-IHE, Delft, Netherlands.

Hoekstra, A.Y. and Mekonnen, M.M. 2012. The water footprint of humanity. *Proceedings of the National Academy of Sciences*, 109, 9, 3232–3237.

Ilic, S., Drechsel, P., Amoah, P. and LeJeune, J.T. 2010. Applying the multiple-barrier approach for microbial risk reduction in the post-harvest sector of wastewater-irrigated vegetables. In Drechsel, P., Scott, C.A., Raschid-Sally, L., Redwood, M and Bahri, A. (eds) Wastewater Irrigation and Health: Assessing and Mitigating Risk in Low-Income Countries. Earthscan IDRC and IWMI, London.

Initiative for Global Environmental Leadership (IGEL). 2013. The Nexus of Food, Energy and Water. IGEL, University of Pennsylvania, Pennsylvania.

Institute of Mechanical Engineers (IME). 2013. Global Food: Waste Not, Want Not. IME, UK.

Intergovernmental Panel on Climate Change (IPCC). 2013. Working Group I Contribution to the IPCC Fifth Assessment Report. Climate Change 2013: The Physical Science Basis – Summary for Policymakers. IPCC, Geneva.

International Food Policy Research Institute (IFPRI). 2002. Green Revolution: Curse or Blessing? IFPRI, Washington, USA.

International Food Policy Research Institute (IFPRI). 2013. Tackling Egypt's Rising Food Insecurity in a Time of Transition. IFPRI, Washington, USA.

IWMI. 2006a. Synthesis paper on irrigation performance in Kenya and lessons learnt from the ten IPIA pilots irrigation schemes. Available at: http://publications.iwmi.org/pdf/H039631.pdf (accessed February 2017).

IWMI. 2006b. Informal Irrigation in Urban West Africa: An Overview. Available at: http://www.iwmi.cgiar.org/Publications/IWMI_Research_Reports/PDF/pub102/RR102.pdf?galog=no (accessed February 2017).

Jimenez, B., Drechsel, P., Kone, D., Bahri, A., Raschid-Sally, L., and Qadir, M. 2010. Wastewater, sludge and excreta use in developing countries: An overview. In Drechsel, P., Scott, C.A., Raschid-Sally, L., Redwood, M and Bahri, A. (eds) Wastewater Irrigation and Health: Assessing and Mitigating Risk in Low-Income Countries. Earthscan, IDRC and IWMI, London.

Kharas, H. 2010. The Emerging Middle Class in Developing Countries. OECD, Paris.

Konar, M., Dalin, C. Hanasaki, N. Rinaldo, A and Rodriguez-Iturbe, I. 2012. Temporal dynamics of blue and green virtual water trade networks. *Water Resources Research*, 48, 1–11.

Kummu, M., de Moel, H., Porkka, M., Siebert, S., Varis, O. and Ward, P.J. 2012. Lost food, wasted resources: global food supply chain losses and their impacts on freshwater, cropland, and fertiliser use. *Science of the Total Environment*, 438, 477–489.

Lal, R. 2011. Sequestering carbon in soils of agro-ecosystems. *Food Policy*, 36(1), S33–S39.

Langford, J., Briscoe, J. and Taylor, N. 2012. The Australian Water Project: Volume 2. The Opportunity of Crisis: a water reform agenda. Committee for Economic Development of Australia, Melbourne, Australia.

Lipinski, B., Hanson, C., Lomax, J., Kitinoja, L., Waite, R. and Searchinger, T. 2013. Reducing Food Loss and Waste. World Resources Institute, Washington DC.

Liu, J. and Yang, W. 2012. Water sustainability for China and beyond. *Science*, 337, 649–650.

Ma, J., Hoekstra, A.Y., Wang, H., Chapagain, A.K. and Wang, D. 2006. Virtual versus real water transfers within China. *Philosophical Transactions of the Royal Society B*, 361, 835–842.

Marshall, F., Waldman, L., MacGregor, H., Mehta, L. and Randhawa, P. 2009. On the Edge of Sustainability: Perspectives on Peri-urban Dynamics. STEPS Working Paper 35, STEPS Centre, Brighton, UK.

Mekonnen, M.M. and Hoekstra, A.Y. 2012. A global assessment of the water footprint of farm animal products. *Ecosystems*, 15, 401–415.

National Intelligence Council. 2012. Global Trends 2030: Alternative Worlds. Available at: https://globaltrends2030.files.wordpress.com/2012/11/global-trends-2030-november2012.pdf (accessed February 2017).

National Water Commission. 2012. Impacts of water trading in the Southern Murray-Darling Basin between 2006–2007 and 2010–2011. Australian Government, Canberra.

Nguyen N. 2002. Global Climate Changes and Rice Food Security. FAO, Rome, Italy.

OECD. 2013. Agricultural Policy Monitoring and Evaluation 2013: OECD Countries and Emerging Economies. OECD Publishing.

Olesen, J.R. and Bindi, M. 2002. Consequences of climate change for European agricultural productivity, land use and policy. *European Journal of Agronomy*, 16, 239–262.

Overseas Development Institute, European Centre for Development Policy Management and German Development Institute. 2012. Confronting Scarcity: Managing water, energy and land for inclusive and sustainable growth. European Union, Belgium.

PepsiCo. 2010. Sustainable agriculture. Available at: http://www.pepsico.com/live/pressrelease/PepsiCo-Unveils-Global-Plans-to-Revolutionize-Farming-with-New-i-crop-Precision-10192010 (accessed February 2017).

Piao, S., Ciais, P., Huang, Y., Shen, Z., Peng, S., Li, J., Zhou, L., Liu, H., Ma, Y., Dins, Y., Friedlingstein, P., Liu, C., Tan, K., Yu, Y., Zhang, T. and Fang, J. 2010. The impact of climate change on water resources and agriculture in China. *Nature*, 467, 43–51.

PricewaterhouseCoopers. 2009. Levering growth in the emerging functional foods industry: trends and market opportunities. PWC, Australia.

Rabobank. 2008. Watering Scarcity – Private Investment Opportunity in Agricultural Water Use Efficiency. Rabobank, Utrecht, Netherlands.

Ridoutt, B.G. and Huang, J. 2012. Environmental Relevance – the key to understanding water footprints. *Proceedings of the National Academy of Sciences*, 109(22), E1424.

Rockstrom, J., Steffen, W., Noone, K., Persson, A., Chapin, S.F., Lambin, E., Lenton, T.M., Scheffer, M., Folke, C., Schellnhuber, H.J., Nykvist, B., de Wit, C.A., Hughes, T., van der Leeuw, S., Rodhe, H., Sorlin, S., Snyder, K., Costanza, R., Svedin, U., Falkenmark, M., Karlberg, L., Corell, R.W., Fabry, V.J., Hanse, J., Walker, B., Liverman, D., Richardson, K., Crutzen, P. and Foley, J. 2009. Planetary boundaries: exploring the safe operating space for humanity. *Ecology and Society*, 14, 32–45.

Rosenzweig, C., Jones, J.W., Hatfield, J.L., Ruane, A.C., Boote, K.J., Thorburn, P., Antle, J.M., Nelson, G.C., Porter, C., Janssen, S., Asseng, S., Basso, B., Ewert, F., Wallach, D., Baigorria, G. and Winter, J.M. 2013. The Agricultural Model Intercomparison and Improvement project (AgMIP): protocols and pilot studies. *Agricultural and Forest Meteorology*, 170, 166–182.

Ruane, J., Sonnino, S., Steduto, P. and Deane, C. 2008. Coping with Water Scarcity: What Role for Biotechnologies? FAO, Rome.

Rygaard, M., Binning, P.J. and Albrechtsen, H-J. 2011. Increasing urban water self-sufficiency: new era, new challenges. *Journal of Environmental Management*, 92, 185–194.

Sadras, V. and McDonald, G. 2012. Water use efficiency of grain crops in Australia: principles, benchmarks and management. GRDC, University of Adelaide, Adelaide.

Steinfeld, H., Gerber, P., Wassenaar, T., Castel, V., Rosales, M. and de Haan, C. 2006. Livestock's long shadow: environmental issues and options. FAO, Rome.

Stockholm International Water Institute (SIWI). 2012. The Water and Food Nexus: Trends and Development of the Research Landscape. SIWI, Stockholm.

Twigg, J. 1984. Vegetarianism and the meanings of meat. In Murcot, A. (ed.) The Sociology of Food and Eating. Gowing Publishing, Aldershot.

Uga, Y., Sugimoto, K., Ogawa, A., Rane, J., Ishitani, M., Hara, N., Kitomi, Y., Inukai, Y., Ono, K., Kanno, N., Inoue, H., Takehisa, H., Motoyama, R., Nagamura, Y., Wu, J., Matsumoto, T., Takai, T., Okuno, K. and Yano, M. 2013. Control of root system architecture by DEEPER ROOTING 1 increases rice yield under drought conditions. *Nature Genetics*, 45, 1097–1102.

UNFAO. 2001. Agricultural Investment and Productivity in Developing Countries, FAO Economic and Social Development Paper No. 148. Available at: http://www.fao.org/docrep/003/x9447e/x9447e00.htm (accessed February 2017).

UNFAO. 2006. Livestock's long shadow: environmental issues and options. Available at: ftp://ftp.fao.org/docrep/fao/010/A0701E/A0701E00.pdf (accessed February 2017).

United Kingdom Environment Agency. 2009. Water for People and the Environment: Water Resources Strategy for England and Wales. Environment Agency, Bristol, UK.

United Kingdom Government Department for Environment and Rural Affairs (Defra). 2011. Guidance on the application of date labels to food. Defra, London.

United Nations. 2005. Good Practices in Agricultural Water Management. United Nations, New York.

United Nations Department of Economic and Social Affairs, Population Division. 2015. World Population Prospects: The 2015 Revision, Key Findings and Advance Tables. Available at: https://esa.un.org/unpd/wpp/Publications (accessed February 2017).

United Nations Development Program. 2006. Human Development Report. Beyond Scarcity: power, poverty and the global water crisis. UNDP, New York.

United Nations Environment Programme and International Water Management Institute. 2012. An ecosystem services approach to water security. UNEP and IWMI.

Vanham, D., Mekonnen, M.M. and Hoekstra, A.Y. 2013. The water footprint of the EU for different diets. *Ecological Indicators*, 32, 1–8.

Verhoeven, H. 2013. The politics of African energy development: Ethiopia's hydro-agricultural state building strategy and clashing paradigms of water security. *Philosophical Transactions of the Royal Society A*, 371, doi: 10.1098/rsta.2012.0411.

Vermeulen, S.J., Aggarwal, P.K., Ainslie, A., Angelone, C., Campbell, B.M., Challinor, A.J., Hansen, J.W., Ingram, J.S.I., Jarvis, A., Kristjanson, P., Lau, C., Nelson, G.C., Thornton, P.K. and Wollenberg, E. 2012. Options for support to agriculture and food security under climate change. *Environmental Science and Policy*, 15(1), 136–144.

Wani, S.P., Rockstrom, J. and Oweis, T. 2009. Rainfed Agriculture: Unlocking the Potential. CABI, Oxfordshire.

Waste and Resources Action Programme (WRAP). 2007. Food Behaviour Consumer Research: Quantitative Phase. WRAP, UK.

Water Footprint Network. 2015. Product gallery. Available at: http://waterfootprint.org/en/resources/interactive-tools/product-gallery/ (accessed February 2017).

Willer, H. and Kilcher, L. (eds) 2012. The World of Organic Agriculture: Statistics and Emerging Trends 2012. Research Institute of Organic Agriculture (FiBL) and International Federation of Organic Agriculture Movements (IFOAM), Bonn.

World Bank. 2009. Climate Change and the Middle East. Available at: http://siteresources.worldbank.org/INTLEBANON/News%20and%20Events/22688823/WB_newsletter_lowres_CC_En.pdf (accessed February 2017).

World Commission on Dams. 2000. Dams and Development: A New Framework for Decision Making. Earthscan Publications Ltd, London.

World Economic Forum. 2010. Realizing a New Vision for Agriculture: a roadmap for stakeholders. World Economic Forum, Switzerland.

World Health Organisation. 2006. WHO Guidelines for the Safe Use Of Wastewater, Excreta and Greywater: Volume 1: Policy and Regulatory Aspects. WHO, Geneva.

6

Consume

6.1 Impact of Consumerism on Water Management

So far, this book has explored how water supports all aspects of our lives by maintaining the biosphere, our essential life support system (Chapter 2), the water fluxes that we are able to exploit (Chapter 3), its role in producing energy ('Live' Chapter 4), and in growing food ('Eat' Chapter 5). The Live and Eat chapters introduce the management concepts required to support the ever-increasing demands for water to meet our basic needs. Most of us live in some form of consumer society and this chapter, 'Consume', explores the third key component driving human demand for water. Here, we look a little more closely at how our consumption of other goods impacts on water, how much of the global demand for water is channelled through large business supply chains, and the critical role of water in supporting national economies.

This chapter examines how much water is used to support our thirst for goods and services, where the industrial powerhouses are, and how industrial use for water compares to water used to provide food, water supply and sanitation. It examines the impacts that global business models and complex value chains have on our ability to manage water resources effectively and fairly. A key element of this issue is the evolution of the concept of 'sustainable business', what this term really means, and how business strategy needs to be far better integrated with water management if population and economic growth are not to lead to environmental decline. If we are to truly change the way water is managed, we need to understand corporations and how they operate.

6.1.1 Water as the Essential Economic Ingredient

Consumption is probably the most complex aspect of the nexus between human activity and water, which also includes Live and Eat. Chapter 5 introduces the concept of food's water footprint, explains the influence of associated virtual water flows on water scarcity and stress, and illustrates the long and complex pathways linking 'food on the plate' to impacted waterbodies. Many of those issues apply similarly to the vast amount of other non-food products that we demand, although they can be much less easy to visualise. The impact of the production of mobile phones on the water environment is, for example, not as immediately obvious as the impact of cropping tomatoes or cotton. Supply and value chains of non-agricultural products (see Breakout Box 6.1) can be much longer and complex than for food and beverage products, and tracing the impacts of individual products to specific waterbodies is exceptionally challenging.

Water Resources: A New Water Architecture, First Edition.
Alexander Lane, Michael Norton and Sandra Ryan.

Breakout Box 6.1 Value chains and supply chains

The concepts of supply chain and value chain are sometimes used synonymously, but the two possess important differences.

The supply chain describes the transfer of goods defined by the required inputs of a given product. For example, a mobile phone requires mined materials and manufacturing to produce the plastics, metals and other components which are combined to form the product. The supply chain includes the raw materials, their manufacturing and the production of packaging and other accessories, and the transport and retail components required to deliver the product to its customer.

The value chain, on the other hand, is defined by the customer requirements and how much each stage in the production process adds to the overall value of the product. *Source:* Feller *et al.* (2006).

6.1.2 Hidden Demand

The average consumer in the USA has a water footprint of 2842 m^3/yr (7790 litres per day), while average citizens in China and India have water footprints of less than half this, around 1080 m^3/yr (2960 litres per day). The numbers vary from person to person depending on their lifestyle, but this consumption is made possible by businesses whose operations we in turn support through our demands. For the majority of people, the things we buy and use make their way to us via a network of predominantly private businesses from the producers, manufacturers, retailers and a whole host of intermediaries.

These businesses and supply chains make it far easier for us to demand water from places and in ways that would otherwise be impossible. They facilitate 'hidden water demand' and make it harder for consumers to make the connection between the products we buy and use, and the very real pressures these activities exert on the water environment. The water supplied to our taps will most likely come from a local waterbody (or at least, typically, from a source within the same region), but what about the water sources affected by our consumption of goods and services? Where are they located? The answer for most people, if the question is considered at all, will be 'No idea. Somewhere far away?' Unless they do some serious investigation, it is likely they will never know which waterbodies their demands impact upon, or how.

We can take this concept one step further and say that, through the global consumption of products, water is being increasingly commoditised, with net movements of water often from developing to developed nations. Research by Lenzen *et al.* (2012) found that the number of water trade connections around the world and the volumes of virtual water traded have more than doubled over the past two decades. The research makes a bold statement that 'developed countries increasingly draw on the rest of the world to alleviate the pressure on their domestic water resources'. This situation has evolved gradually over time as business and trade have become more international and globalised. It has created a situation in which we need to consider not only the ethics of national attitudes (and policies) to water, but also the vulnerability of both virtual water importing and exporting countries to pressures beyond their control (e.g. future climate change, the future trading policies of other nations, and demographic trends).

As the majority of end 'consumers' continue to concentrate in cities in ever greater numbers, the natural resources they depend on are, for many, out of sight and out of mind. Ensuring that the hidden water so essential to our modern consumer lifestyles can sustain this *status quo* requires those with a responsibility to manage water to consider a number of difficult questions:

- Through their lifestyles individual consumers are driving water problems (albeit unintentionally), but to what extent could they be part of the solution?
- Which point in the business supply chain could or should take overall responsibility for tackling the water impacts of consumer lifestyles?
- What is the best approach for galvanising effective action within and across the supply chain? What sort of direction or collaboration is required from government and industry leaders?

6.2 Water Use in Industry: Which Sectors Use the Most?

The UN FAO has developed a global water information system called Aquastat which collates and presents data on 'self-supplied' water abstractions from industry, agriculture and municipal water supplies; the records extend back to 1998 (FAO 2016). 'Self-supplied' refers to the water that users abstract directly from surface or groundwater resources, rather than any water which is supplied to them from a water utility.

Out of 200 countries included in the Aquastat database, agricultural, municipal and industrial abstraction data is currently available for 154 of them. The data show that, globally, 25% of all abstraction is for industrial purposes. However, in nearly three-quarters of these countries, industrial abstraction is much less than this average. Figure 6.1 shows how

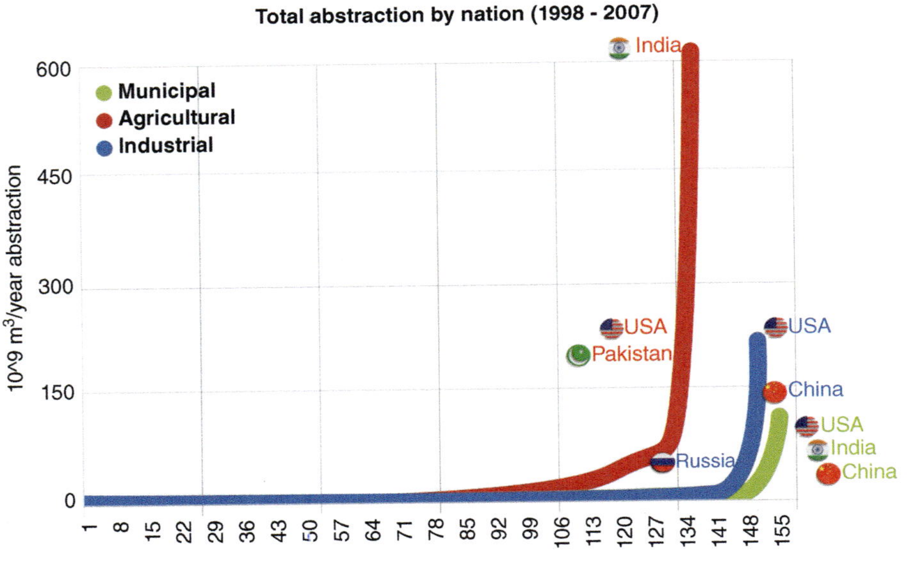

Figure 6.1 Total water abstractions by sector across 154 countries.

Table 6.1 Largest national abstractions for public water supply. Reproduced with permission from Aquastat and World Bank (2014).

Top three abstractors for municipal supplies		Volume (Gm^3/yr)	Population
1.	USA	64	316 million people, 4% global population, 3rd largest population
2.	India	42	1.2 billion people, 18% of global population, 2nd largest population
3.	China	35	1.3 billion people, 19% of global population, 1st largest population.

the largest agricultural abstraction dwarfs the largest industrial abstractions and that both these uses significantly exceed the largest volumes abstracted for municipal supplies. Figure 6.1 also shows how the global data on water use are skewed by a few countries that abstract very large volumes of water.Three countries dominate the Aquastat global abstraction data for industrial use:

- USA: 213 Gm^3/yr;
- China: 128 Gm^3/yr; and
- Russia: 40 Gm^3/yr.

Similarly, total abstraction for agriculture is dominated by a few countries: India, USA and Pakistan. The Aquastat dataset also shows a similar, albeit smaller-scale group of outliers for municipal water abstractions (see Table 6.1). It is no surprise that the countries abstracting the most for public water supply have the highest populations.

The fact that the USA abstracts the most water for municipal supplies despite only having the third-largest population reflects higher rates of per capita consumption. In Chapter 4 we discuss the implications for water management as standards of living, and associated per capita consumption, increase in developing nations. Expertise is not needed to foresee the water problems that increased per capita consumption in China and India will have on demand for municipal water and thus Chinese and Indian water resources.

Before moving on to explore how industrial use of water is split between different products, it is worth exploring further industrial use as a proportion of total abstraction in different countries. Aquastat data illustrated in Figure 6.2 reveals a hockey-stick trend with almost two-thirds of countries using less than 20% of their abstractions for industrial purposes. The global average of 25% of total abstraction being for industrial purposes is skewed by the higher proportions in a few select countries. For example, in Finland, Germany, the Netherlands and Belgium, industrial use is responsible for more than 80% of abstraction, and in Lithuania and Estonia more than 90% (see Table 6.2). These industrial demands for water are dominated by the power sector's need for cooling associated with generating electricity and the manufacturing of refined petroleum products (Eurostat 2014).

The typically inverse relationship between water used for agriculture and water used in industry suggests some form of choice has been made between using water to grow crops or to support non-agricultural industry. Temperate countries such as the UK

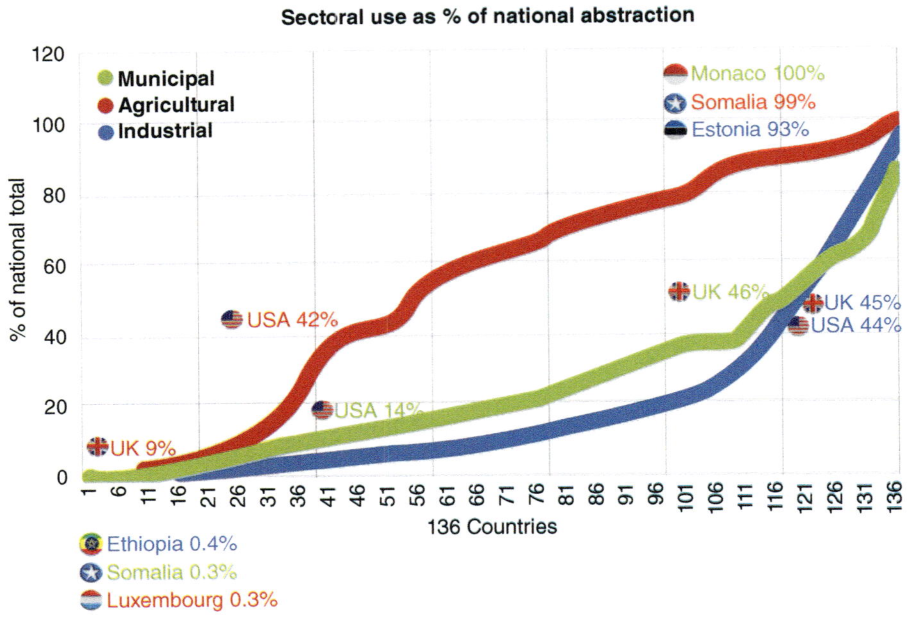

Figure 6.2 Sectoral water abstraction as a percentage of national totals. Adapted from the Aquastat database.

Table 6.2 Countries where total water abstraction is dominated by industrial use.

Country	Proportion of water abstracted that is used to supply industry (%)	Water Risk Indicator	Proportion of water footprint sourced *internally* (%)	Proportion of water footprint sourced *externally* (%)
Estonia	93	2.8	48.3	51.7
Lithuania	92	1.2	73.5	26.5
Belgium	90	3.2	10.9	89.1
Netherlands	85	1.7	5.4	94.6
Germany	83	1.9	31.2	68.8
Finland	80	1.0	52.9	47.1
Austria	79	0.3	31.2	68.8
Hungary	74	0.5	78.7	21.3
France	68	1.8	52.7	47.3
Romania	67	0.8	85.2	14.8
UK	46	2.6	24.8	75.2
USA	45	2.9	79.8	20.2
Australia	11	3.5	88.2	11.8

Water Risk indicator: 4–5: extremely high (>80%); 3–4: high (40–80%); 2–3: medium to high (20–40%); 1–2: low to medium (10–20%); 0–1: low (<10%). National level indicators inevitably limit understanding of local conditions which can vary considerably within a country.
Source: Adapted from the Water Footprint Network and World Resources Institute (2011).

where traditional agriculture is largely rain-fed are more able to use their relatively abundant water resources to support higher-value activities (such as energy production) than countries whose already scarce water resources are currently dominated by the production of agricultural products.

Around the world, water-use decision making is increasingly being influenced by commercial rationale, for example, choosing to allocate water to industrial (or municipal) activities rather than lower-value agricultural uses, a process that has potential impacts for global food production and industrialisation.

6.3 Water Use in Industry: Which Activities Use the Most?

To explore how water is used and managed within the industrial sector (driven by the demand for consumer goods) this section focuses on three core industrial activities:

1) agriculture (non-food products only; food products are addressed in Chapter 5);
2) mining for minerals; and
3) manufacturing.

In many datasets, water utilities (providing mains water supply, sewerage and wastewater services) are included as a fifth industrial subsector but, as the water abstracted is predominantly used in domestic settings, it is also often presented as a separate 'municipal water' sector (as discussed in Section 6.2).

6.3.1 Agriculture: Water to Produce Non-Food Goods

This section focuses on the impact of our demand for cotton (particularly for clothing) and other textiles on the water environment. The use of land and water to produce non-food crops is a double-edged sword. While food crops are generally high volume but low in financial value, they are critical to food security (both locally and globally).

In contrast, non-food crops may ultimately generate higher value, but thereby incentivise increased production to the detriment of food supply. Clothing and textiles, energy, materials used for construction, pharmaceuticals and other more specialised niche products are all outputs of the non-food crop production line that have major implications for the way we manage water. Table 6.3 identifies a few key crops grown for non-food purposes and gives an indication of their water footprints for comparison (noting that these values may vary considerably from region to region). Energy crops are examined in Chapter 4.

6.3.1.1 The Water Footprint of Clothing

Cotton is a major raw material used to produce clothing and fabrics. Around 20 million tonnes of cotton are produced each year; nearly half the fibre is used to make clothes and other textiles worldwide (WWF 2003). The problem for water resources is that cotton is a very thirsty crop, requiring around 20,000 litres of water to produce a single kilo of cotton (approximately the amount needed to produce a T-shirt and a pair of jeans (WWF 1999).

Quantifying the precise volume of water required or the impact of an individual item of clothing is incredibly difficult. The clothing supply chain is highly diverse, complex

Table 6.3 Global average water footprint of non-food agricultural goods (1996–2005).

Good	Typical products	Typical crops	Global average water footprint (m³ /ton)*
Fibres and other textiles	Paper, cloth and fabric (clothes and textiles), string, twine, rope, leather	Cotton (fabric and finished textiles)	9,982
		Flax fibre	3,481
		Sisal fibre (processed)	7,041
		Hemp fibre	2,447
		Jute	2,605
		Leather (bovine)	17,093**
Construction materials	Building materials and insulation made from hemp-lime or straw	Wheat	1,827
Pharmaceuticals	Drugs, herbal medicines and nutritional supplements	Tobacco	2,228
Specialised or niche products	Plastics, paints and inks, essential oils	Maize	947
		Oilseed rape (crude)	3,162
		Palm-oil (crude)	4,787

* Blue, Green and Grey water footprint combined.
** Weighted average of grazing, mixed and industrial farming.
Sources: Crops and derived crop products: Mekonnen and Hoekstra (2010a) and farm animals and animal products: Mekonnen and Hoekstra (2010b).

and lacking in transparency. Nevertheless, a relatively high-level water footprint analysis of clothing was undertaken on behalf of the UK Waste Resource Action Programme (WRAP) to begin to understand overall levels of impact, with the purpose of informing debate on the sustainability of the clothing industry. Table 6.4 is an extract from a summary report reviewing data on the embodied water in clothing (URS 2012). While referring to the turnover of clothing consumed in the UK, it provides an interesting perspective on the size of clothing's water footprint, how it varies depending on material type and the various clothing life-cycle stages.

As shown in Table 6.4, the estimated annual water footprint of clothing is around 6,300 Mm^3 (based on an annual clothing turnover of 2,488,396 tonnes) so the average water footprint is around 2,500 m^3 of water per tonne. By fibre type, the water footprint varies considerably, from as low as 78 m^3 per tonne (86 litres/kg) for synthetic fibres to over 58,000 m^3 water per tonne (approximately 64,000 litres/kg) for silk.

These indicative results show that almost 90% of the total water footprint of the clothing lifecycle is allocated to raw materials (i.e. crop growth: 2,202 m^3 per tonne) and around 10% (318 m^3 per tonne) relates to the processing and manufacturing stage.

Leather is the other extremely water-intensive textile. The water footprint of leather is high for the same reasons as meat, driven by the water requirements of bovine animals over the course of their lifetime. The Water Footprint Network has calculated that a 250 kg bovine cow consumes around 1.9 million litres of water in its lifetime and is ultimately able to produce approximately 6 kg of leather (5% of the total water footprint

Table 6.4 Water footprint of clothing. Cells shaded red indicate the most significant water footprint data. Adapted from URS (2012).

Fibre				UK Water Footprint (WF) of Production and Retail of Clothing in m^3				
Type	**WF (m^3/tonne)**	**as % of UK clothing**	**UK clothing use (tonnes)**	**Raw materials**	**Processing and manufacture**	**Transport**	**UK consumer**	**Total (Mm^3)**
Cotton	3,099	43%	1,070,010	2,806,908,540	493,964,256	5,135	15,611,697	**3,316**
Wool	2,237	9%	223,956	444,834,463	52,975,738	3,161	3,267,565	**501**
Silk	58,153	1%	24,884	1,423,123,949	23,580,664	509	363,063	**1,447**
Flax / linen	2,067	2%	49,768	94,351,009	7,788,438	224	726,125	**103**
Viscose	3,829	9%	223,956	709,611,213	144,552,699	1,069	3,267,565	**857**
Polyester	78	16%	398,143	39,637	25,267,560	1,1766	5,809,004	**31**
Acrylic	128	9%	223,956	22,296	25,357,289	994	3,267,565	**29**
Polyamide	78	8%	199,072	19,819	12,633,780	883	2,904,502	**16**
Polyurethane/ polypropylene	78	3%	74,652	7,432	4,737,668	331	1,089,188	**6**
Total	-	-	**2,488,396**	**5,478,918,358**	**790,858,091**	**14,073**	**36,306,273**	**6,306**
% of WF	-	-	-	**87%**	**13%**	**<0.5%**	**0.5%**	-

is attributed to the 6.1 kg of resultant leather, which equates to approximately 17,000 litres per kg).

6.3.1.2 The Cotton Problem

Cotton is by far the most common fibre used in clothing and textiles and, while per kilogram its production consumes less water than silk or leather, it is still water intensive; furthermore, is often grown in areas where water is already scarce. These production regions include the province of Xinjiang in northern China, Texas in the USA, India, Pakistan, Uzbekistan and West Africa. Some cotton production is rain-fed; however, as for virtually all cropped goods, the yields are much higher when the crop is irrigated (the average yield of cotton is 854 kg per hectare for irrigated cotton and 391 kg per hectare for rain-fed cotton). It is therefore no surprise that almost three-quarters of the total global cotton harvest comes from irrigated land (WWF 1999).

Alternatives to cotton include fabrics made from soy, flax, bamboo and hemp which don't require as much water to grow but remain far less popular as a fibre for clothing.

The environmental impact of intense cotton farming to supply international demand was made abundantly clear in Uzbekistan in the 1980s. Since 1945, four decades of intensive cotton production, augmented by irrigation with water diverted from the Syr Darya River on the east of the Aral Sea and the Amu Darya River to its south, reduced streamflow into the inland sea to a trickle. Simultaneously, cotton fertilisers and other chemicals running-off the land contaminated the watercourses.

Thanks to cotton farming, the 65,000 km^2 area that was once covered by the Aral Sea is almost unrecognisable (see Figure 6.3), with just a small amount remaining. Toxic residues are now found in former coastal regions where receding water levels have exposed sediments, and crop yields have declined in a number of provinces (Columbia University, undated).

As for commercial food-crops, non-food agricultural farming generally depletes local water resources in areas far from where the end-product is received by consumers. The local point of production is rarely where the main financial benefits in the supply chain are to be found (primary production is typically the lowest point in the value chain). This dislocation adds to the problem of water being undervalued by those companies at the head of the supply chain.

As cotton fibres make their way through the various cotton processing and production stages, their financial value increases considerably. At a national level, cotton is critical to the economies of the major producing countries, especially in Uzbekistan where cotton lint generates 75% of the country's export earnings (WWF 1999), and in India and Pakistan where cotton products account for nearly two-thirds of export earnings (IFPRI 2008). In Mali, one-quarter of the population depends on cotton for their livelihoods (OECD 2006).

Cotton 'path analysis' confirms the spatial complexity in cotton-based product supply chains: the paths are long because raw cotton products are processed into fibres, fabrics and garments often with a multiplicity of companies involved. For example, cotton harvested in water-scarce Pakistan is woven into different grades of cloth: low-quality cheap cloth which is then shipped to China to make the 'disposable fashions' that are sold by low-value retailers; high-quality weaves that are used to produce luxury-branded garments; and everything else in between. These observations raise a number of interesting questions for water management and policy.

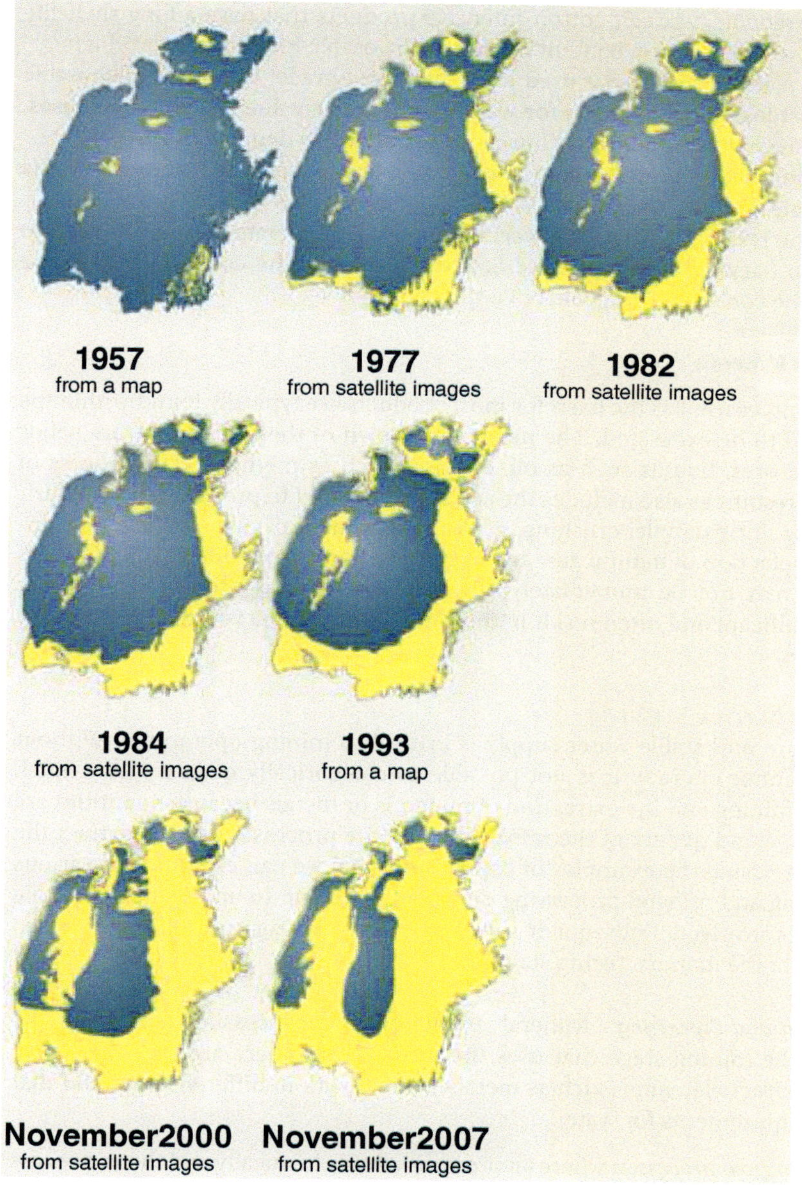

Figure 6.3 A disappearing sea. Reproduced with permission of United Nations Environment Programme (2008).

- Low-grade and luxury cloths essentially use the same raw material, cotton fibre, but fibre length, thread texture and thread count determine the quality of the material. Luxury items designed to have a long shelf-life typically have longer cotton threads, finer textures and higher thread counts than the lower-grade fabrics which are used to produce low-value (often disposable) products. This begs the question, what is the sustainability tradeoff between cotton-intensive products that have a long shelf-life compared to the lower cotton content but more disposable lower-grade products?

- Should scarce water resources be used to produce low-grade, low-value, disposable clothes and textiles? If competition for water from higher-value products increases, what will this mean for the supply of lower-value but highly demanded garments?
- How, if at all, does the value and longevity of the end product reflect attitudes towards the value of water?
- Where does the responsibility (or power) lie to ensure that water resources used to support cotton harvests is sustainable? Is it the producer, the company selling the final project, the consumer, or a collective sponsibility?

6.3.2 Mining for Minerals

The natural resources used as the basis for most products are typically found within the ground and need to be excavated. The most well known of these resources are solids such as coal and ores, liquids such as oil, or gases such as methane. The process of exploiting these resources also includes the activities required to prepare crude materials for marketing, for example, crushing, grinding, cleaning, drying, sorting, concentrating ores, liquefaction of natural gas, and agglomeration of solid fuels (UN Statistics 2010). While it may not be immediately obvious how water is used in mining, the amounts are significant and often result in the generation of large volumes of contaminated wastewater.

6.3.2.1 The Role of Water in Mining

Access to a secure and stable water supply is critical to mining operations. Without water, a mine cannot operate. It is not possible to categorically quantify how much water is used in mining and the extraction of minerals or metals because quantities are subject to the type and quality of the mined ore and the processes used to extract the valuable element. Using the examples of copper and steel we can examine the various impacts of mining and mineral processing on water and begin to understand the scale of water volumes involved. Subsequent stages such as smelting and refining are also considered part of the manufacturing stage.

Mineral Extraction and Processing Mineral extraction and processing represents the activity within the mining stage that uses the most water. There are two alternative approaches to extract elements (such as metals) which work in different ways and that have different requirements for water.

1) *Hydrometallurgical processes*: where an ore is dissolved chemically and then a process such as electrolysis used to recover the dissolved elements (such as gold or copper). Actual volumes of water used per unit of ore vary depending on the specific combination of processes involved but, as a general indication, studies suggest that hydrometallurgical processes can use around 1.6 m^3 of water per ton of copper concentrate (Bruce and Seaman 2014). The water used in hydrometallurgical processes is largely recycled in order to reduce consumption of freshwater and reagents (ANA 2013).
2) *Pyrometallurgical processes*: where ore is incinerated and elements, for example copper, platinum, or iron (used to produce steel) are then recovered from the ashes.

Water is used indirectly to cool equipment such as blast furnaces and, similar to the use of water as a cooling agent in power stations (see Section 4.2.3), the volume of water used varies depending on the cooling system adopted.

Mining operations that include wet separation processes (such as gravitational, magnetic, flotation, flocculation, spherical agglomeration and leaching) use large volumes of water. For example, in the flotation stage (during which sandy particles are removed) water can represent 85% of the ore pulp/water ratio (Levay *et al.* 2001).

Modern processing plants increasingly require high-quality water, and water/ore ratios can range from 0.4 m^3/t to 20 m^3/t (ANA 2013). In 2014, an Australian-based study completed a life cycle assessment of a range of metals with differing levels of ore quality and production routes (Norgate and Lovel 2004). That analysis concluded that the mean value for water consumption (for the mining and processing stage) is 0.7 m^3/t ore.

Example water consumption figures associated with pyrometallurgical processes are as follows:

- To produce copper (mining and flotation), approximately 2 m^3/t is required (Ćirković *et al.* 2014); and
- To produce steel (from iron ore) much larger volumes are required, ranging from 100 m^3/t to 200 m^3/t (ANA 2013).

The same Australian study also concluded that when the full life cycle of metal production (i.e. including the direct and indirect water inputs associated with manufacturing) are taken into account, the mean value for the water consumption (of all the metals considered) is 2.1 m^3/t ore (three times the volume of water used directly). This observation illustrates the importance of accounting for the full water footprint when considering the water management implications of the products that we consume.

Cleaning Ore processing includes cleaning stages which, although in most cases do not have strict quality requirements, do require large amounts of water.

Dust Suppression Large volumes of water are also used during production processes to crush rocks and suppress dust, particularly on haul roads and waste dumps. This water can be lower-quality industrial water or mine water, provided there are no contamination risks.

Transport Using water to create a slurry mix is the primary means of transporting materials in mineral processing. Transporting slurry in pipelines can reduce both costs and energy demands compared with more conventional transport forms such as rail and road. However, significant volumes of water are required to keep the slurry material in liquid form. This type of transport has been used in Brazilian mining since the 1970s, such as with the ore pipeline of Samarco (ANA 2013). Slurried iron ore is piped over 396 km from a mine in Mariana, Minas Gerais, to the plant in Ponta de Ubu, near the city of Guarapari, on the coast of Espírito Santo. In 2004, 15 million tons of pulp containing 70% solids were transported. At this scale of operation, 6 million m^3 of water are used every year to keep the slurry pipe operating.

Other Needs High-quality potable water is also required for domestic purposes in mining company offices and administration buildings, and in camps associated with remote mines. In some cases, and particularly in the case of remote mining operations, the mining industry may also supply water as a secondary activity to nearby towns for use in households or other economic units or to facilities that accommodate mine workers (UN Statistics 2010).

6.3.2.2 Regional Context and Water Management Challenges for Mining

The direct and indirect water use in mining projects must be considered within the context of the regions in which they are located. Mines are found in areas that experience extreme aridity, those that have the highest rainfalls in the world, and those that experience extreme seasonal variations in temperature and rainfall. In the same way that shortages of water have caused power stations to shut down, mines are also at risk. Lack of water can immediately put the brakes on mining, as the following examples illustrate:

- In Mexico in 2011, depleted by its worst drought in seven decades, Goldcorp Inc. was forced to slash planned output at its Penasquito mine when a lack of water made it impossible to operate at full capacity (ICMM 2012).
- Water shortages in South Africa have impacted on power generation with outages causing major knock-on effects for mining operations and, in turn, worker protests in response to mine closures.
- In Peru and Chile, chronic water shortages are forcing operators of metal mines to pump water from desalination plants at the Pacific Ocean, hundreds of kilometres high into the Andes (see Breakout Box 6.2).

Breakout Box 6.2 Water and mining in Chile

The Zaldivar copper mine built in Chile's Atacama desert in the 1990s exemplifies the impact that changes in water availability can have on mining operations. The Atacama, already the driest place on Earth, is experiencing the sharp end of climate change and there is fierce competition for the very limited water resources.

Diego Hernandez, CEO of Antofagasta Plc, was quoted as saying that 'in Northern Chile, there is no underground water for new projects, so any new project will require seawater, desalinated or not' (Reuters 2008). Antofagasta's US$ 1.5 billion Esperanza gold and copper mine was Chile's first mine to be 100% dependent on seawater. Seawater is pumped over 140 km from the coast to the mine at an altitude of 7545 feet.

Desalination does not come cheap. BHP Billiton, which operates the Escondida mine (the world's biggest copper mine), estimates that using desalination triples its water management costs. Despite this, BHP is considering expanding the mine's existing desalination system. At the Zaldivar mine attempts to reduce reliance on desalination and long-distance water transfers are focused on water recycling; the aim is to meet more than 90% of the mine's water demands from recycled water.

Antofagasta also aims to make sure that any water infrastructure put in place supports not only their operations, but also the needs of surrounding residents. To that end, the company runs a desalination plant that provides about 60% of the city of Antofagasta's water.

At the other extreme, too much water can cause flooding, temporary mine closure and the potential release of contaminants to the environment. The mining region of Queensland Australia was hit particularly hard by flooding in 2011. Following years of drought some mines had been designed to catch as much runoff as possible and were completed inundated, with excess water having to be pumped into the mine pits themselves. Huge volumes of water pouring into pits leaked into underground geologies. Eighty-five per cent of Queensland coal mines had to either restrict production or close entirely, with the economic losses estimated at AUD$ 5.7 billion (Queensland Floods Commission of Inquiry 2012). Problems of excess water on mine sites also increase the risk that contaminants seep or leak into the environment from storage lagoons, tailings dams and waste dumps.

Water-related infrastructure now accounts for around 10% of mining capital costs, and this figure is forecast to grow as mines are developed in more remote and inhospitable regions (such as the Andean mountains). In 2009, mining companies spent US$ 3.4 billion on water infrastructure; by 2013 however, this had risen to US$ 11.9 billion (Global Water Intelligence 2011). These increasing costs are driven by:

- the need to respond to water shortages (desalination and long-distance, often high-altitude, pumping);
- elevated standards of wastewater treatment (to meet regulations and to recycle the water);
- increased reliance on low-grade ores demanding more water for each tonne of refined product;and
- decommissioning costs for land and water remediation after a mine closes.

6.3.3 Manufacturing

Once raw materials have been sourced they undergo physical or chemical transformation into new products via manufacturing processes (UN Statistics 2010). Manufacturing industries use significant quantities of water in production processes and for cooling, so also often account for a significant proportion of the water discharged to the sewers. In many places this type of discharge is regulated via permits that control the discharge of trade effluent into the municipal sewerage network. Alternatively (and increasingly), sites may have their own wastewater treatment facilities (which they may also use to provide a wastewater treatment service to neighbouring industries or communities) from which discharges into the water environment may also be regulated via discharge permits. However, this is far from universal and discharge from manufacturing sites is a major cause of environmental and human health problems in many parts of the world.

The largest water uses in the manufacturing sector are typically food and beverage producers. Other manufacturing industries that use a lot of water are textile, paper, petroleum, chemical and fabricated metal production as the follow-on stage from mining and mineral processing.

6.3.3.1 Water Use in Paper Production

The water footprint of paper reveals, rather unsurprisingly, that most of the water in paper comes from the forestry stage. As with metals, water requirements for producing paper differ depending on the types of wood used and the areas of the world where that wood is grown. The water footprint of an A4 sheet of paper (the type typically used for printing and writing) is estimated to be around 2–13 litres (per sheet) or 300–2600 m^3/t (Van Oel and Hoekstra 2012). This is considerably more than the average per ton of metal. The global water footprint of paper can be reduced by choosing production sites and wood types that are more water efficient, and by using recovered paper to produce new paper.

6.3.3.2 Water Use in Fabricated Metal Production

Section 6.3.2 explores the use of water in mining and processing ores. Manufacturing stages following on from mineral processing include smelting and refining, processes which also consume significant amounts of water. Water consumption varies according to the type of metal, and the various processing and manufacturing routes. A life cycle assessment undertaken across a range of metals (Norgate and Lovel 2004) showed water consumption can range from 2.9 m^3/t for steel up to 252,087 m^3/t for gold (results strongly correlated to the grade of the initial ore used to produce each metal).

Steel is produced by injecting pure oxygen into melted iron during the smelting stage, reducing the level of impurities and hardening the metal. Steelmakers use water for various processes and purposes. As per the mineral processing stage, cooling is the dominant use of water:

- Cooling after carbonising in coke ovens: 30,000 to 32,000 L of water per ton;
- Cooling boilers used for converting blast furnace gas to process iron: 75,000 to 227,000 L per ton;
- Cooling boilers used to convert coke oven gas, tars and light oils: 151,000 to 454,000 L of water per ton of coke (US EPA 2008).

In production and finishing, cooling hot strip mills (reheated steel slabs compressed into hot-rolled sheets and coiled through a series of rollers) uses the most water (3,800–7,600 L per ton of hot rolled strip).

Water is also used as a lubricant and a cleansing agent to remove scale from steel products. After iron and energy, water is the steel industry's most important commodity, requiring 284,000 L of water to produce one ton of steel (US EPA 2008). Typically, more than 95% of the water used in steelmaking is recycled.

6.4 Water Risk: Recognising the Magnitude of the Problem

Section 4.2.3 highlights how water management problems can quickly affect business operations and performance. It is often the case that water management incidents create significant detrimental impacts to the environment or social function; however, it is the economic impacts that often stimulate most business action. Rightly or wrongly, when a water management risk manifests itself, it is probably the impact on a businesses' stock price which does most to pique the interest of managers and investors. Any form of under-performance has negative consequences for a business and, increasingly, business leaders and investors have begun to acknowledge water as a key source of business risk.

In 2004, a research paper by the Pacific Institute (Morrison and Gleick 2004) made the statement that 'businesses around the world, from beverage companies to [micro] chip manufacturers, are failing to prepare for the serious economic and political risks posed by growing competition for fresh water, the threat of water contamination, and rising water-related costs. These risks can lead to plant closures, supply-chain disruptions, and public opposition to local business activities' (Pacific Institute 2004).

In the decade that followed, investors of major international and global corporations gradually recognised the vital importance of water to business performance. In 2008, at the World Economic Forum's annual meeting, recognition of the critical issue of water resource scarcity led the United Nations, national governments, major financial institutions and NGOs to form the 2030 Water Resources Group (WRG).

The WRG then began an ambitious project to assess the true scale of the global water management challenge and to identify and begin to explore some of solutions that could be required. The outcome was the ground-breaking *Charting Our Water Future* report (McKinsey & Company 2009) and its sobering picture of an ever-expanding demand for water exacerbated by climate change impacting upon already scarce water resources.

More encouragingly, the report also concluded that future water deficits could be addressed by already proven and available measures. By assessing the viability, deficit-reduction impact and cost of different technologies, the report was able to show how future water crises could be averted. In concluding, the report highlighted the importance of achieving strategies for securing water resources that are joined-up and integrated with broader economic decision making across governments, business sectors, NGOs and water users in agriculture, industry and cities (McKinsey & Company 2009).

A cascade of studies and investigations followed the WRG study, particularly, at least initially, from the individual organisations that had contributed to the original project. One of the most immediate was the 2009 WWF report, *Understanding Water Risks: A primer on the consequences of water scarcity for government and business*, part of a series setting out key concepts in water management in the context of the need for environmental sustainability. Unsurprisingly, the WWF report reiterated how global exploitation had led to significant degradation of ecosystems and the goods and services they provide (WWF 2009).

Societal and business recognition that a healthy and productive environment underpins all aspects of industry and economy is becoming more mainstream; this represents an important step change in the management of water resources, conceptualised in the idea of ecosystem services. This concept and that of water risk formed powerful components of the WWF report, making it clear that as well as being an issue of concern to environmentalists and communities, overexploitation of water has major economic implications for businesses and can adversely affect the ability of governments to meet a whole host of policy goals. In its report, WWF began to explore the different types of risks that water can create for business, highlighting the pervasive impacts associated with impacts on reputation and making the point that the water risks experienced by a particular business often require highly tailored mitigation strategies.

2010 was a major turning point for action on water risk. One of the first significant investigations into how large corporations report on water risk published its findings. *Murky Waters? Corporate Reporting on Water Risk* (Ceres 2010) was in part a response to the global financial downturn of 2008. Ceres stated that 'full corporate disclosure of material business issues [of which water is one] is a core foundation for smart investment decision-making'. The global financial crisis destroyed trillions of dollars of wealth and turned the spotlight back onto material business issues that should not be glossed over or ignored.

Since 2000, the Carbon Disclosure Project (CDP) had been co-ordinating and improving disclosure of corporate risk related to climate change; partly due to investor demand (banks, pension funds, large insurance companies) it expanded its remit to include water in 2010 (CDP 2010). At first, very few companies disclosed their water performance data (150 in 2010); of those that did, the responses were dominated by water uses by the business at the head of the supply chain only. By 2014, the number of disclosing companies had increased to 1064. Of these, 68% reported that water posed a substantive risk to their business (CDP 2014a). The 2014 CDP responses indicated that awareness of water risk had risen in all business sectors (not just those with the most obvious reliance of water, such as food and beverage companies) and that rhetoric had moved on from simple water reduction commitments to the business value associated with improving water stewardship. The reporting companies were collectively responsible for abstracting 912,000 Gm^3 of water and consuming 11,200 Gm^3 of this amount over the course of a single year (CDP 2014a).

Figure 6.4 shows the breadth of water risk awareness across business sectors, but also indicates how short-term considerations dominate current business thinking.

Recognition that water has the power to undermine industrial productivity and even derail economies has pushed water up the business agenda arguably more so than any 'green' lobbying or conservation concerns have ever done. Just seven years passed from 2008, when the World Economic Forum identified the need to raise the profile of water

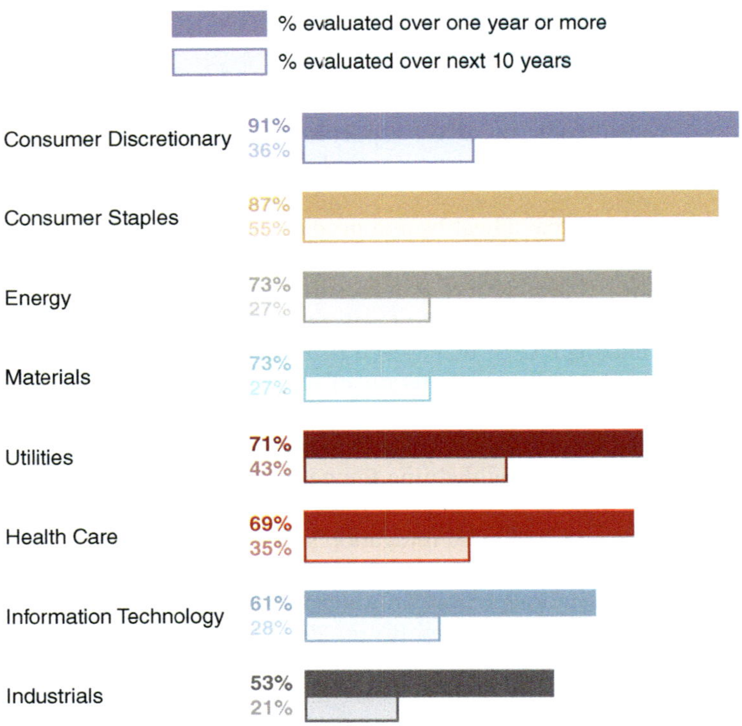

Figure 6.4 Sectoral assessments of exposure to water risk.
Reproduced with permission of Carbon Disclosure Project (2014).

security as a risk issue, to 2015 when water crises were identified as 'the biggest threat to global security over the next decade' (WEF 2015). This rapid increase in business recognition has however in some ways created a disconnect between the traditional 'water' authorities (governments and river basin authorities) and the newly focused business groups. This relationship is discussed further in Chapter 7. For now, let us concentrate on how the water risk concept has evolved.

6.5 Water Risk: Defining and Quantifying the Risk

Once piqued, interest in the concept of water risk was explored by investors, businesses and consultants and eventually disaggregated into different categories. We can illustrate these categories as a water risk model, as shown in Figure 6.5.

Figure 6.5 shows the relationships between the various types of water risk (the rectangles) and the consequent potential impacts for business (the circles). The CEO Water Mandate (Global Compact 2015) focuses on three risk categories – physical, reputational and regulatory – but in Figure 6.5 we define six.

6.5.1 Physical Risks

Physical risks are defined as 'those which directly relate to water shortage, flooding, or poor water quality – all situations that can impact on the ability of a company to operate.'

The most water-intensive parts of many businesses are often located in regions where water management issues are the most pressing. 'Physical risks' are defined as those which can directly impact on the ability of a business to operate. For example, too

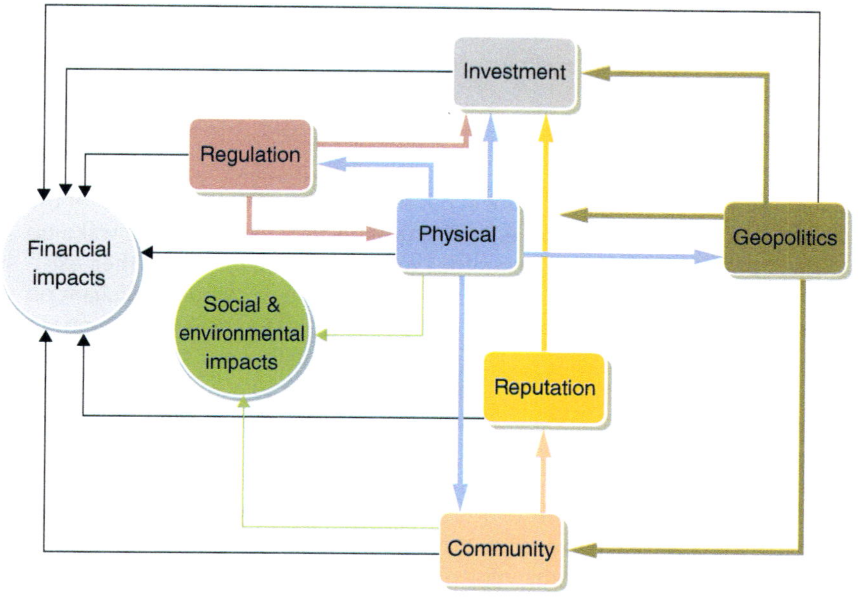

Figure 6.5 Water risk model.
Source: Amec Foster Wheeler (2013).

little or too much water, or water of the wrong quality could result in a power station having to shut down, a situation that may become more likely under climate change.

Too little water is the classic scarcity problem, and could occur when there is insufficient surface water available to abstract (or to discharge effluent into) or where groundwater bores run dry as water levels are progressively depleted. There may be too little water available to abstract directly, or there may be restrictions imposed on the use of mains supply by the utility.

The problems associated with too much water are obvious and much more visible. Flooding devastates people's lives, businesses and economies. Unlike water scarcity, which is generally a long-term and gradually increasing problem, flooding can occur unexpectedly and catastrophically (see Breakout Box 6.3 for example), and it can happen anywhere and from a variety of different sources. Flooding can have severe impacts on business wherever it hits in the supply chain, destroying raw materials, damaging factories and goods, and creating logistical problems.

Governmental approaches to flood risk management vary around the world and this impacts on the ability of business to adapt to and mitigate risks. In many developed nations, flood risk modelling has improved understanding of the interactions between rainfall, geology and land use so that areas at risk of flooding can be defined and different types of flood risk response developed as a result, such as land-use planning to restrict activities in the floodplain. Elsewhere, a lack of data leaves gaps in the knowledge of flood processes and therefore in the ability of a business to prepare and adapt.

As well as causing business production shutdowns, reduced water availability and flooding can also degrade water quality leading to longer-term risks. Issues such as depleted oxygen concentrations, increased temperature, algal growth, reduced dissolution capacity and elevated sediment concentrations all contribute to diminished water quality. The level of impact on a specific business varies according to the production process in question, however; reduced source water quality (often combined with reduced volume) invariably leads to reduced business productivity and lost revenue.

Breakout Box 6.3 Fukushima power plant disaster

For power plants, reduced output or shutdown has major financial consequences for the operating companies and their share value. The Fukushima disaster was a major wake-up call to the nuclear power industry; many countries subsequently re-evaluated their existing nuclear power programs and the stock prices of many energy companies reliant on nuclear sources fell. The greatest losses were felt by the Tokyo Electric Power Company whose stock price crashed due to the direct damage of its nuclear plants in Fukushima (Serita and Xu 2012). The impact on energy company stock prices across Europe varied significantly between Member States, and this could be linked to their differing nuclear energy policies. Generally, what followed the disaster was a wealth transfer from nuclear energy companies to renewable energies companies (Mama and Bassen 2011), although the shareholder wealth in nuclear and conventional energy companies in the United States seemed to be unaffected (Betzer *et al.* 2011).

These examples serve to highlight the vulnerability of the power sector to water risks, not just due to their dependency on water for cooling but also through the reputational risks associated with using, consuming and polluting water to produce energy.

Notwithstanding the improvements in our current understanding of the patterns of and processes behind physical water scarcity and flooding, these risks are subject to constant flux. Climate change and new patterns of population growth mean that the past is no longer an acceptable proxy of the future. This uncertainty adds further complexity to the risks which business and investors are beginning to grapple with. Businesses may find that site operations or supply chains which have previously operated successfully start to experience problems relating to reduced water availability, flood risk and/or diminished water quality. In such a scenario, a reactionary approach is the only available response to mitigate the risks. Alternatively, increasing the breadth of our water resource data and investigating future water availability now represents a more proactive and efficient means for businesses and investors to mitigate risks, while also identifying opportunities for growth, investment and innovation.

6.5.2 Geopolitical Risks

Interstate conflict has re-emerged as one of the world's major concerns after years in which financial crises and environmental issues dominated the risk agenda. According to the World Economic Forum in 2015, geopolitical risks are taking centre stage. Geopolitics is essentially the relationship between the politics of people in different geographic locations. As such, geopolitical relations are typically the outcome of government activity. Businesses are typically less interested in becoming involved in geopolitics, instead preferring to focus on doing good business. Unfortunately, geopolitics has the ability to derail business by incentivising or discouraging certain practices.

For example, business can find itself facing water problems related to the geopolitics of cross-border water issues. Watercourses and river basins generally do not fit neatly within the political boundaries delineating nations and territories. Approximately 60% of global freshwater is within the 276 lakes and river basins which cross national boundaries. This book says many times over that water is a shared resource, but trading systems and conflict over other resources such as land and oil demonstrate that sharing is not a concept that people find easy.

Transboundary rivers cross at least one political border and some of the largest shared water resource systems flow through multiple nations, often between which many geopolitical tensions exist as the following examples demonstrate:

- **Mekong River**: The Mekong flows 4909 km through six countries: China, Myanmar, Thailand, Laos, Cambodia and Vietnam. Notwithstanding the signing of a treaty in 1995 to establish the Mekong River Commission with a mission to promote collaborative management of the river, the six governments continue to struggle to share the resource. For example, in 2010 Laos proposed to build the Xayaburi Dam. Subsequent consultations did not resolve disagreement. In 2012 Laos and Thailand decided to proceed with the dam, despite ongoing opposition from Cambodia and Vietnam.
- **River Jordan**: A conflict hotspot in the Middle East, Israel and Jordan have argued over the River Jordan since 1955. Tensions peaked in 1967 and moves to divert the River Jordan, Israel's main source of drinking water, were at least a contributing factor behind the Six-Day War. The war ended with Israel quadrupling the territory it controlled, including exclusive control of the waters of the West Bank and the Sea of Galilee; these resources collectively provide Israel with around 60% of its fresh water.

- **Indus River**: The signing of the Indus Waters Treaty between India and Pakistan was facilitated by the World Bank in 1960. The Indus River is a vital water resource for Pakistan but, with its source waters in India, Pakistan feared India could use the water resource of the Indus as a political lever, especially during times of war. Since its signing, it should be noted that India has not revoked or reneged on the Indus Waters Treaty, even during three subsequent Indo-Pakistani wars (1965, 1971 and 1999).
- **River Nile**: Home to 160 million people and with its water resources shared between 10 countries, the Nile is vital in the fight against poverty and in achieving economic development. Treaties and agreements have been tabled, agreed, ratified and rejected for over a century. Disagreements exist between the lower riparian countries of Egypt and Sudan and the upper riparian nations of Ethiopia, Tanzania, Uganda, Rwanda, Kenya, Burundi, Democratic Republic of Congo and South Sudan. Arguably, at the core of the problem are the 1902 and 1929 treaties that give Egypt and Sudan veto power over upstream activities, and concerns by upper riparian countries that subsequent draft agreements would maintain the provision of this power to Egypt and Sudan (Stein and Mackenzie 2014). Backed by recent strong political commitments and external funding, Ethiopia is now in the process of initiating an ambitious dam-building program (Verhoeven 2013). Eight major dams are under construction, supporting a program with the aims of improving domestic energy supply, exporting electricity to other east African nations and overcoming Ethiopia's climatic variability to increase irrigation and in turn provide crops for domestic consumption and export markets (Verhoeven 2013). The potential for diplomatic ramifications is however significant and the downstream Nile Basin countries, particularly Egypt, have been vociferous in their objections to the schemes. How Ethiopia chooses to manage the water resources that pass through its borders will have considerable influence on regional politics.

Competition between water users in river basins that cross political boundaries can be particularly destabilising. The Euphrates–Tigris River Basin is one such transboundary basin (shared between Iraq, Turkey, Iran, Syria, Saudi Arabia and Jordan) and has been the focus of numerous attempts to draft and agree treaty agreements on the shared use of water. Unfortunately, to date none of these attempts have been successfully agreed or implemented, largely because of regular conflict between the riparian nations. Water in this region is critical to development and stability and so the continued lack of agreement (and lack of trust) between the basin countries will continue to hinder business growth and economic development. This in turn has the potential to drive further conflict as the population of the region increases, water demand for agriculture and other sectors grows, and as water infrastructure projects (such as the Ataturk dams in southeast Anatolia, Turkey) spark controversy.

While water may not be the primary cause of geopolitical tensions or the trigger for domestic or international conflict, perceived or real instances of water mismanagement by riparian neighbours frequently act to exacerbate political relationships, making conflict far more likely. Sharing of an essential resource such as water, one that provides so many benefits to so many activities, is not a concept we find easy to embrace. The UN recognises that while transboundary waterbodies do create potential for discourse and conflict, they also provide opportunities for cooperation and promotion of regional peace and security as well as economic growth. UN records also highlight how water disputes can be successfully handled through diplomatic means. Almost 450 agreements concerning international waters were signed between 1820 and 2007; of the 150 treaties that have been signed, only 37 disputes were recorded in the last 50 years. Conflicting interests can best be solved by cooperation, adequate legal and institutional frameworks, joint approaches to planning, and sharing of benefits and related costs. Transboundary agreements are valued by all parties because they make international relations over water more stable and predictable, see Breakout Box 6.4 (United Nations 2014).

Breakout Box 6.4 The River Danube: a model of transboundary threats and action

The River Danube is the longest river in the European Union and its tributaries flow through 19 countries. The main river stem flows 2,800 km from its headwaters in Germany's Black Forest through Austria, Czech Republic, Slovakia, Hungary, Slovenia, Croatia, Bosnia and Herzegovina, Serbia, Montenegro, Bulgaria, Moldova and Ukraine before discharging into the Black Sea in Romania (ICPDR undated). The International Commission for the Protection of the Danube River (ICPDR) is an international organisation that coordinates actions across Member State boundaries to benefit the Danube, including ongoing water quality assessments.

Threats to the Danube. Industry, agriculture and tourism all depend on the Danube and its tributary network as a resource yet, as these industries prosper, they threaten the very water resource on which they depend. The Danube is intercepted by dams from its source to its mouth as people across Europe harness its water and inherent energy resource. The dams interrupt the continuity of water flow, fragment wetlands, block the transfer of sediments from upstream to downstream (which contributes to the erosion of the delta beaches), and disrupt the migration and spawning of fish species. The most critical fish species in the Danube is the endangered sturgeon, the fish that produces lucrative black caviar. A combination of dams, pollution and illegal fishing has driven this 200 million-year-old fish to the brink of extinction.

As the Danube flows through its riparian countries, the list of its entrained contaminants grows; industrial effluents (from smelters, paper mills, chemical plants and tanneries) give way to agricultural pollutants (fertilisers, farm pesticides and manure) as the land-use changes. From the upper to the lower reaches of the Danube, water quality monitoring (ICPDR undated) shows significant overall increases in: suspended solids; organic pollution; organochlorine pesticides; concentrations of heavy metals (especially cadmium) peaking in the middle reaches; concentrations of nitrite, ammonium and phosphorus; conductivity (caused by dissolved salts); and alkalinity.

Nutrient levels are high throughout the whole basin, with legal limits in groundwater often exceeded (although recent trends indicate this situation is improving). The main sources of nutrients are agriculture (50%), municipal wastewater (25%) and industry (25%).

Heavy metals and other hazardous substances discharged in industrial effluents and municipal wastewater (microbiological issues) are also a problem. The Baia Mare cyanide spill in 2000 and Ajka red sludge spill in 2010 both received worldwide media attention. The red sludge incident was caused when a dam at a Hungarian aluminium plant failed, releasing 700,000 m^3 of sludge. As a result, 10 people died and about 1,100 ha of land were contaminated.

Addressing the threats. While the environmental situation within the Danube Basin remains pressing, it is much improved from the 1980s. The collapse of communism in the region also led to a collapse of much of the heavily polluting industry (ICPDR undated). However, as the economy in Central and Eastern Europe continues to bounce back, so too does the risk of former problems re-emerging. The regulatory environment across Europe has improved significantly since the 1980s and, from a water perspective, arguably the single biggest driver for change has been the Water Framework Directive (WFD), a programme of measures in place since 2009 which member states of the EU have objectives to meet. The riparian nations have also taken on board one of the fundamental principles of the WFD: to encourage the active involvement of all interested parties in decision making. Joint working is also ongoing to restore natural watercourses and develop green corridors and fish migration aids in order to combat some of the dysfunctional habitat situations created by the numerous artificial influences in the river (particularly the dams and major water abstractions).

Water quality in the Danube has also benefited from the legislative requirements of the Urban Wastewater Treatment Directive, adopted in 1991, which has increased the level of treatment at municipal wastewater treatment works. The Directive requires that all European agglomerations with a population of more than 2,000 are served with sewerage and wastewater treatment to at least secondary treatment level, in order to significantly reduce biological components in effluent. The Danube Delta was identified as a sensitive catchment area (due to eutrophication), thus requiring the relevant Member States to apply more stringent treatment via nitrogen and phosphorus removal (European Commission 2012). Compliance against the directive continues to increase, to the benefit of water quality throughout the basin. While most stretches of the Danube are still moderately polluted, this river is an example of how cooperation and collaboration can yield improvements.

The ICPDR has also established the Danube Accident Emergency Warning System, activated whenever there is a risk of transboundary water pollution, to help authorities initiate the necessary environmental protection and public safety measures.

For businesses operating in environments with geopolitical tension, the volatility of political relationships acts as a key source of risk, and one over which they have very little control. Geopolitical water risks also make other water risks more likely, for example through physical changes to water availability or through regulatory change.

Businesses need to identify and plan for geopolitical water risks by, for example, consulting with regulators and governments over how virtual water policy could be used to alleviate physical water scarcity. Geopolitical risks for business are likely to increase as the implications of virtual water trading become more widely understood. Business and governments need to prepare for the potential conflicts that this could trigger.

6.5.3 Reputational Risks

Reputational risks are defined as those which impact on a company's brand. These can relate to real or perceived physical outcomes and social or environmental injustices, particularly where business activities are believed to impact on ecosystems or community access to clean water. A brand's reputation can be damaged exceptionally quickly.

Risks to the reputation of a business are closely and directly linked to the other risk categories, particularly the physical and social/community risks. Images or articles highlighting dried-up river beds, contaminated watercourses, dead fish and damaged habitats or depleted drinking water sources are highly emotive. Brand value is one of the most valuable assets a company possesses, and one which can take a long time and huge investment to develop and nurture.

Strong brands create competitive advantages by commanding a price premium. They also decrease the cost of entry into new markets and can help attract and retain talented staff (Pasquali 2015). Brand value can however be damaged extremely quickly and sometimes irreparably, especially in a modern consumer era of rapid communications and social media. Even in instances of conjecture, where no evidence of business culpability is present, erosion of brand value can quickly translate into falls in market share or stock price.

As the significance of brand value to a company's market capitalisation continues to grow, brands are increasingly targeted by campaign groups. The more successful the brand, the higher the risk of it being challenged on environmental, social and human rights issues.

6.5.4 Social and Community Risks and Impacts

The success of business operations can be affected by local community attitudes and the relationship the company fosters with stakeholders in its local area. Risks may be linked to concerns raised directly by people who are affected (or perceive themselves to be affected) by the activities of the business. Businesses are also at risk from campaigns by protest and lobby groups recognising real (or again perceived) negative impacts of company activities on community groups less able to generate their own campaigns. Water-related issues that have the potential to draw business criticism and opposition from stakeholders include degraded local watercourses and habitats due to over-abstraction and/or water quality discharges, depleted water resources such as groundwater aquifers, reduced access to safe and reliable drinking water supplies, or public health incidents thought to be due to pollutants released by the company.

Social opposition to the presence and activities of a company can significantly impact on its social licence to operate, and even lead to conflict (see Breakout Box 6.5). As instances of water scarcity increase, as the number of people affected by this scarcity grows and as education leads to greater awareness of the value of water, more and more businesses are likely to experience social water risks.

6.5.5 Regulatory Risks

Regulatory risks are defined as 'those which restrict water use either due to governmental controls, or a lack of control. These relate to restrictions on water abstraction and disposal and can include volume and quality permitting regimes, water allocation, water pricing and controls on water infrastructure'.

Water risks associated with regulation relate to the following:

- The stability or instability of the regulatory regime: businesses need stable regulations in order to plan how their business will grow. When regulations are changed frequently, with little consultation, the risk that business will suffer increases;

- Constraints on water use or discharge: Chapter 4 explains how for water the tendency to over-exploit is strong, and that one response could be to control access to the resource. This type of control creates both risks and opportunities; and
- Failure of water management policies: these may cause immediate business impacts or lead to knock-on effects as other water risks manifest (e.g. water scarcity).

Stability and predictability of regulation is critical for effective business planning. In an effectively regulated environment, activities are investigated and understood, and environmental priorities and objectives are likely to be in place. While these may impose some regulatory burden on business, the processes are understood and can therefore be planned for. In a regime such as this, the water resources on which a business relies are more likely to be effectively protected from external influence, a huge advantage to the business in question. In many developed countries, policies and procedures for regulating water resources have been in place for many years and are well understood by most

Breakout Box 6.5 Social opposition

The mining sector is particularly vulnerable to social and community risks. Communities in many mining regions are well aware that mines and the rich jobs they bring are only temporary and are increasingly demanding benefits that provide longer-term compensation. Facilities that can provide fresh water for decades to come are emerging as a major bargaining chip for miners looking to secure an all-important social licence to operate. Still, for many local communities and stakeholders, the potential damage to the quality and quantity of a local water supply outweigh any benefits a mining project can bring.

Poor communities, often the most at risk from the impacts of mining projects and typically lacking capacity to respond, are increasingly fighting back. For example, residents in Peruvian mountain towns, afraid of losing access to fresh water, delayed Zijin Mining Group of China's US$ 1.4 billion Rio Blanco copper project and Anglo American's Quellaveco copper project. Similarly, environmental protests in Cajamarca, Peru, mainly over water issues, led to the mining company Newmont suspending the US$ 4.8 billion Minas Conga gold and copper mining project in 2011. In an effort to resume the project, Newmont considered increasing the storage capacity of reservoirs used by the mine and exploring opportunities to provide water supplies to local communities (Newmont 2013).

The mining industry is not the only sector vulnerable to social and community risks. Despite its many and various water stewardship initiatives, Coca-Cola (and its various subsidiaries) have been the subject of strong community opposition to factories and bottling plants. In 2000, Coca-Cola opened a new bottling facility in Kerala, Southern India. Local communities complained that the company's abstractions from the shared groundwater resource were excessive, had reduced their shares and that the limited sources remaining were of very poor quality (unsuitable for bathing and drinking). The community protests led to the local village council refusing to renew Coca-Cola's abstraction licence when it was due in 2004, a ruling which was upheld by the regional council. Coca-Cola was subsequently forced to suspend production and, despite temporarily re-opening, the site has been shut down since 2007. Local people remain unhappy and continue to call on the State and National Government for compensation for the damage caused to their health and to the environment. *Source:* Righttowater (undated).

sectors of the economy. While legislation and regulations are updated and changed, this usually takes place gradually and in consultation with stakeholders.

Riskier regulatory environments are typically those which are newly emerging, or where oversight is limited, poorly structured or absent. For example, a lack of regulation means there is likely to be little or no control over water abstractions or discharges. Data describing water resource availability or water quality are also likely to be limited. Businesses operating in this type of environment are vulnerable to a variety of unknowns. In emerging regulatory environments, regulations may be introduced or changed rapidly, possibly without consultation. Businesses will find it far easier to operate under a highly strict yet stable regulatory regime as opposed to one where water resource management is weak or where the regulatory and political landscape is volatile. The influences of water policy and regulation are discussed further in Chapter 7.

6.5.6 Financial Implications of Water Risks

For a long time the price of water has simply meant the cost of a water bill or a fee for an abstraction licence. In the UK for example, most businesses are used to being presented with a bill at the end of every month or quarter for the amount of water they have consumed. When compared to the cost of other utilities and overheads, the number in that bill is often too small to create any incentive to reduce consumption. The low cost reinforces a perception that water is a resource with little value (see also Chapter 10).

By looking at water differently however, and by considering the implications to business if that water is not available, the value of that same resource and the potential financial impacts of its impairment become far more significant. As an example, General Motors has modelled the implications of having to shut down one of its facilities in Northern Mexico in the event of a drought and concluded that it would cost the company in excess of US$ 25 million every month (Balch 2013).

This shift in business mindset is beginning to gain traction but is in no way a mainstream practice. Increasingly, the drive for change is being led by investors that are increasingly aware of the potential for water risks to damage their profits. As a result, the pressure they exert on businesses to take action is growing. No longer will they accept a business that has no risk management plan, no comprehension of the water risks that it faces and no strategy for reducing their potential impacts. While mitigating risks will always incur some costs, these are almost always dwarfed by financial consequences experienced if a water risk is realised.

6.6 Managing Risks and Seizing Opportunities: The Path to Maturity

Many commentators still claim that despite evident water stresses, few organisations prioritise water as a management priority, instead taking its availability for granted and reacting to water shortages, flooding or quality issues as and when they arise. They claim that few organisations actively measure their water use or manage the risks associated with depleting and degrading water resources. They say 'you can't manage what you don't measure' and, in many respects, this assertion is very true (at least in respect to hindering informed management). While companies may claim to have an understanding of the importance of sound water management, at least publically, in reality traditional responses to water risks are often reactive rather than proactive. Invariably, this means that the root causes of particular risks are much less likely to be addressed.

Thankfully, progress is being made and measuring water use (at least primary inflows and outflows) is now fairly commonplace. Many companies, particularly the larger international brands, now also realise the commercial significance of water risks. Some companies are forging ahead, developing pioneering approaches to reduce their risk and, in turn, increase their competitive advantage. Such companies are often identifiable through their involvement with initiatives such as the Aqueduct project, the Alliance for Water Stewardship and the Global Environmental Management Initiative.

The path to mature water risk management evolves through a variety of means including:

- business leadership;
- collaboration with academia and researchers who can help companies quantify their water risk profile;
- public pressure; and
- investor requirements for more information on the risks that water poses to their business interests.

As the number of companies reporting and disclosing on water, such as through the CDP annual water report, continues to increase, this in turn raises expectations on companies that have not yet joined the party.

Figure 6.6 illustrates the phases that companies pass through on their way to a more astute management of water risk. We have defined these phases as:

- the age of taking water for granted;
- the age of water reduction; and
- the age of water stewardship.

The key characteristics of each 'age' are discussed in turn in the following subsections.

6.6.1 The Age of Taking Water for Granted

Doing good business is fraught with challenges: keeping one step ahead of the competition; bringing in new customers and maintaining existing ones; developing new product lines; balancing sticking with what works with exploring new business ideas; improving profitability; and keeping shareholders and investors satisfied. For any business, there are a million and one issues that could create problems but traditionally water has not been considered one of them.

Who worries about water supply when drinking water is always available, safe and reliably supplied by either the state or some other water service provider? If a direct supply is required and your business has access to a borehole to tap into a groundwater resource, then what's the concern? Past weather patterns may have fluctuated here and there from season to season with the odd El Niño/La Niña but, on the whole, regional climates have been nice and stable, allowing people to adopt misguided expectations that all is and will continue to be well.

Water services engineering and management has allowed itself to become invisible. Water pipes are laid underground and, unless there is a mains leak, are generally forgotten

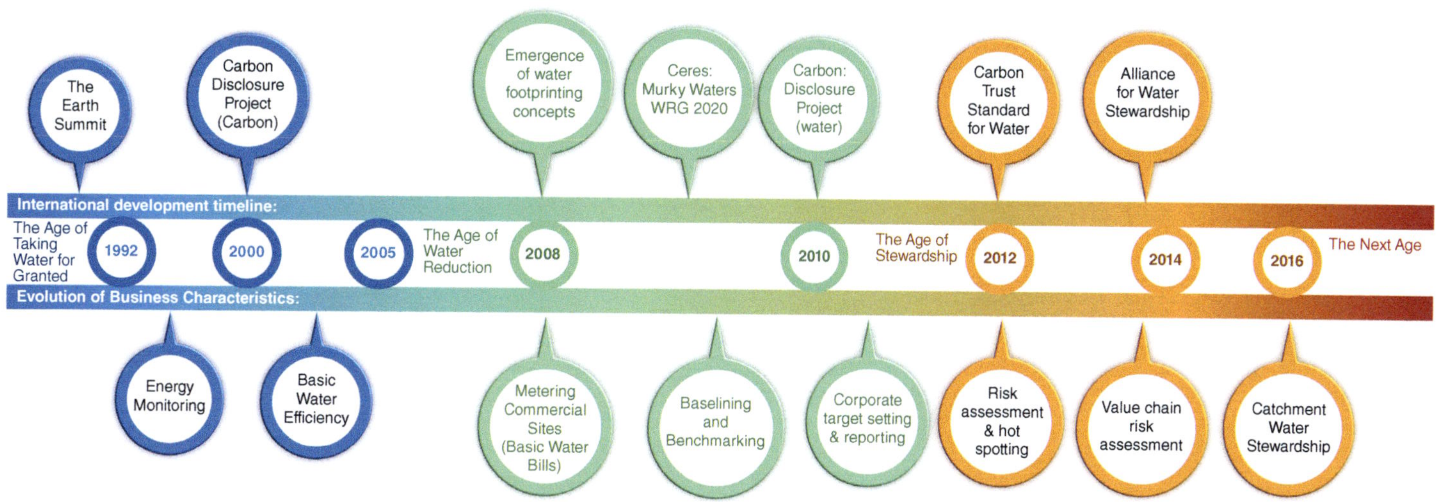

Figure 6.6 Simplified illustration of path to corporate water management maturity.

about. Most people probably couldn't tell you where the treatment works is that produces their drinking water supply, and even fewer like to think about where their wastewater goes. For non-water scientists, engineers and managers, this is not exciting stuff. It is certainly not the kind of information that people running businesses generally choose to distract themselves with.

In the Age of Taking Water for Granted, a state in which many people and businesses still exist, water is just not on people's radar. At best, there may be someone with a site management role who is aware of the water that a company uses, or who occasionally has to make sure the water bill (if there is one) gets paid. In the Age of Taking Water for Granted, it is very possible that no one even bothers to look at the bill. In comparison to all the other overheads a company or family has to deal with, water is cheap, reliable and invisible. As a result, individuals and companies are sheltered from the inherent scarcity of water and, as such, take it for granted. More recently however, as research has allowed water risks to be more accurately defined and as a greater number of business impacts have been shown to be linked to these risks, the transition away from the Age of Taking Water for Granted has gathered momentum.

6.6.2 The Age of Water Reduction

Once the penny begins to drop that maybe water should be thought about in some way the immediate response is typically to save water, which is invariably a good thing. Every drop not used is a drop that has not been abstracted from the environment, not pumped through a supply system and not treated. For hot water the implications are even better; hot water not used is energy not used to heat it. These are all beneficial outcomes.

The problem with the Age of Water Reduction is that water suddenly becomes very site-specific. While there will have been some trigger to generate the transition from taking water for granted to reducing water use, the links between water use and water in the environment can remain quite abstract. The trigger for saving water in most cases is a realisation that saving water can save money (even if just a small amount). Governance may also play a role. In the UK for example, changes to regulations in the 1990s and 2000s imposed restrictions on the type of water fittings (in terms of their water consumption) that could be installed in new building developments. This spawned a proliferation of dual flush toilets and spray taps in public buildings. Ask people at the time what they thought about the changes and you would typically get a blank response or maybe some recognition that it was 'environmentally friendly'. The Age of Water Reduction can probably be best described as 'being green – and saving a bit of cash'. It certainly does not venture into considering the broader water environment as something to be wary of and proactive about.

As concerns over global warming and the impact of greenhouse gas emissions have galvanised action to measure and monitor carbon emissions and energy use, interest has also gradually grown into monitoring of other key resources, including water. The foundations for the international community's Age of Water Reduction can be argued to have spawned from the global climate change and environment summits of the 1990s.

In 1992, 109 heads of state attended the popularly named 'Earth Summit', the UN Framework Convention on Climate Change (UNFCCC), from which emerged the Kyoto protocol on carbon emissions. Governments also came together at the United Nations Conference on Environment and Development in Rio de Janeiro that same year, and there have been many others since. These summits initially helped to create an accepted recognition that our desire and needs to live, eat and consume are causing environmental problems, but also a sense of optimism that solutions could be found.

In reality however, good international intentions have often failed to result in clear progress on the ground, with vital actions held back by bureaucracy and by the difficulty of securing the necessary multilateral agreements. With governments slow to commit to targets and action, it has been left to individual businesses and industry sectors to take the lead. Often, they are the parties identifying and managing the water risks that governments have failed to take real action on. The disconnection between governments and industry with regards to the management of water and water risk is explored further in Chapter 7.

Government inertia and the knock-on effects of uncertainty for business meant that it took many years following Kyoto for large corporations to start reporting their carbon emissions. In 2008 the CDP was formed to provide a global platform for individual companies to report and catalogue their carbon emissions. The 2014 CDP 'Carbon Action Report' received responses from 225 companies (CDP 2014b).

The early stages of the CDP's water project, and of water disclosure more broadly, loosely align with the relatively limited objectives of what we term here the Age of Water Reduction. The CDP and investor signatories had their eye on the wider issues of water risk when it launched the water report, but many of those early disclosing companies were probably very much still operating within their own Ages of Water Reduction.

The term 'water efficiency' (using water wisely and reducing unnecessary use and consumption) is popular within the Age of Water Reduction. This is often accompanied by more widespread uptake of water meters, water efficient equipment, development of more astute water-use strategies and the adoption of explicit water-use targets. Examples of the flavour of the steps taken by select global corporations operating within the Age of Water Reduction are outlined below:

- **Coca-Cola** (2008): to improve water efficiency system-wide by 20% by 2012, compared with a 2004 baseline;
- **Sainsbury's** (2008): to achieve a 50% relative reduction in water use compared to the 2005/06 baseline, saving 1 billion litres of water each year;
- **Morrisons** (2008): to achieve a 15% reduction in water use against 2005 levels by 2010;
- **General Mills** (2006): to reduce water use by 20% by continuing to monitor areas of high use and by identifying opportunities for water conservation; and
- **Royal Dutch Shell** (2011): to ensure that in water-scare areas, operations have water management plans that set out what the company monitors and how to reduce water use.

While water-use reduction targets are often a key component of improved business management, they do represent a rather blunt metric and one that may fail to address the complexity of influences on water risk. As our knowledge of this complexity has grown, those companies that continue to focus solely on water reduction are increasingly perceived as being behind the curve, perhaps even lacking the commitment or ambition to truly tackle our collective water crises.

Many companies at the very start of the water risk management journey don't even specify water reduction targets, relying instead on generalised commitments to reduce their use. In 2012, the UK-based Carbon Trust found that just one in seven of 475 companies from the UK, USA, China, South Korea and Brazil set water reduction targets or disclosed their water performance (Carbon Trust 2012). Such companies are still operating within the Age of Taking Water for Granted.

Of those companies that have set water management targets, the justification for specific targets is often unclear. For example, a target to achieve a 20% reduction in consumption begs the question: does 20% represent a meaningful threshold in relation to specific data and information, or has a figure of 20% been put forward because 20 is a 'good' number? How can a business genuinely promote the sustainability of its water reduction plan without any reference to the water situation in which its business is located, the waterbodies that are actually under pressure, and the ways that its business activities influence and are influenced by them? Furthermore, as the concepts of water footprint and virtual water have grown in acceptance in use, it is now common knowledge that the water risks of a businesses extend far beyond the farm or factory gate.

Many corporations, and even broad industry sectors, remain very much rooted in the Age of Water Reduction. What is more surprising, and worrying, is the seemingly persistent reluctance of a number of other business sectors to engage in disclosure initiatives, even when many of the supply chains that sustain them appear susceptible to a broad variety of water risks. In researching for its 2014 report, the CDP found that 58% of companies in the Energy sector and more than half of companies in the Consumer Discretionary and Industrials sectors did not respond to requests to disclose on their water management (CDP 2014b). Given the critical value of water to the process of extracting primary energy sources and generating final energy products (as explored in depth in Chapter 4), this reluctance to engage appears illogical. CDP even goes so far as to state that the Energy sector is the most at-risk industry (CDP 2014b). In contrast, the Consumer Staples (food and drink), Consumer Discretionary and IT sectors are moving forward, seeking to reduce their exposure to the multiple risks described in Section 6.5 and to capitalise on the competitive advantage available to those companies that can maintain their operations within a safe water envelope.

Considering the CDP water risk database in more detail allows us to identify those business sectors taking the lead on incorporating water risk management into their business strategies. Since its inaugural reporting year, responses to the CDP water project have been and continue to be high among the food and drink sector (69% response rate in 2014). This may come as no surprise considering the reliance of the companies in this sector on water as a raw ingredient. The food and beverage sector is also often acutely exposed to environmental campaigns and so, by taking a leading role in disclosure, their 'social license to operate' is enhanced. Of perhaps greater interest is the increase in response rate in the Minerals sector (i.e. mining) where response rates have increased from 16 out of 27 companies contacted (59%) in 2010 to 22 out of 30 companies (73%) in 2014 (the highest sectoral response that year).

6.6.3 The Age of Water Stewardship

Recognising the significance of the global water crisis and understanding that reducing water use will help to alleviate water stress is an important first step that a number of businesses have taken. To begin to make a real difference, however, this step needs to precede a longer and more complex journey that moves businesses beyond a focus on internal water use towards a more comprehensive understanding of how internal water risks interact with the plethora of water issues beyond the factory fence.

It is the supply chain where the largest component of the water footprint and water risk portfolio of most businesses is located, and therefore where the root cause of the majority of water risks lie. For the business at the head of the supply chain, and therefore that which is ultimately responsible for and impacted by its activities in the supply chain, managing supply chain risks is often complicated by a lack of direct control. Companies often talk of needing to 'first get their own house in order' before trying to grapple with what may be murky and unwieldy supply chains. The potential impacts of such a blinkered approach can however be devastating. In 2011 for example, Toshiba, Honda, Toyota and several other companies were forced to shut down hundreds of factories when the worst flooding in Thailand in almost seven decades hit several major industrial parks. The effects of this shutdown were felt much more widely however, extending to companies operating in countries many thousands of kilometres away due to a shortage of key parts that were otherwise produced in the Thai factories.

For those companies that do take the path towards the Age of Water Stewardship, a number of best-practice tools are available to help; see Table 6.5.

Once a business's supply chain risks have been examined, the next logical and necessary step is to decide what to do to minimise the risks that are identified. Of course,

Table 6.5 Tools supporting water risk assessment

Tool	Description
Aqueduct, by the World Resources Institute	The World Resources Institute's Aqueduct tool is an online platform showing water risks by country and river basin. It contains a variety of datasets and illustrates water stress as defined by a range of indicators.
Water Footprint methodology, by the Water Footprint Network	The Water Footprint Network has developed a methodology for calculating the water footprint of individual products, companies and even nations. The footprint reveals the point in the supply chain where water issues are most significant and therefore where most of the water risk lies.
Water Risk Filter, by WWF	The Water Risk Filter is a tool that helps companies and investors ask questions about water. It improves awareness about water risks in relation to specific water regions and catchments.
The Global Water Tool, by the World Business Council for Sustainable Development	The Global Water Tool takes site location and water use data to generate a water inventory, reporting indicators and other risk and performance metrics, and includes an online mapping system.
The Local Water Tool, the GEMI	The Local Water Tool was developed by GEMI in cooperation with the WBCSD to expand on the functionality of the Global Water Tool. It evaluates the external impacts, business risks, opportunities and management plans related to water use and discharge at a specific site or operation.
Aqua Gauge, by Ceres	Developed by Ceres and other partners, Aqua Gauge focuses on management practices and business approaches to addressing water risks.
The Alliance for Water Stewardship's Water Standard	Once risks have been identified, the Water Standard provides advice on what actions a business could take to mitigate the risks and how to galvanise collective action.

there is the option to do nothing (and hope for the best); however, climate change, competition for water resources and investor confidence are likely to inhibit any company considering the continued pursuit of a 'businesses as usual' route.

Another option is to 'cut out' the risk. This might be achieved by switching suppliers, closing down a site or choosing not to develop in a certain geography or sector. This approach on its own may avoid specifically identified risks but is likely to generate a new range of risk-related decisions, regarding the performance of new suppliers and locations and shareholder responses for example.

A third option is to look for ways to identify and help to tackle the root cause of the risk. The word 'help' is used very deliberately in this respect as it is almost always the case that no single company, organisation or stakeholder can solve water management problems on their own. Collaboration is key, and astute businesses that recognise this imperative are beginning to encourage and incentivise collective action in two important ways:

1) by imposing targets and embedding efficiency requirements in contractual agreements; and
2) by influencing water use within their catchments of operation through, for example, financial incentives or direct support that allows collaborative stakeholder engagement to flourish (see Breakout Box 6.6).

Tackling the root cause of water risks will often require a business to become involved with people and undertake actions well outside its areas of core expertise and activity. For example, by taking action to protect regional water supplies, improve water quality or increase access to clean water and sanitation, companies will provide benefits to the broader community while simultaneously reducing many types of water risk (physical, social, possibly regulatory and most definitely reputational).

It is often possible to identify when a company has transitioned from the Age of Water Reduction into the Age of Water Stewardship through the change in tone, content and language in their corporate sustainability reporting. Some examples include:

- **Coca-Cola**: a quote from Coca-Cola in 2014 (by which time the job title Director of Global Water Stewardship had emerged): 'Coca-Cola meeting its goals isn't going to solve the water crises that exist in all these places ... you may have made your impact sustainable or positive, but that hasn't mitigated the overall stress in a given

Breakout Box 6.6 Collective water stewardship action

Marks and Spencer together with Woolworths South Africa used the WWF Water Risk Filter and identified the Western Cape of South Africa as a major water risk hot spot in their respective supply chains, particularly because of the high water consumption per unit weight of stone fruit crops such as nectarines.

Working with their suppliers and other farmers in the catchments, the retailers applied the Alliance for Water Stewardship Water Standard to identify shared problems and inefficiencies and implement actions across farms in the regions to improve agricultural practice. This process achieved water savings across the region, improved catchment water quality and generated other benefits including increased crop productivity, reduced loss of applied fertilisers and improved soil conditions.

watershed… The big challenge will be how to build a consensus from a broader set of actors, industry as well as government and society to amplify the work we are doing' (Perella 2014).

- **Sainsbury's**: the supermarket chain became one of the first organisations to achieve the Carbon Trust Standard for Water, helping to pilot the methodology. Water stewardship has become one of Sainsbury's key environmental targets and it forms part of its 20 by 20 Sustainability Plan which, the company says, is the cornerstone of its business strategy (Carbon Trust 2013).
- **General Mills**: According to its Water Policy, since 2006 the company has realised that 'General Mills has assessed that 99% of the water use associated with our value chain occurs upstream of our direct operations in agriculture, ingredient production, and packaging. For this reason we have committed to sustainably sourcing 10 priority ingredients by 2020, representing over 50% of our total ingredient buy.' The company is now 'pursuing a long term, multi-stakeholder water stewardship strategy' (General Mills 2014).
- **Walmart**: The company has not set water consumption reduction targets. Analysts suggest that the purpose of this stance is to start conversations with suppliers about increasing the sustainability of their processes and products, rather than imposing arbitrary standards suppliers must comply with (Bloomberg 2013). The Walmart-owned UK supermarket Asda has mapped its entire global fresh produce supply chain and found that a staggering 95% of its fresh produce category is under threat from the impacts of a changing climate (2degreesnetwork.com 2014).

The stewardship approach to water management forms the backbone of the Alliance for Water Stewardship (AWS) partnership, a coordinated call for action from a broad variety of companies, organisation and NGOs that see the benefit and need for a radical overhaul to the way in which most businesses consider water risk management (see Breakout Box 6.7).

Breakout Box 6.7 The Alliance for Water Stewardship

The 2008 founding members of the AWS come from a broad variety of sectors and stakeholder categories, and include a number of global corporations already feeling the impacts of water risks.

Since 2008, the AWS has formed a Global Water Roundtable and an International Standard Development Committee, and has developed the International Water Stewardship Standard. The Standard is a framework to help any organisation understand how they use water and how they can reach out to involve other stakeholders to manage their shared water resources. The Standard 'defines a set of water stewardship criteria and indicators for how water should be stewarded at a site and catchment level in a way that is environmentally, socially, and economically beneficial' (AWS 2014).

The Standard goes beyond validating a business's approach to water management to provide a clear roadmap for collective action. It enables companies to identify and evaluate the impacts and risks from water abstraction, discharge, other water users and communities as well as on-site processes. One of the key principles of the Standard is that the solutions it promotes will be grounded in collective action.

The standard also provides guidance on how to overcome the inertia that can inhibit engagement, empowering individual companies to take the lead to invite and inspire water stewardship partners.

Definitions, instructions, tools and guidance are essential in the world of water stewardship, particularly because even the biggest companies often struggle to understand what water stewardship really means; and they often know even less about what actions constitute appropriate and effective responses. The structure and content of the Standard is likely to evolve as more companies seek AWS accreditation, a process that will no doubt lead to the development of other accreditation standards.

Water stewardship's core principle of collective action brings us neatly back to one of the underlying themes of this book: that despite much talk, the relationships between the companies, institutions, organisations and societies with a stake in water management need a radical overhaul in order to convert inward-looking approaches into outward-looking problem-solving. We need to create a future in which private sector companies, governments, regulators, water utility companies and communities support each other so that water problems are more accurately understood, solutions more integrated and sustainable, and delivery processes far more effective.

Whether the concepts characterising the Age of Water Stewardship are sufficient to address this complex web of concerns remains to be seen, and it may be that a further step is required to place water management on a truly sustainable footing. The remaining chapters of this book explore some of the concepts and ideas that the authors believe could enable water management to progress along such a path.

References

2degreesnetwork. 2014. Supply Chain: 95% of Asda's fresh produce is at risk from climate change. So, what's it going to do about it? Available at: https://www.2degreesnetwork.com/groups/2degrees-community/resources/95-asdas-fresh-produce-at-risk-climate-change-so-whats-it-going-do-about-it? (accessed January 2017).

Agência Nacional de Águas (Brazil) (ANA). 2013. Water resource management and the mining industry/National Water Agency, Brazilian Mining Association. Antônio Félix Domingues, A.F., Boson, P.H.G., Allpaz, S. Translated from: A gestão dos recursos hldricos e a mineração, Portuguese edition, 2006.

Alliance for Water Stewardship. 2014. The AWS International Water Stewardship Standard Version 1.0 (April 8th 2014). Report for Waste Resources Action Programme.

Balch, A. 2013. CDP report reveals corporations starting to take water scarcity seriously. The Guardian. Available at: http://www.theguardian.com/sustainable-business/cdp-report-corporations-water-scarcity (accessed January 2017).

Betzer, A., Doumet, M. and Rinne, U. 2011. How policy changes affect shareholder wealth: The case of the Fukushima Daiichi nuclear disaster. University of Wuppertal. Available at: http://ftp.iza.org/dp5896.pdf (accessed January 2017).

Bloomberg. 2013. Climate Blog: Walmart Sets New Energy Goals, Lacks Water Targets and Comprehensive Electronics Recycling. Available at: http://www.bna.com/walmart-sets-new-b17179873594/ (accessed January 2017).

Bruce, R. and Seaman, T. 2014. Reducing fresh water use in the production of metals. CESL Limited. Available at: http://www.cesl.com (accessed January 2017).

Carbon Disclosure Project. 2010. CDP Water Disclosure 2010 Global Report. Available at: http://www.circleofblue.org/wp-content/uploads/2010/11/CDP-2010-Water-Disclosure-Global-Report.pdf (accessed February 2017).

Carbon Disclosure Project. 2014a. From water risk to value creation. CDP Global Water Report 2014. Available at: https://b8f65cb373b1b7b15feb-c70d8ead6ced550b4d987d7c03fcdd1d.ssl.cf3.rackcdn.com/cms/reports/documents/000/000/646/original/CDP-Global-Water-Report-2014.pdf?1470394078 (accessed February 2017).

Carbon Disclosure Project. 2014b. Why companies need emissions reduction targets. The key to a low-carbon economy. Available at: https://b8f65cb373b1b7b15feb-c70d8ead6ced550b4d987d7c03fcdd1d.ssl.cf3.rackcdn.com/cms/reports/documents/000/000/848/original/Carbon-action-report-2014.pdf?1472035953 (accessed February 2017).

Carbon Trust. 2012. Businesses sleepwalking into a Resource Crunch. Available at: http://www.carbontrust.com/about-us/press/2012/12/businesses-sleepwalking-into-a-resource-crunch (accessed January 2017).

Ceres. 2010. Murky Waters? Corporate Reporting on Water Risk. Available at: https://www.ceres.org/resources/reports/corporate-reporting-on-water-risk-2010 (accessed February 2017).

Ćirković, M., Trujić, V. and Bugarin, M. 2014. Synergy of Energy Resources of Copper Pyrometallugy in RTB Bor-Serbia. Association of Metallurgical Engineers of Serbia. Available at: http://metalurgija.org.rs/mjom/vol20/No4/5_Cirkovic_MME-2004.pdf (accessed January 2017).

Coca-Cola. 2008. 2008 Annual Review. Available at: http://www.coca-colacompany.com/content/dam/journey/us/en/private/fileassets/pdf/2012/12/2008_annual_review.pdf (accessed February 2017).

European Commission. 2012. Technical assessment of the implementation of Council Directive concerning Urban Waste Water Treatment (91/271/EEC) Situation as of 31 December 2009 or 31 December 2010. Final report as of 12 December 2012. Available at: http://ec.europa.eu/environment/water/water-urbanwaste/implementation/pdf/Technical%20assessment%20UWWTD.pdf (accessed January 2017).

Eurostat. 2014. Water use in industry. Available at: http://ec.europa.eu/eurostat/statistics-explained/index.php/Water_use_in_industry (accessed January 2017).

FAO. 2016. Aquastat database. Food and Agriculture Organization of the United Nations (FAO). Available at: http://www.fao.org/nr/water/aquastat/data/query/index.html?lang=en (accessed February 2017).

Feller, A., Shunk, D. and Callarman, T. 2006. Value Chains Versus Supply Chains. BP Trends. Available at: http://www.bptrends.com/value-chains-versus-supply-chains/ (accessed February 2017).

General Mills. 2006. General Mills: Environmental Sustainability. Available on request from: https://www.generalmills.com (accessed February 2017).

General Mills. 2014. General Mills Water Policy. Available at: http://www.generalmills.com/en/News/Issues/water-policy (accessed February 2017).

Global Compact. 2015. The CEO Water Mandate. Water related business risks. Available at: http://ceowatermandate.org/business-case/water-related-business-risks (accessed January 2017).

Global Water Intelligence. 2011. Water for mining. Opportunities in scarcity and environmental regulation. Available at: http://www.globalwaterintel.com/client_media/uploaded/Water%20for%20Mining/GWI_Water_for_Mining_sample_pages-.pdf (accessed January 2017).

International Commission for the Protection of the Danube River (ICPDR) Undated. The Danube River Basin. Facts and Figures. Available at: https://www.icpdr.org/main/danube-basin-facts-and-figures-brochure (accessed February 2017).

International Council on Mining and Metals. 2012. Water management in mining: a selection of case studies. Available at: https://www.icmm.com/document/3660 (accessed January 2017).

International Food Policy Research Institute (IFPRI). 2008. Cotton-Textile-Apparel Sectors of Pakistan. Situations and Challenges Faced. Available at: http://www.ifpri.org/sites/default/files/publications/ifpridp00800.pdf (accessed January 2017).

Lenzen, Manfred, Bhaduri, A., Moran, D., Kenmoto, K., Bekchanov, Geschke, A. and Foran, B. 2012. The role of scarcity in global virtual water flows. ZEF – Discussion Papers on Development Policy No. 169. Centre for Development Research, University of Bonn.

Levay, G., Smart, R.St.C. and Skinner, W.M. 2001. The impact of water quality on flotation performance. *The Journal of The South African Institute of Mining and Metallurgy*, March/April 2001, 69–76.

Mama, H. B. and Bassen, A. 2011. Contagion effects in the electric utility industry following the Fukushima nuclear accident. Available at: https://papers.ssrn.com/sol3/papers.cfm?abstract_id=2084267 (accessed February 2017).

McKinsey & Company. 2009. Charting our Water Future. Economic frameworks to inform decision-making. 2030 Water Resources Group. Available at: http://www.mckinsey.com/business-functions/sustainability-and-resource-productivity/our-insights/charting-our-water-future (accessed February 2017).

Mekonnen, M.M. and Hoekstra, A.Y. 2010a. The green, blue and grey water footprint of crops and derived crop products, Value of Water Research Report Series No. 47, UNESCO-IHE, Delft, the Netherlands.

Mekonnen, M.M. and Hoekstra, A.Y. 2010b. The green, blue and grey water footprint of farm animals and animal products. Value of Water Research Report Series No. 48, UNESCO-IHE, Delft, the Netherlands.

Mekonnen, M.M. and Hoekstra, A.Y. 2011 National water footprint accounts: the green, blue and grey water footprint of production and consumption. Value of Water Research Report Series No. 50, UNESCO-IHE, Delft, the Netherlands.

Morrisons. 2008. Taking Good Care. Our Corporate Social Responsibility Report. Available at: https://your.morrisons.com/Documents/Corporate_Social_Res.pdf (accessed February 2017).

Morrison, J. and Gleick, P. 2004. Freshwater Resources: Managing the Risks Facing the Private Sector. Available at: http://pacinst.org/wp-content/uploads/sites/21/2013/02/business_risks_of_water3.pdf (accessed January 2017).

Newmont. 2013. Conga Fact Sheet. June 2013. Available at: http://www.yanacocha.com/wp-content/uploads/2014/04/Conga-Fact-Sheet.pdf (accessed February 2017).

Norgate, T.E. and Lovel, R.R. 2004. Water use in metal production: a life cycle perspective. CSIRO Minerals. Available at: http://masters.donntu.org/2012/feht/ponyka/library/ist.pdf (accessed February 2017).

OECD Development Centre. 2006. Policy Insights, No. 30. The Cotton Sector in Mali: Realising its Growth Potential. Available at: http://www.oecd.org/countries/burkinafaso/38145914.pdf (accessed January 2017).

Pasquali, V. 2015. Best Global Brands 2015. Global Finance. Available at: https://www.gfmag.com/global-data/economic-data/best-global-brands (accessed February 2017).

Perella, M. 2014. Cola-Cola looks beyond water targets to wider policy challenge. Edie Newsroom. Available at: http://www.edie.net/news/4/Cola-Cola-looks-beyond-water-targets-to-wider-policy-challenge-/ (accessed February 2017).

Queensland Floods Commission of Inquiry. 2012. Final Report. Chapter 13: Mining. Available at: http://www.floodcommission.qld.gov.au/publications/final-report (accessed January 2017).

Reuters. 2008. As South American rivers dry up, miners tap ocean. Available at: http://www.reuters.com/article/2008/02/21/environment-latam-environment-mines-dc-idUSN2138359120080221 (accessed January 2017).

Righttowater. Undated. Case against Coca-Cola Kerala State: India. Available at: http://www.righttowater.info/rights-in-practice/legal-approach-case-studies/case-against-coca-cola-kerala-state-india (accessed January 2017).

Royal Dutch Shell. 2011. Shell Sustainability Report 2011. Available at: http://reports.shell.com/sustainability-report/2011/servicepages/downloads/files/entire_shell_sr11.pdf (accessed January 2017).

Sainsbury's. 2008. Our Values Make Us Different. Corporate Responsibility Report 2008. Available at: https://www.j-sainsbury.co.uk/media/171780/cr2008_report.pdf (accessed February 2017).

Serita, T. and Xu, P. 2012. Energy stock returns and the Fukushima Nuclear Accidents. Available at: http://www.akes.or.kr/eng/papers(2012)/14.full.pdf (accessed January 2017).

Stein, R. and Mackenzie, G. 2014. Implication of the Entry Into Force of the UN Watercourses Convention for Southern African States. International Water Law Project. Available at: http://www.internationalwaterlaw.org/blog/category/africa/ (accessed January 2017).

United Nations. 2014. Transboundary boundaries. Available at: http://www.un.org/waterforlifedecade/transboundary_waters.shtml (accessed January 2017).

United Nations Statistics Division. 2010. International Recommendations for Water Statistics. A background document for the 41st session of the United Nations Statistical Commission. Available at: http://unstats.un.org/unsd/statcom/doc10/BG-WaterStats.pdf (accessed February 2017).

URS. 2012. Review of Data on Embodied Water in Clothing Summary Report. Available at: http://www.wrap.org.uk/sites/files/wrap/Appendix%20V%20-%20Water%20footprint%20report.pdf (accessed January 2017).

US Environmental Protection Agency. 2008. 2008 Sector Performance Report. Available at: https://nepis.epa.gov/ (accessed February 2017).

Van Oel, P.R. and Hoeskstra, A. Y. 2012. Towards quantification of the water footprint of paper: a first estimate of its consumptive component. *Water Resource Management*, 26, 733–749.

Verhoeven H. 2013 The politics of African energy development: Ethiopia's hydro-agricultural state-building strategy and clashing paradigms of water security. *Philosophical Transactions of Royal Society, A*, 371, 20120411, doi: 10.1098/rsta.2012.041.

World Bank. 2014. Population ranking. Available at: http://data.worldbank.org/data-catalog/Population-ranking-table (accessed January 2017).
World Economic Forum. 2015. Global Risks 2015. https://agenda.weforum.org/2015/01/why-world-water-crises-are-a-top-global-risk (accessed January 2017).
World Resources Institute. 2011. Q&A What is a Green Economy? Available at: http://www.wri.org/blog/2011/04/qa-what-green-economy-0 (accessed January 2017).
WWF. 1999. The impact of Cotton on Freshwater Resources and Ecosystems. A preliminary synthesis. Available at: http://wwf.panda.org/?3686/The-impact-of-cotton-on-fresh-water-resources-and-ecosystems (accessed January 2017).
WWF. 2003. Living Waters conserving the Source of Life. Thirsty Crops. Our food and clothes: eating up nature and wearing out the environment? Available at: http://www.worldwildlife.org/publications/thirsty-crops-our-food-and-clothes-eating-up-nature-and-wearing-out-the-environment (accessed February 2017).
WWF. 2009. Understanding Water Risks: A primer on the consequences of water scarcity for government and business. Available at: https://d2ouvy59p0dg6k.cloudfront.net/downloads/understanding_water_risk_iv.pdf (accessed February 2017).

Part III

Existing Water Architecture

7

Existing Management of Water Resources

One of the first lessons that any new water professional learns when embarking on their career is that the solutions to most water problems are constrained by a complex framework of governance structures, water ownership philosophies and investment models. As far as possible, we try to ensure that decisions and investments are based on evidence and the principles of sound science. Unfortunately, political strategies and corporate agendas often override logic and science. Policy and legislation implicitly define how water is valued and prioritised, so water professionals must be bold and engage confidently with governments in order to assert the importance of science and scientific evidence within water management decision making.

So far, much of this book has explored the physical nature of water management, how we use water and the consequences of this use on the environment, as well as some of the technical solutions that are available. We have examined how hydro-climates and landscape features combine to form the water resource systems that societies have developed into water supply systems. We have explored how urbanisation is changing the way we live and use water, and how globalisation is increasingly disconnecting point of consumption from point of production. The changing climate, mass urbanisation and globalisation of food and other goods are forcing us to urgently rethink water management systems and attitudes that previously developed slowly over time. Before we present our vision of the New Water Architecture that the world so urgently needs, let us examine the existing management system in closer detail.

7.1 Governance

The World Economic Forum (WEF) has examined the causes and drivers of a multitude of risks affecting global populations and identifies the failure (or lack) of water governance as one of the primary factors contributing to (water) crises (World Economic Forum 2014). The OECD agrees, claiming that 'the water crisis that many countries face is essentially a crisis of governance' (Charbit 2011).

Governance is important because it defines how we organise ourselves. It is the overarching administrative system that enables and directs how organisations and stakeholders operate and interact with each other to manage water resources and deliver water services. Governance is not about scientific or technical mastery; it is a political, social and economic framework. Good governance creates an enabling environment, whereas poor governance can lead to inertia or even create barriers to good practice. As

Water Resources: A New Water Architecture, First Edition.
Alexander Lane, Michael Norton and Sandra Ryan.

observed by the likes of the WEF, unfortunately it is all too often the case that institutional inertia, administrative bureaucracy, corruption and financial mismanagement compound the challenges associated with physical water scarcity.

Governance in water is complex, not least governance in the water sector itself, but also integrating the water sector with the wider context of national or international governance (see Chapter 11 on institutional integration). Within the water sector, governance defines the roles of water supply and sewerage authorities, environmental and economic regulators, customer groups, trade bodies and academia. Each individual group also has its own internal governance structure that sets out decision-making hierarchies and operational management procedures.

As in any business, 'good' governance is essential to the successful management of water resources and the successful delivery of water services. The complexity of managing a shared resource such as water demands good leadership that provides a clear vision and strategy. The institutions involved need to be strong in their own right, they need to be technically robust, effective and financially stable, but they also need to be well coordinated. Good governance recognises that everyone has a stake in ensuring sound water management and that it takes participation and inclusion to ensure transparent and accountable decision making. Good governance recognises that all water management activities (e.g. policy making, resource assessments, planning and consultations, and not just service delivery) need to be adequately funded.

On the other hand, an apparent lack of accountability or transparency is a tell-tale sign of poor governance and these conditions help corruption to thrive at the expense of sound water management. Poor governance is often driven by highly politicised agendas and frequently leads to questionable decision making. Poorly designed policies, inappropriate or unsustainable projects, ongoing inequalities in water distribution, inadequate mechanisms for public participation, poor financing and limited if any monitoring and evaluation create water crises that need not exist. Plummer and Slaymaker's 2007 review of the Millennium Development Goals concluded that the evidence suggests there is a direct correlation between those countries most lacking water services and those with the weakest governance. Factors that can lead to poor governance include a lack of political will and fierce competition for resources, especially between two or more countries (see Section 6.5.2 on transboundary water management).

International investment organisations such as the World Bank and the Canadian International Development Agency recognise the risks of investing money to improve water resource management in countries with poor governance. These organisations have developed tools to assess the quality of governance within individual countries. Such indicators generally focus on specific subsets of governance that relate to democracy, human rights, policies, public sector management, accountability, legislation, corruption, financial management and internal conflict. Readers with an interest in this topic are encouraged to explore the issue further; Water Governance in Africa (WPP 2010) is a good starting point.

7.2 Structure of Water Management

Chapter 2 and 3 explore how longstanding climatic and cultural conditions have shaped our relationships with water and therefore the approaches to how we manage it. As our societies have grown and our relationships with water have become increasingly demanding and complex, we have responded by creating legal arrangements to control our use of water. Water management legislation is usually implemented at national or state level and is typically designed to ensure preservation of water resources, maintain certain water quality standards, clarify and protect water rights, manage conflicts, and specify the roles and responsibilities of those involved in water supply and wastewater management activities. Policies and guidelines are typically developed to implement the requirements and expectations set out by the legislation.

Inevitably, the priorities and requirements for water management vary between different legislative regions; there are often significant differences in both the issues addressed and in the level of stringency prescribed from location to location. Nevertheless, there is a general structure in which water management principles and activities are formulated and implemented, and the three main components are policy, legislation (which includes regulation) and guidance:

- Policy: a course of action, a principle or a statement of intent.
- Legislation (and regulation): the mechanism that transfers policy into law. There are different facets and legislative components.
- Guidance: legislation and regulations can be supported by tools such as statutory and other formal explanatory guidance. The regulatory tools are cascaded between administrative scales and hierarchical organisations.

Figure 7.1 illustrates the differences between these three components, how they interact and how they are administered at different scales within governance systems.

Legislation is legally binding and typically authorises, mandates, prohibits or restricts a particular activity. When we talk of legislation, we often mean instruments such as Directives or Acts implemented at national and sometimes international scales. In Europe for example, Directives are tools applied by the European Union to decree an outcome that Member States must achieve, without dictating the means of achieving that outcome. Individual Member States are then required to transpose the Directive into their own national laws. Acts are typically national- or state-level pieces of legislation that may or may not be driven by higher-level requirements. These types of laws are often referred to as primary legislation.

The Water Partnership Program (WPP 2010) argues that where legislation is in place, the institutional roles associated with water governance can be at their strongest. For example, the role of regulators is vital for ensuring compliance with legislation. The way that water regulators are organised varies significantly between countries, as does the level of independence that regulators have from the operational side of the water sector. However, their objectives are typically to protect environmental water needs, safeguard drinking water, ensure affordability for water consumers and support the sound economic performance of water utilities. Regulators also serve to 'level the playing field' between user and provider in what may otherwise be a monopolistic environment. In some instances this role extends to explicitly safeguarding the public or customer consultative interests. A functioning regulator is therefore a critical feature of good governance.

In developing countries and emerging economies, specific 'roles' for regulators may not yet exist or may only have been recently created by legislation. The difficulty of establishing regulators with the 'right' powers and remits is a challenging one and even in many developed nations regulators roles' change frequently. In the UK

Regulatory tool:	Federal / multi-state Government	Administrative hierarchy State / Country	Implementing body responsible authority)
POLICY	Federal policy (2-way process)	Government policy	(governmental policy not made at this level)
LEGISLATION	Directives e.g. Water Framework Directive (European Union) Transposition (1-way process)	Statutes, regulations, directions	(legislative tools not developed at this level)
	Regulations Decisions	Transfer directly to implementing body for action and enforcement	
GUIDANCE	Guidance documents supporting Policy/Directives/Decision e.g. Common Implementation Strategy (supporting the water Framework Directive)	Governmental guidance Statutory Codes of Practice Guidance endorsed by administrative groups, e.g. UK Technical Advisory Group (to the Water Framework Directive) Increasing levels of detail	Position statements Regulatory and operational guidance
Hierarchy of consistency	Between states / countries	Within state / country	Across the implementing body

Figure 7.1 Policy, legislation and guidance.
Source: Reproduced with permission from UKWIR (2015).

for example, the Water Services Act 1991 introduced an environmental regulator to oversee the newly privatised water utilities (privatised under the Water Act 1989). Later, as lessons were learned and needs changed, the Water Act 2003 and subsequent Water Act 2014 altered the roles and requirements of many of the stakeholders. This type of legislative evolution is a common characteristic of water management in almost all countries.

Before an item of legislation becomes law it is often referred to as a 'bill', after which, if it is passed by the government, it becomes law. Bills are typically developed within a government department (ideally, although not necessarily, through a consultative process with a broad group of stakeholders) and are raised by a minister, possibly in response to high-profile or influential policies that call for change. Bills may also be raised by non-ministerial politicians in an attempt to try to change legislation. Lobbying of those who ultimately vote on whether to pass a bill is a powerful tool to bring about legislative change; in democratic societies, this activity is itself heavily regulated (for example by the *Transparency of Lobbying, Non-Party Campaigning and Trade Union Administration Act 2014* in the UK).

7.3 The Role of Policy in Decision Making

Policy is a key instrument in the governance of water, often formed in response to an existing or emerging problem; policy makers intend to create a positive outcome or avoid a negative impact. A specific policy usually includes a set of actions or expectations depending on what is being managed and the authority of the policy makers. Unlike legislation, policies cannot compel or enforce, and so policies are often implemented through legislation.

Policies governing the management of water sources and their exploitation for human gain have evolved in response to the needs deemed to be imperative by the societies of the time. As those needs invariably vary from place to place, a complex pattern of policy can often emerge. In Europe for example, supra-national (e.g. EU), national, regional and local government policies all influence water management.

Typically, policies for the management of natural resources tend to involve one or a combination of governmental command-and-control, market-based tools and community-based, informal arrangements. Whatever its type, whether principally developed by government, business, professional institutions, voluntary bodies or community-based organisations, 'good policy' must involve broad stakeholder participation. It is this process of engagement and collective learning that makes the difference between a powerful and successful policy, supported by all and which achieves its aims, and a weak or inappropriate policy that lacks broad awareness or support and therefore either fails to meet its aims or generates unintended consequences.

Early examples of water supply policy can be seen in the informal arrangements or customary laws that governed water sharing prior to the industrialisation and urbanisation of communities (see Breakout Box 7.1). At this point in time, it was typical that the competing demands from largely agricultural and domestic users could comfortably be met by local water sources. The informal arrangements used to control allocations to these demands remain common in developing countries, although they now increasingly exist alongside more formal mechanisms within hybrid frameworks (Butterworth *et al.* 2010). Customary laws also continue to exist and exert an important influence at a local scale in many developed nations.

Because customary laws are agreed and overseen by local stakeholders that typically possess an intimate understanding of their local hydrological environment, they can often be highly effective at managing water supplies within sustainable limits as long as the demands upon them are relatively straightforward to understand (and assuming that the rate of change in these demands is relatively slow). However, when competition for water increases rapidly, the lack of formal and explicit policy can pose difficulties for

Breakout Box 7.1 Customary laws for managing water

Customary laws for managing water relate to those informal arrangements and practices, often implemented by indigenous groups and passed down from generation to generation, that have evolved over time on the basis of accepted moral norms. In close-knit communities, resource management may be achieved entirely through customary arrangements. Appropriately accounting for such arrangements is therefore critical when developing formal water management policies and legislation.

the management of common resource pools such as water. It is in such scenarios that formalised policies governing water supply are often required, typically prepared and implemented by government authorities.

During the period when now-developed countries industrialised, formalised water supply policies tended to focus on the exploitation of a limited number of then-abundant water sources. Such an approach lent itself to exploiting economies of scale by developing water supply networks, piping water from a source (usually a river, lake or aquifer) to a centralised water treatment facility and then on to its various customers. In turn, these centralised networks allowed people to live in locations comparatively far from water sources and at densities that would otherwise have been infeasible. This policy approach was also adopted for the provision of other resources (such as electricity) but is predicated on the assumption that the original source of the resource remains abundant. While this may have been the case when centralised schemes were first planned and implemented, it is now increasingly apparent that many of the water sources on which large cities rely are dangerously fragile. For example, a punishing drought throughout 2014 in the southeast of Brazil caused the volume stored in the main reservoir system of the city of São Paulo to fall to less than 10% of its capacity. By the end of 2014, it was estimated that the 6.5 million residents reliant on the reservoir system might have as little as 2 months of water supply remaining.

The contemporary challenge for developed countries and cities is therefore to manage and supplement existing infrastructure and its enabling policy in order to restore the exploitation of water sources to within sustainable limits while continuing to provide the necessary social foundation for its citizens.

7.4 Types of Policy and their Development

Water management practitioners tend to group policies into one of two main categories:

1) policies that focus on *identifying* high-level goals, such as setting targets to increase access to water and sanitation (CEO 2005); and
2) policies that focus on *achieving* goals by seeking to influence certain types of behaviour (these policies typically attempt to define or infer 'good behaviour' and/or regulate 'bad behaviour', even though policy itself does not compel or enforce).

Policies can also be categorised according to the nature of their overall aim, for example whether or not they seek to mitigate a risk or adapt to a change (NWC 2012).

Once established, a policy may remain a standalone reference point, intended to act as a guide to stakeholders, or it may be supported by legislation in order to enforce those practices that will allow its aims to be achieved. It is not uncommon to find policy that explicitly sets out to either change a piece of legislation or create a new one.

Alternatively, policy may be formed in response to legislation or regulation, particularly where there is a need to interpret and implement the requirements set out at the legislative level. For example the UK Government's policy on 'Reforming the water industry to increase competition and protect the environment' (Defra 2015) provides focus on how the water utilities can become more efficient and better able to meet customer needs, in response to the Water Act 2014.

All organisations, not just those with a regulatory role, are free to develop policy. However, a policy's ability to stimulate action (its ultimate aim) will vary depending on the status of the policy-making organisation within the broader governance framework.

Opinion varies significantly between organisations within the water sector; while there may be consensus on broad issues, such as increasing access to safe water supply and sanitation, there is often argument over more localised goals or how broad goals should be achieved (e.g. whether to meter domestic water supplies or not). The OECD specifically highlights that understanding multi-level governance challenges in water policy requires a holistic approach to coordination (Charbit 2011).

Changes to water management policies are often instigated in response to specific, catastrophic events. For example, severe droughts can change the emphasis in policies concerned with water metering, demand management or attitudes to resource development. Flood events often drive changes in policies on investment and approaches to both short- and long-term flood risk management and response. Policy is generally developed by people in positions of power and influence within an organisation and, because of this, policy can change suddenly. Changes in leadership also typically lead to changes in policy, driven by new ways of thinking or new opinions on which issues are most important.

A common cause of disjointed or suboptimal water management within the water sector is the difficulty in reconciling the often conflicting messages of different policies within the governance framework (e.g. those coming from central versus local government bodies). Where poorly thought out or single-minded policy is implemented, the risks of ineffective or even damaging water management decisions are high. Confusing policy often leads to management approaches that address only a single issue, and often create new risks as a result. Debates surrounding the chlorination and fluoridation of potable water supplies typify this issue. Chlorine reduces the spread of a wide variety of diseases but its by-products are potentially carcinogenic (Viscusi 1994), while fluoridation is beneficial at low concentrations but toxic at higher levels (Tiemann 2011).

Even well-thought-out policy can be weak. Lack of monitoring, poor enforcement, corruption, lack of political will, differences in national policy implementation locally and other more urgent priorities for action by government all serve to undermine the objectives of policy.

The processes of developing and then implementing policy should be regarded as a continuous cycle from which lessons are regularly learned and acted upon (see Figure 7.2). The following subsections explore how policies related to our dominant uses of water have evolved over time.

7.4.1 Water Policy for Domestic Supply

In industrialised and industrialising nations, the provision of a domestic water supply is almost universally achieved by constructing and operating a pipeline network that exploits a limited number of large blue water sources via a centralised water treatment plant. Water is abstracted from the blue water source, pumped or gravitated through pipelines to a centralised water treatment facility, treated to a high water quality (invariably potable standard) and then distributed to homes and commercial properties.

The water utilities, governments or other governance bodies generally set policies on how the supply system should be managed, for example defining what level of leakage is

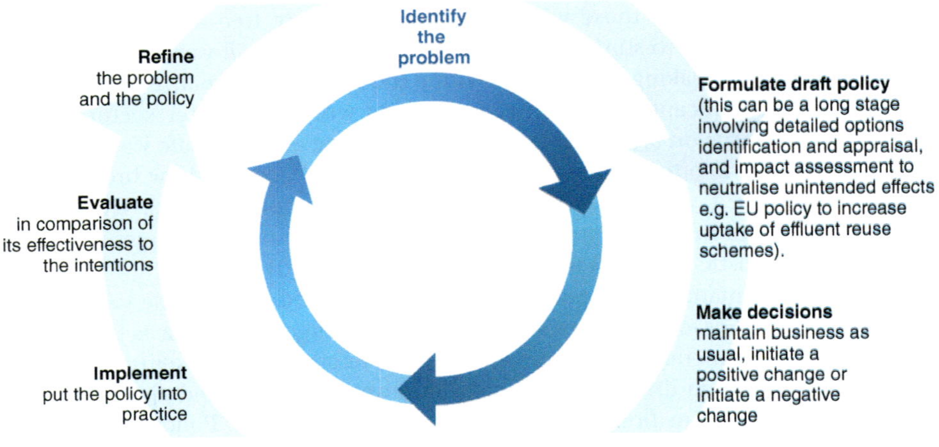

Figure 7.2 Continuous cycle of policy development and implementation.

tolerable and what approaches should be employed to control leakage. Other supply policies may relate to the maintenance and replacement of pipes, metering supplies, water efficiency expectations, and how best to mitigate and adapt supply systems to climate change.

As explained in Chapter 4, the success of structured water policies to protect public health and provide reliable service cannot be disputed. However, the centralised approach to providing water supply is rigid and therefore vulnerable to the growing uncertainties that now characterise urban water supply, for example those associated with climate change, population growth and political cycles. Alternative or hybrid systems that give a greater role to decentralised or satellite components and that exploit more localised water sources and treatment options have the potential to increase the resilience of the overall system (see further discussion in Chapter 4 and Section 7.8). However, adoption and implementation of such decentralised solutions is often held back, discouraged by a number of factors including the following:

- Policies that often place a legal obligation on water utilities to achieve a certain level of service and price controls. For example, Ofwat, the economic water regulator in the UK, has policies setting out expectations on minimal services levels that water companies are required to meet. This can encourage risk-averse decision making and incentivises maintenance of the *status quo*.
- The failure of many national policies to successfully incentivise investment in the research, development and demonstration of emerging water supply options.
- The fact that individual components of pipeline networks require maintenance at different times. This means that, in many cases, like-for-like replacement is the only feasible means of maintaining continuity of supply. As such, infrastructure lock-in is a common problem.

Despite these constraints, examples of new approaches to domestic water supply are increasing in number. Gradual changes in policy are supporting this process. However in many cases, the switch to or integration of a new water supply option often reflects a reactive government response to a severe shock, a major drought or system failure for example, rather than a coherent strategy. In such cases, it's far less likely that the decision will lead to optimum outcomes over the long term. Alternatives to the centralised water supply network include the following:

- Expanding the water source base by adopting options such as indirect and direct potable reuse (see case studies in Chapter 4) as well as establishing other drought-response contingencies such as desalination and managed aquifer recharge. Reuse options help water supply systems to become less linear and more circular, thereby reducing dependencies on traditional water sources that are likely to become less capable of meeting total demand in future. For example, a study in Australia found that recycling stormwater in the city of Adelaide could deliver 60 GL of water a year, a similar volume to that which could be provided by a recently constructed desalination plant (Page et al. 2014). Desalination may be too costly to rely on as a primary component of the water supply portfolio, but can provide vital resilience to the broader network during times of acute scarcity.
- Satellite systems within which reliance on the centralised pipeline network is reduced by using locally abundant water sources (such as rainwater, domestic grey water and stormwater) to meet the demands of a small population (e.g. household, apartment block or local community).
- Taking a hybrid approach to water treatment so that users (or at least a greater proportion of users) receive water that is fit for purpose, rather than receiving water that is fit for the purposes of all the users of the network.

Reducing the amount of water that is lost during its transfer through the pipeline network (leakage, one component of which is termed non-revenue water) also provides a means of increasing water supply. In many countries, developing and developed, the proportion of non-revenue water can exceed 30%, an issue that is often the focus of anger from customers. Reducing volumes of non-revenue water requires utilities to develop effective maintenance strategies and, crucially, that the revenues the utility generates meet the associated investment needs. Even in the USA, only one-third of utilities earn enough through fees to operate a financially sustainable business (Black and Veatch 2014). While domestic customers often feel that water supply is a right that should be provided at very little cost (e.g. 2014 protests in Ireland over the imposition of water fees; Galbraith 2014) it needs to be realised that water supply is a service that incurs a significant cost. These costs must be recovered in order to support a viable and sustainable supply system.

Briscoe (2011) argues that the cause of poorly performing systems is most strongly influenced by the seriousness and capability of the prevailing government. Briscoe highlights that, in many developing country cities, utility performance is weak. As a consequence, it is often the poorest in society who lose out; a lack of connections (particularly in rapidly expanding peri-urban areas where property rights may be ill-defined) forces the poor to buy water from vendors for several dollars a cubic metre while the rich, connected to the network, pay just a few cents. The latest example of this in 2016 was the water crisis in Pakistan. Karachi, the largest city in Pakistan, needs 4 billion litres per day to meet the needs of more than 20 million people. However, only half that volume is delivered, leaving people without water even for basic needs such as drinking, sanitation, and laundry. Once the Hub Dam emptied, it left Karachi completely dependent on transfers from the Indus River, more than 120 km away. A leaky system, poorly performing pumping stations and water being illegally siphoned off to sell on the black market are all blamed on the mismanagement of water by the State.

It is not always the case that developed-country cities demonstrate best-practice urban water governance. Briscoe (2011) gives examples of a range of alternative and successful approaches to delivering urban water supply (e.g. privatised utilities in Manila, hybrid models in Brazil and public models in Phnom Penh). In the latter case study, over a 10 year period the utility was able to transform into a financially sustainable business while also significantly reducing non-revenue water and increasing the reach of its distribution network by more than 500% (Biswas and Tortajada 2010b).

7.4.2 Water Policy for Agriculture

While sound domestic water supply policy is critical for public health, policy on water use in agriculture has a much greater influence on total water abstraction and water resource management. This is because the production of crops and livestock is responsible for 86% of our global water footprint (Hoekstra and Chapagain 2006) and agricultural activities almost always makes up the largest share of national water footprints (Hoekstra and Mekonnen 2011).

Globally, the vast majority of agricultural systems are rain-fed, that is, they rely on green water, and this was especially true of early farming systems. With populations small and the opportunity costs of green water use low, rain-fed agriculture supplemented by limited blue water use and governed by customary arrangements was largely sufficient to meet demands for food, despite the impacts of occasional drought. As populations grew however, the variability of rain-fed agricultural production began to exert greater influence and led to a more systematic and sustained exploitation of blue water sources. The mechanisms of this early exploitation were, and in some cases remain, governed by customary law. However, as competition for water has continued to increase, the use of formalised policies has become more prevalent.

Agriculture is a vital component of virtually every national economy, especially those in the early stages of economic development and where livelihoods are often directly reliant on the sector. The socio-economic importance of agriculture has in turn, strongly influenced the evolution of policy governing the supply of water to its users. This has frequently resulted in the allocation of large surface water and groundwater entitlements at zero or very low cost to farmers. This highly incentivised scenario for the exploitation of water has also, often inadvertently, been supported by the policies of other resource sectors, most notably energy (heavily subsidised fuel supplies have often further encouraged the pumping of water from watercourses and aquifers).

While successful in supporting the growth of the agricultural sector without any perception of the inherent scarcity of water, policies such as these have ensured that many farmers feel no need to consider the efficiency of their water use. As a result, simple but highly inefficient forms of irrigation tend to dominate agricultural systems, even in developed countries today. For example, flood irrigation has very low efficiency rates, but globally is the dominant form of irrigation (FAO 2002); around half of irrigated land in the USA employs this method (Porter 2014).

The impacts of growing demands for water for irrigation are now increasingly borne out in the depletion of those environments that would otherwise be supported by blue water sources. The Aral Sea is given as an example in Section 6.3.1. As another, in Gujarat State (India), free groundwater and the free electricity to pump it contributed to severe groundwater depletion, the near bankruptcy of the state electricity board, and unreliable power supply to farmers and other rural residents (Giordano and Shah 2014).

As demands on blue water sources have grown, open basins (those where RWR exceeds total demand) are increasingly transitioning to closing and closed catchments. In these scenarios, it has most typically been the environment (that water use whose benefit to humanity is most crucial and yet most difficult to value) which has lost out in the competition for water. Awareness of the importance of ecosystem services is now growing and, as a result, agricultural water supply policies are beginning to evolve. Most countries now at least recognise the need to reduce the perverse water-related subsidies they provide to farmers, albeit if the lure of appeasing this stronghold of public support remains powerful in practice (Biswas and Tortajada 2010a).

Efforts to achieve or to work towards cost recovery naturally incentivise improvements in the efficiency of agricultural water use. The 'more crop per drop' paradigm is explored in Chapter 5 and includes initiatives that focus on closing the yield gap in rain-fed agricultural systems by integrating land and water planning through means such as the reuse of crop residues and improved soil management. These initiatives help to ensure that infiltration of rainfall is increased and that surface runoff is reduced. Some authors continue to argue however that too much focus remains directed towards the construction of new water supply projects and that not enough is directed to the management of the resource (Tortajada 2010). It should also be remembered that any change in green water use will impact blue water availability. A holistic approach to managing all water supplies must therefore be the primary aim of water supply policy.

In many countries with limited financial capital, and especially where cost recovery in water supply remains elusive, government incentives such as low-cost loans and tax breaks are required to fuel growth in agricultural innovation. Several national and supra-national governments provide these mechanisms; however, in most global regions, investment in agricultural research and development is falling.

As an alternative to the incentive-based approach, agricultural water-use efficiency improvements could be mandated by, for example, reducing water entitlements in order to force a change in on-farm water management. While these policies can be effective, they must consider the financial and human resource capacity of farmers to respond. They are therefore more likely to succeed where farmer's access to knowledge, capital and appropriate agrarian water management expertise is high.

Encouraging greater recycling and reuse of water in agriculture is also receiving renewed attention. Interestingly, this process often requires primary producers to revisit the frugal approaches to resource management that allowed early rain-fed agricultural systems to account for the vagaries of drought and flood. Treated wastewater provides a reliable and climate-independent source of water, while the use of wastewater itself (although requiring careful management to minimise human health risks) can also be hugely beneficial, potentially providing most of the nutrients required for crop growth that would otherwise be supplied by fertilisers. In Israel, almost three-quarters of urban wastewater is reused and most of this goes to support peri-urban agriculture (Rygaard *et al.* 2011). With the correct safeguards in place, policies that encourage wastewater reuse in the agricultural sector can provide an important ameliorating effect on overall water demand. Even black water (that containing human faeces) can be used to produce solid fertilisers suitable for use in urban horticulture not intended for human consumption (Schuetze *et al.* 2013).

The role of land-use management to mitigate flood risks is clearly understood and it is typically agricultural land in upper- and mid-level catchments that has the greatest potential to control the flow of water further downstream. Farming 'ecosystem services', in particular cultivating and maintaining natural water retention systems, as an alternative to producing food and milk (often in saturated markets) is a policy that the authors believe should be given more attention.

7.4.3 Water Policy for Industry

Depending on their physical location relative to a blue water source, industrial water users are typically supplied with water via either licences granting them permission to abstract surface or groundwater directly, or through a centralised distribution network. While historically subordinate to agriculture in terms of national economic importance, industrial business has always held significant political clout; as a result, it is often afforded generous water entitlements. For example, the framework for granting surface water licences in Texas, USA means that Dow Chemical, one of the largest manufacturers of chemicals and plastics in the world, is afforded priority supply of water from the Brazos River. As the river's oldest water user the company has priority over all others, allowing it to become the largest abstractor and pitting it against farmers, cities, power plants and other stakeholders, particularly during times of water scarcity.

Increasingly, businesses are becoming more aware of the risks that water supply poses to their profitability. Manifestations of these risks might include:

- failure to secure sufficient water to meet the needs of their factories, which could have a direct impact on operations or production output;
- public perceptions of improper water use (whether proven or not), which could create broad and damaging impacts on brand perception and customer loyalty; and
- a change in the policies that apply to a business's water use, which could significantly impair how operations can function.

Chapter 4 highlights a number of water-related impacts recently experienced by a range of companies and industries. It is the increased recognition and quantification of these risks rather than a change in government policy *per se* that is increasingly driving business to adopt new approaches to water supply and use, including the following:

- reusing water though operations to reduce the volume of water taken from blue water sources. For example, it is claimed that Coca-Cola and its bottlers have spent nearly US$ 2 billion since 2003 to recycle and reuse water at 863 plants around the world (Clark 2014);
- investing in dedicated supply systems independent of local sources. For example, in order to support copper mining in Chile's Atacama Desert, BHP Billiton is investing US$ 3 billion in a seawater desalination plant at a port over 150 km away, as well as the pipeline (and its associated ongoing operating costs) to pump the water to a mine at an elevation more than 3000 m above sea level (Clark 2014); and
- collaborating with NGOs and engaging in multilateral sustainability initiatives as well as investing in self-run social welfare initiatives.

Government policy could do more to support these trends. For example, by mandating that companies report on their water use through disclosure pathways easily accessible to the public (e.g. CDP see Chapter 6), the pressure to implement improved water management practices would increase. The benefits of this change would not just be realised as a reduction in demands on the natural water source, they would also fundamentally improve the performance and long-term viability of the business in question.

7.5 The Rise of Decentralisation and Consultation

The centralised approach to water management described in Section 7.4 is typically implemented through a top-down approach to governance, one that often fails to extensively engage the public in the planning process. As the limitations of centralised systems are increasingly borne out, and as more and more stakeholders seek to engage in water management decision making, an alternative approach based on participatory planning and the more widespread adoption of decentralised management options has become increasingly prevalent.

Centralised infrastructure implemented through top-down governance is often the default model of countries in the early stages of developing formalised approaches to the governance of water resources. While such an approach may allow policies and infrastructure to be implemented relatively quickly (by dodging the potential delay that can occur when participatory models are adopted), it also foregoes the significant knowledge and experience of those local groups directly involved in the day-to-day management of local water resources. Conversely, decentralised models enable local governments and water users to play a much larger role in the policy-making process, thereby often resulting in mutual buy-in to the process that can help support its implementation.

Interestingly, it is frequently developing countries and emerging economies that are taking the participatory approach to water resource management, essentially reinforcing customary arrangements that were often the backbone of natural resource management in small rural communities. For example, a large number of African nations are decentralising their water governance systems, with responsibility for service provision being passed from central to local governments. Although this transition brings its own challenges, particularly when local authorities are not staffed with the skilled personnel required, examples of the benefits of decentralised and consultation-rich policy making are evident in many regions. Two excellent examples are Mali's approach to developing its National Sanitation Policy, and Uganda's preparation of its National Water Action Plan (see Breakout Box 7.2).

Notwithstanding the benefits of a more inclusive and participatory approach to the governance of water resources, there is a danger that centralised governments devolve power to local actors too quickly, before the necessary enabling policies and legislation are in place, and before local capacity and competence are sufficient. The Water Partnership Programme (2010) suggests that this risk is most commonly realised in the areas of procurement, project management and financial management.

Breakout Box 7.2 Participatory water management in Mali and Uganda

In Mali, water resources have been managed via a decentralised government approach since the 1990s. Approximately 700 administrative districts, or communal councils, control water and sanitation services (US AID 2010). The challenges of adopting a decentralised approach to water management have been seen as an opportunity to develop skills and build local capacity, and the country has not rushed the preparation of policy. It took two years between 2005 and 2007 to develop its National Sanitation Policy, including defining the goals to achieve between 2015 and 2020, preparing guiding principles for affected stakeholders and identifying the responsibilities of each stakeholder (AMCOW 2008). The policy is supported by a mechanism to coordinate implementation activities and by guidelines to ensure sustainable financing. The policy also defines and describes initiatives intended to build local capacity and to continually monitor the performance of the water management system (AMCOW 2008).

Uganda's journey towards a more integrated approach to water resource management is set out in its Water Action Plan, developed through a consultative and participatory process which defines the short- and long-term water management roles and responsibilities of its various stakeholders. The process of developing the plan involved assessing the human resource capabilities and management tools available to the various local stakeholders and, in turn, their needs for capacity building. Central government representatives worked alongside those from the local districts, public sector water services providers, and partners from the private sector to agree and implement the action plan. The plan was developed over a 10 year period to 2004 and has enabled Uganda to assert its important role in the management of the Nile Basin. The process has fostered more consistent policy and legislation and provided clearer guidance on priorities for water use allocation and the management of wastewater. Stakeholder participation in the process has also fostered local-level involvement and empowered more widespread engagement on water management locally, regionally and internationally (Jønch-Clausen 2004).

7.6 Regulation of Water Management

Regulation is the mechanism that transposes law into something workable in practice. The primary purpose of regulation is often to protect against risk: protecting the environment from damage; protecting human health; and protecting the operating requirements of abstractors and dischargers within catchments. While it is clear that the governance structure and regulatory framework differs from country to country, the core issues of concern are generally the same:

- regulating sources of water (traditional and alternative);
- regulating drinking water and non-potable quality;
- managing different and competing demands for water;
- treating and disposing of wastewater; and
- regulating environmental conditions (maintaining environmental water flow or level requirements).

Table 7.1 Typical contents of a regulatory toolbox.

Hard tools	Soft tools
Price controls	Operational guidance
Statutory requirements	Best-practice evidence
Licensing and permitting	Auditing checklists
Compliance reporting	Evaluation tools
Technical methodologies (mandatory)	Assessment guidance
	Stakeholder engagement guidance
	Planning guidance

Regulations are statutory instruments (in the same way as Directives or Acts) and are typically prepared by government, with the aid of lawyers, interpreting the often-wordy requirements of a Directive or an Act into documents more focused on implementation. It is slightly misleading to refer to legislation and regulation as two separate entities, as regulations are legislative tools. Typically, the contents of a regulatory toolbox can be grouped into 'hard' and 'soft' tools (see Table 7.1).

While the process of changing the law is complex and often difficult, regulations are amended comparatively frequently. In fact, in some cases the regulations themselves may specify that they are reviewed at regular intervals. For example, the US Safe Drinking Water Act (originally passed by US Congress in 1974) requires the US Environmental Protection Agency to review each piece of national primary drinking water regulation at least once every six years and revise the piece of regulation as appropriate, so that it can continue to achieve its objectives. This requirement is important for helping to ensure that regulation remains fit for purpose; however, it can also make it difficult for practitioners to stay up to date with their legal obligations.

7.6.1 Regulating Sources of Water

People have always abstracted water from the environment and, for much of human history, the scale and intensity of this process has grown without too much impact. The problem now is that 7.5 billion people need water from what are essentially the same sources and volumes of available supply.

Wherever an abstraction takes place, it has the potential to place the immediate environment under stress by reducing available water volumes and/or flows. Most hydroecology (aquatic-dependent habitat, flora and fauna) is highly sensitive to flow and/or water volume (although the specific requirements of particular species vary significantly and are not always fully understood). For example, littoral environments in streams and lakes that are otherwise adapted to regularly changing water regimes are particularly susceptible to sustained drops in water level (Ryan and Griffiths 2001; Reed *et al.* 2008). Groundwater-dependent ecosystems are susceptible to the declining water tables that accompany unsustainable groundwater abstraction, a process that in coastal areas can also cause saline water to move into previously fresh groundwater environments. As population growth, economic advancements and lifestyle changes continue to require more and more water abstraction, effective regulation has never been more important.

Many countries attempt to regulate water abstractions by administering a system of water rights or entitlements. Unfortunately, however, the administering of such systems has often failed to appropriately consider the complexity of the water resource environment. In many surface water catchments, the management of these rights is often facilitated by the direct regulation of river flows using physical infrastructure (such as dams and weirs) to store and release water. The flows in heavily regulated catchments are therefore relatively reliable (see Breakout Box 7.3). In contrast, in physically unregulated catchments, water flows reflect seasonal rainfall patterns and are not supported by the managed release of stored water. Such rivers are now very rare but still persist in some remote regions, for example the rivers of the Northern Territory of Australia.

Breakout Box 7.3 Twentieth-century abstraction licensing and the need for reform

UK: The water abstraction licensing system was introduced in the 1960s, more as a process of registering abstractors than with any significant sustainability principles. The majority of licences did not include expiration dates, so abstractors expect to hold the licence in perpetuity. Significant increases in population, the reality of climate change and modern environmental obligations (i.e. implementing the European Union Water Framework Directive's key objective of achieving a good chemical and ecological status in all surface and groundwater bodies, interim targets to be set for 2015 and 2021 with full compliance by 2027) forced the UK regulators to re-examine the system. Provisions were added into revisions to the Water Act enabling authorities to amend or remove licences without compensation if investigations prove the licence is causing serious damage to the environment, and to introduce time limits on licences.

USA: Across the USA water abstraction systems are managed at State level. While there are differences in detail, these typically adopt either a riparian landowner's right to water or an application system, both of which are usually on a first-come-first-served basis. As in the UK, the pressure on water resources now draws attention to how unfit for purpose these approaches are in the twenty-first century, but by no means has this led to universal reform.

In Texas for example, surface water is owned by the state and held in trust for its citizens. The state grants the right to use this water to different parties, such as farmers or ranchers, cities, industries, business and other public and private interests, and during drought water abstraction is prioritised purely on the basis of the date on which permits were issued (Texas Commission on Environmental Quality 2015). In contrast, groundwater belongs to the landowner and abstraction is subject to the Rule of Capture. Under that rule, landowners do not own the water but have a right to pump and capture whatever water is available, regardless of the effects of that pumping on neighbouring wells (Texas A&M University 2014). Concerned that such unregulated abstractions could lead to over-abstraction, the Texas legislature has authorised the establishment of groundwater conservation districts to manage significant aquifers (KBH Energy Centre 2015).

Australia: Australia provides a model of why and how pre-existing licensing regimes can be reformed. The riparian doctrine (which gives landholders constitutional rights to access and use waters adjoining their land) was introduced to Australia during the period of European settlement. However, it was not long before the Australian climate highlighted the inadequacies of an approach based on European conditions where there was little competition for relatively abundant water resources. State governments soon took control of water resources from landowners and, in the early twentieth century, each state established statutory licensing systems granting rights to water based on the area of irrigable land, rather than property or proprietary rights. Although water rights were allocated on a first-come-first-served basis, once granted there were no 'seniority' rights (as in the western US), meaning the reliability of existing licence holders' rights was compromised as more rights were granted. Over a period of 25 years, abstraction management in Australia has been reformed to manage increasing scarcity and consequent competition for resources. The reformed Australian system now defines existing rights as a share of the water available (Piure 2014).

The first, relatively rudimentary, abstraction licensing systems were established during the twentieth century, in many cases simply allocating water on a first-come-first-served basis or on the basis of land ownership (see Breakout Box 7.3). The precise details of abstraction licensing systems differ between countries; however, common elements include:

- definition of the specific location or multiple locations at which abstraction is authorised;
- a volumetric abstraction limit (usually a daily maximum amount, but often also with an annual or other time-defined cap); or
- a defined purpose of use (e.g. public water supply); typically, the licence holder cannot use the water for a purpose other than that stated on the licence.

The licensing system usually requires abstractors to record their abstractions and submit those records to the regulator on an annual basis. This creates two datasets – a register of licensed abstractions and a record of actual abstractions – which the regulators can then use to analyse trends in water abstraction within catchments. The level of analyses that can be undertaken is dependent on how robustly the licensing system is administered. In countries with extensive datasets on past and existing abstractions as well as quantified information on the RWR available at the catchment scale, useful determinations of the volume of water that can be abstracted without detriment to the environment can be made. Increasingly, these analyses are revealing catchments which are over-abstracted or over-licensed (where over-abstraction would occur if all abstraction licences were utilised to their maximum allowable limit). This knowledge, combined with increasingly frequent observations of the impact of over-abstraction on the environment, are leading to calls to reform outdated abstraction licensing systems (see Breakout Box 7.4). However, when it comes to changing licensing systems the power and lobbying ability of some existing abstractors, as well as a perception held by some of a right to abstract (often stemming from the old first-come-first-served approaches to licensing) frequently act as strong brakes on change.

Breakout Box 7.4 Regulating flows in the River Severn catchment

The River Severn is a large catchment in the UK which supports abstractions supplying several cities. The environmental regulator uses a variety of regulatory tools, including abstraction licences, to manage groundwater and surface water resources to achieve a prescribed minimum flow and thereby to protect both environmental requirements and abstractors' needs. However, the combined stress of increasing demand and the impact of climate change on the water available in reservoirs and groundwater is raising questions about the feasibility of maintaining the prescribed flow. Given the growing pressures, in future it may be necessary for the environment and/or abstractors to accept a reduction in the minimum flow.

Traditional abstraction licensing systems typically focus on ensuring compliance with abstraction limits set many years ago and that often pre-date more recent work that better understands environmental water requirements. Regulatory authorities also find themselves grappling with the difficulties of controlling potentially unsustainable, but nonetheless licensed, abstractions on the one hand, and implementing national and international legislation or agreements to protect the environment on the other. Regulators may also be tasked with protecting the operational rights of licensees, for example by maintaining minimum flows necessary for abstraction. Implementing new rules to regulate groundwater abstraction in California (Breakout Box 7.5) and the emergence of water trading in Australia (Breakout Box 7.3) are enhancements of the traditional water licensing approach which demonstrates the need to continually refine and improve the regulatory environment.

7.6.2 Regulating Drinking Water and Non-Potable Quality

Once water has been abstracted, regulations also govern the means by which it can be supplied to various uses. These regulations are principally driven by the aim of safeguarding human health; however, variation in the stringency of the controls imposed by regulations in different jurisdictions and different countries is vast, particularly with regards to non-potable uses.

The World Health Organization provides a range of guidance materials intended to ensure that the quality of drinking water supplies meets human health requirements. It advises that the key principle underlying the legislative structure of the drinking water sector should be to protect and improve public health through the sustainable provision of drinking water of adequate quality, in sufficient quantities to all the population continuously, at a price which is affordable (WHO 2011).

The US provides a good example of how different regulatory instruments can be used in unison to protect drinking water. The US Environmental Protection Agency (EPA) sets highly prescriptive legal limits on the levels of certain contaminants in drinking water, recognising human health requirements and the quality levels that can be achieved using best available technologies. These quality regulations are supported by additional regulations, also set by the EPA, on drinking-water testing methods and frequency. These regulations all sit beneath the US Safe Drinking Water Act 1974, a national piece of legislation that requires individual states to set and enforce their own

Breakout Box 7.5 Groundwater abstraction in California

Probably one of the highest-profile problems arising from a lack of regulation is the depletion of groundwater reserves in California. It is also an example of how opposition can make it difficult to impose regulations. In October 2014 more than 80% of the state was in extreme or exceptional drought, yet it remained the only western US state without groundwater regulations.

Despite regulating how much people can abstract being the seemingly obvious response to plummeting groundwater levels, plans to regulate groundwater abstraction faced major opposition, particularly from farmers arguing that regulation would infringe their property rights. As well as having no control over a resource on which more than half of public water supply depends, unregulated systems mean there is no information on how much water is pumped or how much deeper boreholes are being sunk. The situation is a prime example of water laws which were never designed to deal with modern challenges. Linking water rights to land ownership, unmeasured usage with no disclosure and perverse 'use it or lose it' incentives have created a system which encourages waste.

Despite the opposition the state of California has now passed a bill requiring each groundwater basin to report everyone's pumping quantities and the depth of groundwater each year, in line with specific water level objectives. The goal is to stop groundwater depletion but also to create incentives to replenish aquifers and use them sensibly as a water management tool. The bill aims to put an end to the profligate use and abuse of groundwater.

The problem of opaque 'water rights' is not unique to California; establishing a transparent system of publicly registering abstraction licences is a common solution to the unsustainable use of scarce 'shared' water resources. Transparency also helps to emphasise the interdependency of different uses of finite water resources (UNESCO 2014).

drinking water supply controls in order to achieve a prescribed set of standards (US EPA 2013).

Approaches to regulating non-potable water uses such as irrigation vary wildly from none at all in many countries to a default requirement to ensure that all water supply meets drinking water standards. This latter approach typically reflects the perceived risks associated with the water source in question, the purpose of use and the likely levels of human exposure. In developed countries with a history of water policy, agricultural use of non-potable water is more likely to be subject to regulations. In many developing countries however, regulations are often non-existent. The WHO has issued guidance on how to regulate the risks and opportunities of using untreated or partially treated wastewater to irrigate agriculture (WHO 2006).

7.6.3 Managing Demands for Water and Enforcing Best Practice

As water resources come under increasing pressure, the currently widespread expectations of unlimited access to water have to be changed. In many ways, the concept of managing and controlling demands for water is largely a policy issue (it is usually up to regulators, utilities and other authorities within governance systems to agree on what demand management stance to take and then seek to achieve). Demand management measures such as compulsory metering of households and the creation of volumetric tariff structures are controversial, divisive and therefore highly politicised issues. This means that while particular water authorities or utilities may have their own demand management policies, standardised regulations are rare.

Occasionally regulators may impose a specific requirement on an individual water supply provider. For example, following Thames Water's breach of its leakage targets in 2004–05 and 2005–06, the economic regulator Ofwat (2006) secured a legally binding commitment from Thames Water to meet future leakage targets and replace £150 million of additional leaking water mains at the expense of its shareholders. Thames Water completed all of the actions required by the commitment, meeting its leakage targets in every year from 2006–07 to 2009–10, reducing total leakage from 860 ML/day in 2005–06 to 670 ML/day in 2009–10.

7.6.4 Regulating Wastewater Treatment and Disposal

Wastewater regulation is often secondary to water supply regulation, as the need for safe drinking water takes priority. In many developing countries there is no wastewater regulation at all, while in developed countries wastewater regulations followed after regulation of water supply. Until the mid-nineteenth century, sewage treatment was often little more than dilution by receiving waters. As populations boomed however, particularly in major cities such as London and New York, widespread human health epidemics drove more advanced sewerage and treatment systems.

The pioneering biological treatment methods in the early twentieth century and continued advancements in treatment technologies since then have also been accompanied by evolution in the nature of regulation, focusing both on the level of treatment required and the permitted quality of the effluent to be discharged into the environment.

Reflecting the risks to human health (and the environment) from wastewater disposal, wastewater regulations are now among the most prescriptive and detailed of the water regulations in developed nations. The established regulatory frameworks leave very little room for interpretation. Discharge permits are used to limit the volume of water and/or concentrations of specified substances permitted to be released into receiving waters by a particular permit holder. In Europe, Directives are driving increased scrutiny of the wastewater disposal regulations of Member States. The pressure is on to ensure that waterbodies achieve the 'good ecological status' objectives of the Water Framework Directive.

While discharge consents are often crucial in ensuring water quality objectives can be met, they can also create barriers to residential or commercial development in areas where inadequate wastewater treatment means that consent obligations will not be achieved if development goes ahead. Nutrients are a major issue in wastewater disposal, as discharge of phosphates and nitrates into waterbodies provide the ingredients for algal growth and eutrophication. Discharge consents typically set strict limits on the concentrations of substances such as these. However, in response to increased inflows to wastewater treatment works (e.g. generated by increased population), permitted concentration limits are reduced by the regulators to protect the receiving waters. As concentration limits are driven down, more advanced treatment is usually required to meet the consent. Problems arise when treatment works are already operating with the best available technology, meaning their scope to improve performance becomes more and more limited. In cases such as these, it is crucial that local planners, developers, water utilities and regulators work together to plan and invest appropriately. Ultimately, there are often difficult tradeoffs to be made between increasing the level of treatment required to comply with discharge consents, and the carbon footprint and financial cost of implementing such upgrades.

7.6.5 Regulating Environmental Conditions

All human activities are ultimately sustained by the regulating, provisioning and cultural services provided by the environment. By regulating water flows and by purifying water, ecosystems play an essential and irreplaceable role in the global hydrological cycle. Although the importance of ecosystem services and the links between ecosystem services and human benefit is indisputable, in modern societies the increasingly long and complex pathways through which an ecosystem service is realised as a human benefit means that we frequently forget they exist.

Before industrialisation and rapid urbanisation, the links between ecosystem services and the communities reliant upon them were direct and immediate. If a crop failed due to drought, people went hungry and livelihoods were jeopardised. As a result, customary laws accounted for the needs of the environment for water. Today, local environmental impacts associated with a lack of water of an adequate quality hardly ever translate to human impact in the distant cities and industries at the top of the supply chain. For example, should a particular crop fail, supermarkets simply switch suppliers and the customer experiences no loss of service. It is because of this lack of a direct link that ecosystem services are rarely ascribed their true value in decision making, ensuring they are effectively provided free of charge. Because modern globalised systems act to shelter society from the impairment of localised ecosystem services which their activities cause, irreversible environmental degradation can result over time. Attempts to replicate the lost ecosystem services through alternative, artificial means may then require huge financial investment. Unfortunately, it is often only when environmental thresholds are approached, or passed, that policy responses are initiated. For example, it took one of the most severe droughts on record and the alarming depletion of its aquifers for the US state of California to enact laws in 2014 that regulate the pumping of groundwater for the first time (see Breakout Box 7.5).

While there is now at least a broad recognition of the importance of ecosystem services to human wellbeing, frameworks for their valuation, considered by many commentators to be a prerequisite for sustainable ecosystem management, remain a work in progress. A focus on creating protected areas to limit environmental impact remains the dominant regulatory response (Muradian and Rival 2012). However, such an approach fails to explicitly recognise and communicate the links between the environment and the beneficiaries of ecosystem services. In contrast, if ecosystem services can be accurately valued, these links would become clear; in turn, this would establish a real incentive for their preservation that may ultimately help to break the link between economic development and environmental degradation.

The Economics of Ecosystems and Biodiversity (TEEB) initiative aims to incorporate the economics of nature into mainstream environmental management, demonstrating the value of ecosystem services in economic terms and, where appropriate, suggesting means to capture those values in decision making (Sukhdev *et al.* 2014). The TEEB initiative is outlined in a series of reports that identify the economic principles of measuring and valuing ecosystem services. Those reports guide policy makers in appropriate means of investing in natural capital, and describe the risks and opportunities that ecosystem decline presents to businesses. Although argued by some to unnecessarily commoditise nature and subject it to a number of challenges associated with identifying accurate and comparable valuation methods, TEEB is increasingly being considered and applied by governments and industry. For example, the application of TEEB in Norway has resulted in biodiversity being reflected in sustainable development indicators in the national budget, the use of ecosystem services to support climate change adaptation and mitigation, and the recognition and application of these benefits in efforts to improve public health (UNEP 2013). A variety of comparable valuation initiatives are also underway, such as the Economics of Land Degradation scheme and the UNEP Green Economy Initiative (UNEP 2013).

7.7 Regulatory Models

It is easy to spot the different models of regulation being applied in different regions or to specific activities. Regulatory approaches do not simply evolve organically. Although acute problems (or recognition of chronic problems) may trigger calls for an immediate change in approach, the nature and speed of change is subject to the core regulatory principles of the model in question. The range of regulatory models includes:

- self-regulation: most often seen in water systems managed by the public sector;
- regulation by performance monitoring: based on the concept that a requirement for disclosure will ensure continual improvement in performance;
- regulation by performance contract: such as a requirement to report to a network of stakeholders; and
- independent autonomous regulation: this is arguably the most appropriate method whereby a regulator, free of political interests or pressures, is able to objectively assess performance.

Under the latter model (independent autonomous regulation), the World Bank (2006) identifies three criteria against which performance should be assessed:

1) legitimacy: the regulatory system should protect consumers from the exercise of monopoly power (for example through high prices and/or poor quality of service);
2) credibility: investors must have confidence that the regulatory system will honour its commitments (such as maintaining agreed minimum tariff levels); and
3) transparency: regulation and related information are available to all.

Analyses of the costs and benefits of different approaches to regulation are often undertaken by the administrations responsible for the delivery of legislation, for example the European Commission (2013). Where such analyses exist, they provide transparency as to the impacts of policy.

Some advocates for 'deregulation', such as the American economist Milton Friedman, argue that even if an industry does do business in an undesirable way, such as polluting watercourses, regulation is unnecessary because injured parties could simply sue offending companies for damages. His argument is that the threat would serve as a sufficient disincentive to corporate misbehaviour. Counter to this assertion is the argument that such injury is unacceptable and that individual parties should not have to launch law suits in such a way.

Another alternative to imposed regulation is self-regulation, whereby industry is expected to regulate its own activities in an environment perceived to be free from bureaucracy. The concept of self-regulation is believed to be flawed in several ways, not least because of the need to check that the industry is indeed regulating itself and achieving the types of desired outcomes that it may express in high-level statements.

Despite the range of different policy models, some form of imposed regulation is invariably necessary to achieve appropriate water management. Sensible regulation looks to risk-based approaches to develop imposed measures without creating unnecessary bureaucracy. For many long-established regulatory systems however, the process of identifying and removing unnecessary red tape can itself be beset by administrative challenges, not least that of achieving agreement between all parties on where and how to reduce regulatory burdens.

Irrespective of the regulatory model applied, monitoring and evaluation create the transparency essential to good governance. This means that implementation can however be resource intensive and expensive; in governance systems where money is scarce, monitoring and evaluation is therefore often one of the first activities to be scaled back. Invariably, such a decision acts only to the long-term detriment of the regulatory model itself. Policies need to recognise the value of monitoring and evaluation in order to provide the regulatory evidence that supports good governance and decision making (see Section 7.4).

7.8 Regulatory Phases: Unregulated versus Highly Regulated

Notwithstanding the power of law to force action, sometimes the decision is made not to legislate on a particular issue. This may be because non-mandatory tools are available which sufficiently address the issue, or because it would be too impractical or expensive to regulate. It is also important to remember that governments are frequently criticised for creating unnecessary or unenforceable legislation.

Regulation works well for activities that have clear authorised and prohibited actions which can be easily monitored and evaluated, and where determination of compliance is not subjective. Mandatory monitoring programmes or prescribed operational methods help to remove ambiguity from these relatively routine processes.

In contrast however, some parts of the anthropogenic water cycle are less well suited to imposed regulation and can often be better managed via non-mandatory 'best practice' approaches. Potential examples include the management of diffuse pollution, which could be achieved by encouraging farmers and other landholders to adopt land management practices that minimise erosion and leaching of nutrients and other chemicals into watercourses.

Identifying which issues are better managed through non-mandatory approaches and incentives is challenging and can involve the governance system passing, via trial and error, through three regulatory phases: an unregulated or lightly regulated phase;

an over-regulated phase; and finally a mature phase of regulation, as explored in the following subsections.

7.8.1 The Unregulated or Lightly Regulated Phase

In its most basic form an unregulated system is a 'free for all' although, where water resources are under pressure and stressed, this more commonly translates as a 'free for a small number of privileged users' situation, typically those with the greatest ability to access the resource. Until a problem emerges there is often no regulatory driver, but lack of regulation can lead to and exacerbate water shortages and water access inequality. Any perception that unregulated (and unsustainable) water use is predominantly a problem confined to developing countries is misplaced. There are plenty of examples of richer developed nations experiencing water problems due to unregulated use; see Breakout Box 7.5 for example.

Spain is a developed nation with a highly stressed water environment not helped by the limited RWR (60% less per capita than average European levels) and high rates of abstraction (62% higher than the European average; Global Water Forum 2015). The Tagus-Segura water transfer was designed and built to relieve water pressure by transferring water 286 km from the Tagus River and its reservoirs (in the northern province of Guadalajara) and from Buendía (in the province of Cuenca) to the Talave Reservoir on the Segura River. Despite the implementation of this grand scheme in 1978, water shortages remain common in the Alicante, Murcia and Almeria provinces. The scheme is said to be inherently flawed, partly because the recipients in the receiving Segura Basin view the significant water infrastructure as the solution in itself and their subsequent abstractions remain unregulated. Similarly, the engineering scheme is not supported by regulations to ensure minimal flows are maintained in the donor Tagus Basin (WWF 2003). As well as intensifying water shortages, the lack of regulation has led to increased pollution and an expansion of irrigation. The consequences are not just environmental. An illegal water market has emerged which discriminates against traditional land uses, and has dramatically worsened illegal immigration and labour exploitation associated with farm workers (WWF 2006).

In addition to the obvious risk of exacerbating water scarcity, unregulated systems also increase the human health risks incurred when water sources of unsuitable quality are used for certain activities. Water pollution in the environment and in drinking water creates environmental, social and economic problems. In developing and transitional countries, wastewater treatment is often still inadequate for most people and regulatory frameworks either do not legislate the issue (they are essentially unregulated) or do not enforce regulatory requirements. UNESCO claims that global focus should be on improving wastewater management in developing countries where wastewater collection and treatment is still very poor (UNESCO 2014). Increased levels of treatment combined with a regulatory system that specifies water quality objectives and enforces water quality monitoring would help to reduce pollution.

The response to problems that emerge in unregulated systems is often to create and impose regulations. Where the benefits of regulation are realised, the temptation can be to develop more and more control. Over-regulation can be a difficult temptation to resist, and one which can pose just as many detrimental consequences as no regulation at all.

7.8.2 The Over-Regulated Phase

Too much regulation can create systems which are inflexible, burdensome and over-complicated, involving multiple organisations across several levels of governance that frequently have conflicting agendas. The system for regulating wetlands in the US is a good example of the problems that can arise when organisations within a governance system have different roles and agendas with regards to a single issue, in this case wetland protection. Breakout Box 7.6 explores the case study.

Depending on the speed with which its constraints are realised, and the ease with which regulations can be changed, the over-regulated phase may persist for many decades, particularly if its current form supports the interests of a powerful group of stakeholders. Such a situation has the potential to create a regulatory framework that

Breakout Box 7.6 Wetland protection in the US: issues of over-regulation

In the USA, five federal agencies share primary responsibility for protecting wetlands but each has a different mission that influences its focus and priorities.

1) Navigation and water supply are controlled by the US Army Corps of Engineers.
2) The Environmental Protection Agency focuses on the contribution of wetlands to the chemical, physical and biological integrity of the US water resources.
3) Management of fish, wildlife-game species, and threatened and endangered species is the responsibility of the US Fish and Wildlife Service.
4) The National Oceanic and Atmospheric Administration takes charge of wetlands as a coastal resource.
5) Wetlands affected by agricultural activities are managed by the Department of Agriculture, Natural Resources Conservation Service.

The value of wetlands has taken a long time to be recognised, and disagreement on how to protect these resources has led to discrepancies in local, State and Federal guidelines. This problem of conflicting policies has led to programs that both encourage and discourage the conversion of wetlands. Programs that indirectly facilitate wetland degradation include the following:

- The National Flood Insurance Program actively encourages development in floodplains by providing low-cost Federal Insurance.
- The Payment-in-Kind Program indirectly encourages farmers to place previously unfarmed areas, including wetlands, into production.
- The Water Resources Development Act encourages water development projects that have the potential to damage wetlands.

At the same time:

- The Coastal Zone Management Act provides Federal funding for wetlands protection programs in most coastal States.
- The Comprehensive Environmental Response Compensation and Liability Act establishes liability of the US Government for damages to natural resources.
- The US is a signatory to the Ramsar Convention, which maintains a list of wetlands of international importance and encourages the wise use of these natural assets.

> Despite widespread recognition of many and varied benefits that wetlands provide, the conflicting interests of landowners, the general public, developers and conservationists (served by different government agencies, each with their own priorities) are the source of much tension and controversy in US wetland protection policy. Attempts to reconcile some of these differences are being made, but many policies will have to be modified to achieve consistency.
>
> This case study also highlights the critical need for the successful implementation of regulation, not just its development. Wetlands will never be protected unless regulations can be adopted and/or enforced. A critical element of the successful implementation of regulation is the need to educate the public in order to stimulate public-led conservation efforts. Source: United States Geological Survey (2002).

ties itself in knots and rather than providing clarity on how to manage water resources for their safe and effective use, creates confusion, frustration and inertia in the face of evolving and growing challenges.

Where over-regulation persists, usually what we often eventually see happening is a phase of deregulation, typically initiated because the administrative burdens of excessive regulation become too great and too expensive to maintain. Deregulation may however also be stimulated by a realisation that non-mandatory regulatory approaches can be more flexible, more inclusive, more engaging and therefore ultimately more successful then rigid, imposed structures.

7.8.3 The Mature Phase

The regulation of water abstraction is one area of water resource management regularly cited as ripe for reform. These calls for change are often strongest in countries where the traditional first-come-first-served, fixed and permanent licence system is increasingly seen as inadequate to address growing demands and more variable supplies.

Traditional approaches to abstraction regulation have, in many countries, cultivated attitudes of water entitlement and led to the unhelpful use of terminology such as 'water rights'. The need to abstract water and the consequences of competition and over-abstraction are high-profile issues affecting large and powerful stakeholder organisations. As such, there are often strongly opposing views on reform. Notwithstanding the need to engage all stakeholders when planning appropriate water management, it is our firm belief that, while all people require access to clean water and sanitation, water remains a shared resource and the concept of entitlement to large volumes of water regardless of the environmental situation is not one that has a place in a future of effective and fair water management. Modern abstraction licensing has to be more flexible to cope with increasingly extreme rainfall patterns and a wider range of abstraction needs. The emergence of mechanisms to facilitate trading of water licences therefore marks a welcome addition to the portfolio of tools available to planners to support appropriate water resource management.

In addition to controlling the quantities of water we abstract, adherence to acceptable water quality levels is vitally important to the environment, human health and ultimately, the economy. Control of water quality can very easily become entrenched in the

Breakout Box 7.7 Lighter-touch regulatory models: the Dutch administrative burden reduction programme

In 1994, the Dutch government initiated a programme to drastically reduce the regulatory burden impacting on businesses. A target was set to cut the cost of red tape by 25% and, between 2003 (when the programme was significantly enhanced) and 2007, this ambition target was achieved. Over 190 simplification measures successfully eliminated €4 billion of administrative burdens. The Netherlands was the first country in the world to set and achieve such an ambitious set of reforms (World Bank Group 2007).

This success is attributed to four factors:

1) the 25% target attracted significant attention;
2) regulatory reforms were linked to the government's budget cycle, thereby attracting political leverage and making the reforms feasible;
3) a Dutch Advisory Board on Administrative Burden was established as an independent watchdog of the reforms which helped to build and maintain momentum; and
4) there was commitment across all major political parties of Parliament to reduce business costs.

These features are now being adopted by other countries around the world.

over-regulation phase. It is easy for regulations to become highly detailed and prescriptive, based on exact concentrations of chemical substances or biological contaminants, or precise treatment requirements for certain types of water. However, the range of substances is ever-changing and the growing number of 'unknown unknowns' in raw water and wastewater quickly becomes too difficult to regulate in this way. Alternative regulatory systems are moving away from focusing on prescriptive water quality levels and, instead, turning towards outcome and risk-based principles that often require collaborative catchment management activities and pollutant characterisation assessments.

Adding additional layers of regulation is often an inappropriate response to a problem as it can impose counterproductive, inflexible burdens on both the industries being regulated and the regulators themselves. In light of this constraint, various regulatory models have been developed that attempt to measure and reduce the regulatory burden, such as the Dutch Standard Cost Model (see Breakout Box 7.7).

7.9 Governance Silos

To fully understand the roles of different levels of government in water resource management, it is important to recognise the interests and mandates of the numerous departments involved in delivering a water programme and their unique, as well as overlapping, responsibilities. Governments like to organise themselves along departmental lines. Exactly how this is done varies from place to place, but generally we see individual departments taking charge of environment, trade and industry, energy, health, education, etc. Each department then typically sets out its goals according to its own focus and priorities.

The same departmental approach happens within water education and research, with issues being taught and studied in distinct and often siloed disciplines within geography, biology, chemistry, engineering, and social science departments, for example. It is no surprise then that there is also a tendency in water utilities to reduce problems and challenges into their component parts: hydrology, water quality, clean water treatment and supply, wastewater treatment and disposal and hydroecology for example. Unfortunately, the issues afflicting the water sector can rarely be isolated into such discrete boxes. Broader perspectives are required to recognise the relationships to other aspects of water management, as well as to the management of other resources (e.g. food and energy). Because the water sector tends to split itself along these scientific, engineering and policy lines, the siloed mentality that results often fails to grasp the root causes of water management problems.

Adequate integration of governance and knowledge all too often eludes us because of conflicting and entrenched perspectives, experiences and agendas which lead to clashes and conflicts when attempts at integration are made. When integration becomes a process and an end goal in itself it can become part of the problem, especially if the task is to integrate solutions that have been developed in isolation. True and effective integration within the water sector would have the overall function of a catchment ecosystem at the heart of all water management priorities, with all stakeholders engaged and acknowledging the broader water resource management issues that ultimately affect their own specialist areas.

Governance within any sector will always be weakened if the linkages between it and the wider governance context are overlooked (Water Partnership Programme 2010). In developed nations, the influence of the water sector in governance may be clouded by long supply chains and the disconnection between customers and water sources. In many developing nations however, there can often be very strong links between how water resources are managed and how water services are delivered, and wider governance issues of political stability, economic development, social justice and gender equality. Integrating water into these issues in an efficient and effective manner therefore has the potential to provide significant benefits to the country in question.

7.10 Breaking the Silos and Integrating Water Supply Policy

With the growing demand for water from expanding and increasingly affluent populations coupled with uncertainties over the potential impacts of climate change, stakeholder conflicts over water and scenarios requiring the better accounting of tradeoffs in water policy are increasing.

While the traditional response to shortages of adequate water has been to increase supply (e.g. through construction of new dams, groundwater bores and pipelines), such approaches have often failed to account for the root causes of the supply failure as well as the numerous knock-on effects that result from any water supply decision. To ensure that the tradeoffs associated with increasing supply are managed in the most effective manner possible, water policy needs to evolve to better account for several key factors as follows:

- The number one requirement is for society, and in turn policy, to address the current undervaluation of the role of water in the provision of ecosystem services. In our modern, consumerist societies, the most viable means of achieving this aim is to place an accurate financial value on the benefit that water supply provides to its various consumers. Accurate valuation and pricing arrangements that achieve full cost recovery would secure the funding necessary to invest in the initiatives and infrastructure that can sustain our exploitation of water sources within safe limits. Without this accurate valuation, consumers will continue to be sheltered from the true scarcity of water supplies, therefore making the overexploitation of water sources more likely to continue. As has been the case to date, in such a scenario, environmental requirements will lose out in the competition for water, undermining the ecosystem services on which all human activities ultimately depend.

- Valuing ecosystem services achieves a fundamental prerequisite that enables market processes to allocate water to those demands that use it most efficiently. Market-based policies have become increasingly popular; they are one of the means by which countries are addressing the Kyoto agreement on climate change, for example (Dietz *et al.* 2003). These policies are often more attractive to governments than command and control measures because the change they promote is driven by market participants making choices in a context created by governments, rather than by governments making decisions directly (Kiem 2013). For example, through a cap and trade policy, a market can be created for the trade of a particular environmental allowance, such as water. A limit is placed on the take of water from a given source and then entitlements for water up to this cap are created. Entitlements can then be traded in the market between competing users. Breakout Box 7.8 explores the water market of the Murray-Darling Basin of Australia.

Breakout Box 7.8 The water market of the Murray-Darling Basin, Australia

Probably the best-known example of a water market is that of the Murray-Darling Basin in Australia, an area home to more than 2 million people and one that produces more than one-third of Australia's food. In 1995, a cap was imposed on abstracting water from the system and, in 2004, the National Water Initiative replaced the existing system of water licences with tradable water rights (Quiggin 2007). Although the market-based approach has experienced a number of challenges, the cap was broadly successful in halting the unsustainable growth in abstractions of water and there is now a significant volume of temporary water trade (Quiggin 2007). It is estimated that in 2008–09, part of a period of prolonged drought in Australia, water trading and the reallocation of water used in agriculture in the southern Murray-Darling Basin increased Australia's GDP by AUS$ 220 million (Commonwealth of Australia 2010). Furthermore, the same report argues that trading has helped individual irrigators to better manage and respond to external drivers (such as drought, commodity prices and policy changes) by allowing more flexible product decisions (such as shifting between crops).

A number of other, smaller water markets also exist in rural Australia, their proliferation being indicative of a transition away from a siloed, engineering-focused approach to water supply planning to one that increasingly incorporates economic mechanisms within a broader planning framework (Horne 2013).

An important factor to consider when devising market-based polices for a given good is their potential to create inadvertent tradeoffs. For example, assume that the most efficient user of water in a market is the petroleum industry. Through the effective functioning of the market, water would be allocated to this user. However, while perhaps being efficient in water use, the petroleum industry is responsible for a significant volume of greenhouse gas emissions. Should a water policy be effectively incentivising this process? For complex commodities, Muradian and Rival (2012) argue that markets perform less well. Perhaps water is too complex a commodity to be managed in this way. It is clear at least that market-based policies relating to water use have to be very carefully thought out in order to avoid detrimental impacts to other natural resources and ecosystem services.

Successful markets also require strong institutions, good data and appropriate information about values and uncertainty that can be readily assimilated by users, as well as active market participation based on a thorough knowledge of its performance. They are therefore better suited to those nations that have the capacity to invest in building this capability. In developing countries, an alternative and potentially more successful means of improving water supply management would be to focus on enhancing the customary laws that have traditionally governed the management of environmental assets. Through incentivising inclusive and genuine stakeholder participation, and buffered from outside forces, these customary frameworks have successfully sustained resources for centuries (Dietz *et al.* 2003). Furthermore, where participants genuinely buy in to the approach, the sense of community management provides a strong bind and makes sustainable resource management readily achievable. Indeed, several characteristics of successful customary frameworks, particularly those related to securing and sustaining stakeholder engagement, provide valuable examples from which decision-making practices in major cities and developed nations could learn.

A key feature of successful future water supply policy will be its ability to encourage flexibility in water management in order to ensure that services continue to be provided in spite of future uncertainties. Decisions should also be taken independently of extreme events when public and political emotions are otherwise elevated. These aims require incentivising the adoption of broader water source portfolios; utilising smaller but locally abundant blue water sources in satellite networks; prioritising the capture of green water and the reuse of household grey water and stormwater; and especially (given its climate independence) encouraging the reuse of treated wastewater.

It is also clear that water supply policy needs to be formulated with adequate representation from all those stakeholders that rely on it. While the links between water and other key human resources are increasingly apparent, it is important to remember that this coupling occurs at multiple scales; for example, augmenting energy (and food) requirements will require rational water policies and integrated institutions.

With the number of closed and closing river and groundwater basins continuing to rise, the lack of spare water source capacity means that no stakeholder can take their continued water supply for granted. All stakeholders therefore need to be present at the decision-making table and, crucially, this requires that the environment is also afforded fair representation. Failure to achieve this diversity in stakeholder participation will increase the likelihood that initiatives and policies in one sector inadvertently cause detrimental impacts in another. The impact on water consumption from the growth of the biofuel industry (in many nations supported by mandates for biofuel use) is one example. The power sector is another. In this industry, demands for water can vary significantly depending on the technologies used. For example, power plants that adopt air-based cooling systems can require as little as 10% of the water required by plants that use a once-pass flow of water (Rodriguez *et al.* 2013). Furthermore, as the power generation sector looks to reduce its greenhouse gas emissions, carbon capture technologies are increasingly touted as a potential solution, even though they may increase the water use of a power plant by as much as 90% (WEC 2010).

Ultimately, water supply policy should seek to enable adaptive responses by local stakeholders that are guided by overarching fundamental principles, but based on local conditions and implemented through local institutions.

7.11 Evolution of Integrated Water Resource Management

The concept of Integrated Water Resource Management (IWRM) has become synonymous with efforts to break down the research and policy-making silos described above. IWRM has been widely embraced by the research community and represents the dominant paradigm of the current time. The Global Water Partnership (GWP) defines IWRM as: 'a process which promotes the coordinated development and management of water, land and related resources, in order to maximise the resultant economic and social welfare in an equitable manner without compromising the sustainability of vital ecosystems and the environment' (GWP 2000).

The benefits of managing water use and catchment activity as a network of interacting processes have been recognised for many decades. Competing uses of water, interactions between land use and water, and the tradeoffs between the use of different resources led mid-twentieth-century water practitioners to begin considering what was later to become encapsulated by IWRM. The idea is simple: by looking at the whole water cycle rather than just discrete components, and by working with natural ecosystem processes, more sustainable and enduring solutions can be identified, often with multiple benefits across a given catchment (CIWEM 2011). The idea may be simple, and with hindsight exceptionally obvious, but it has taken decades to define the concept, draw out the underpinning principles and develop practical implementation methods (see Figure 7.3). Because of its breadth however, the implementation of IWRM continues to suffer teething problems.

Realisation of the need for more integrated management of water resources and an appropriate framework for its delivery has grown slowly, with roots that can be traced back through various major research conferences as far as the 1940s. IWRM first became a mainstream concept at the 1977 UN Conference on Water held in Marta del Plata, Argentina. At this event, IWRM was showcased as a means to 'incorporate the multiple competing uses of water resources' (Rahaman and Varis 2005). It took until the 1992 UN International Conference on Water and Environment, held in Dublin, Ireland and its preparations for the UN Conference on Environment and Development in Rio de Janeiro, Brazil later that year for the concept to be refined into four guiding principles (World Meteorological Organization 1992):

1) recognition that water is a finite resource and that it should be managed in an integrated manner;

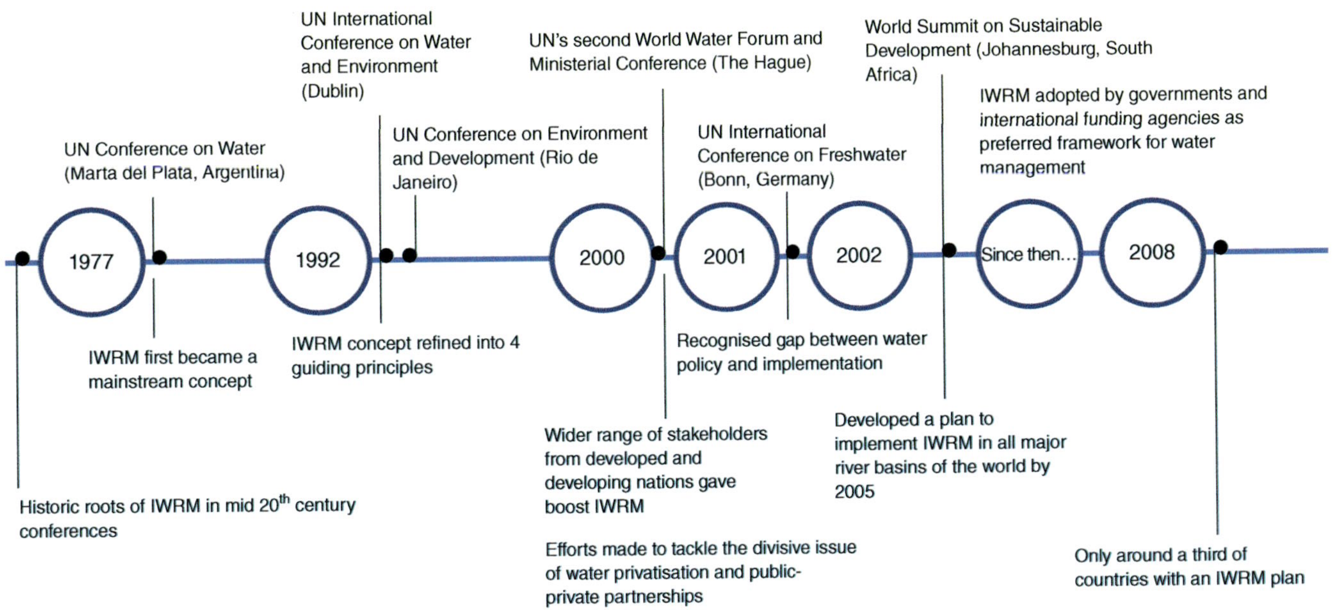

Figure 7.3 IWRM at a glance: from concept to implementation.

2) the requirement for this to take a participatory approach (incorporating users, planners and policy makers) at all levels of governance;
3) the central role played by women in IWRM (particularly in developing nations); and
4) the need for water to be considered as an economic good.

While Dublin achieved much in elevating the profile of IWRM, the fourth of its principles was widely criticised by developing nations as failing to consider issues of equity and poverty (Rahaman and Varis 2005). As a result, active participation by many developing countries in IWRM was constrained until the UN's second World Water Forum and Ministerial Conference in 2000, held in The Hague. The event's inclusion of a wide range of stakeholders from developed and developing nations gave IWRM the boost it needed. The forum acknowledged that 'food security, ecosystem protection, empowerment of people, risk management from water-related hazards, peaceful boundary and transboundary river basin management, basic water demands, and wise water management are all achievable through IWRM' (Shen and Varis 2009).

Importantly, The Hague forum made efforts to tackle the often divisive issue of the role of water privatisation and public-private partnerships in achieving successful outcomes. After several countries had experienced privatisations that had gone badly wrong, many water professionals opposed the argument that privatisation could contribute to achieving IWRM objectives. They contested that 'the water sector is interrelated to many functions that demand government presence, i.e. flood control, drought alleviation, water supply, and ecosystem conservation' (Shen and Varis 2000; Rahaman and Varis 2005).

The 2001 UN International Conference on Freshwater held in Bonn, Germany recognised the gap that often exists between the development of water policy and its implementation. A lot of time and effort has been invested into developing the simple concept of IWRM into arguably quite complex policy (albeit often necessarily so), without due consideration of how such policies will be implemented in practice. The 2002 World Summit on Sustainable Development held in Johannesburg, South Africa made steps towards rectifying this situation, guiding the development of a plan to implement IWRM in all major river basins of the world by 2005 (WSSD 2002). Although the global community fell a long way short of this target (6 out of 27 developed countries and 20 out of 53 developing countries claimed to have completed or initiated national IWRM plans by 2008; UN-Water 2008), the intention served as a clear attempt to translate IWRM from concept to tool.

IWRM has since been adopted by the majority of governments and international funding agencies as their preferred framework for water management; however, the vagaries of its definition have led many authors to criticise its use. For example, in a by no means exhaustive review of the literature, Biswas (2008) found 41 sets of issues that the respective authors argued should be addressed by IWRM. This suggests that the paradigm lacks a clear operational definition, allowing it to mean many things to many people; it therefore enables decision makers to claim most water management outcomes as examples of IWRM and consequently indicative of a successful project. The vagaries of IWRM also make the development of indicators against which to measure the true performance of specific schemes problematic. It is also argued that as the concept becomes more established there is a risk, or a tendency, for practitioners to lose sight of the core principles through a preoccupation with procedure, dogma and compliance (IWMI 2012). Water practitioners should re-embrace the simplicity of the philosophy of IWRM.

Despite its limitations, IWRM has been widely adopted as best practice by most major international funding agencies. This means that developing countries that currently lack water supply infrastructure must align their water management proposals with the IWRM ethos in order to optimise their chances of securing funding. Since the IWRM paradigm grew primarily from the developed world, with only limited involvement from developing nations, there exists an implicit assumption that developed country solutions will succeed in developing country contexts. However, because the needs and therefore water management practices of developed and developing countries differ, this assumption is rarely true in practice. In developing countries, the pressing need is to construct and manage infrastructure (Briscoe 2011), rather than just to manage it (see Breakout Box 7.9).

Critiques such as those presented in Breakout Box 7.9 have led several authors to claim that, while useful at framing broad and high-level thinking, IWRM fails to effectively find solutions to site-specific challenges. More pragmatic problem solving, based on an evaluation of specific local needs and guided rather than constrained by the principles of IWRM, represents a potentially more efficient means of identifying and implementing those water resource management options that are most sustainable in the local context.

Breakout Box 7.9 IWRM in developing countries

IWRM has been the prevailing paradigm of the water management sector for the last three decades, most dominantly between 1980 and 2000 (Biswas and Tortejada 2010a). Its implementation is characterised by basin-scale water management measures, support for the development of catchment management authorities, and initiatives such as water entitlement reform and those intended to ensure full cost recovery. Investment in infrastructure is typically not a primary focus of IWRM (Butterworth *et al.* 2010).

In practice, the implementation of IWRM has often proved problematic, especially in developing countries where the primary need is often for the physical infrastructure that can provide the foundation for water supply systems. The long-term success of typical IWRM measures may also be less likely in developing countries where the power of customary rights remain strong (Butterworth *et al.* 2010). To illustrate this issue, Giordano and Shah (2014) considered the evolution of water supply policy in Tanzania.

Tanzania is faced with a rapidly expanding population and a climate that tends to provide unreliable levels of rainfall that vary significantly with geography. The nation's RWR is 2020 m^3/cap/yr (UN-Water 2013) and, while it is not currently classified as water scarce, a lack of water storage infrastructure close to its major cities does restrict water access (Noel 2010). Recognising this constraint, Tanzania's 1991 water policy called for improved water storage (Giordano and Shah 2014). As is the case for many developing countries, domestic finance streams were insufficient to fund the policy's implementation; investment therefore had to be sought from international funding agencies, funding agencies whose approach to water supply management centred on IWRM ideals that often discouraged infrastructure construction. Forced to align with the requirements of the funding organisations, the Tanzanian government compromised on its water policy and passed regulation that reformed the allocation of water-withdrawal permits and water taxes and that supported the development of river basin organisations (Giordano and Shah 2014), all potentially effective water management measures in their own right but perhaps not those best aligned to the most pressing needs of the country's citizens.

Pakistan is another country that could benefit from the careful development of additional water storage capacity. Its alternating exposure to drought and monsoonal-driven flood suggests that efforts to slow the movement of water across the landscape could help provide more water to human users without detriment to environmental needs (by capturing high flows), while also protecting human assets. Briscoe (2011) argues that the reluctance of large funding institutions to support the construction of storage infrastructure is impeding the nation's development.

7.12 Traditional Water Planning Responsibilities versus a Corporate-Driven 'Water Risk' Agenda

Section 6.6 highlights the dominance of corporate sector influence on approaches to water management and Section 3.4 explores the extent of globalised virtual water trading associated with the products that businesses produce. From the perspective of a water manager, increased business awareness of water risks is welcome if it leads to more sustainable attitudes and management systems. However, it also raises concerns for the future of water management policy in locations where there is little governance or a weak governance framework. Here, powerful organisations could seek to create policy and implement water management systems which focus solely or predominantly on their own vested interests in securing water supply.

What is apparent is that much of the evolution of the water risk concept and its associated corporate assessment tools has predominantly been business or investor driven, albeit sometimes with a partnership influence from other stakeholders and non-governmental organisations (NGOs). The key absentees from these partnerships are governments and water utilities, exactly those authorities that have traditionally held the responsibility for managing water resources and supplies. While these stakeholders frequently express concerns about water scarcity, for example through the recent and continued proliferation of reports on water stress, they rarely engage in the water risk debate. This then points to an important question, whether the 'business driven' agenda (which is more dynamic, more flexible and arguably more urgent than that of governments and utilities) is accelerating beyond the governing capacity of those traditional authorities? The remit of water managers in the traditional authorities generally has a fixed (and relatively small) geographical scope, whereas that of the 'new' water managers within international companies with supply chains crossing the globe is on a completely different scale.

7.13 Summary

This chapter has focused on the fundamental principles that affect how water management decisions are made and ultimately how such systems control which water projects are implemented. One of the authors recalls their early years as a junior water professional when a relatively senior former boss said: 'this game that we play involves a lot of very clever words and some very clever science, but it also involves a lot of stupid people in influential positions who control the rules of the game'. We are reminded of those words every time a good solution based on sound evidence is rejected because it doesn't fit within a predetermined agenda. The seriousness of the problems we need to resolve make the stakes very high in this 'game', but the takeaway point from that statement was the need to understand the rules of the game. Understanding how the rules are defined is especially important when you are faced with a situation where you need to question the validity of those rules and how to effect a change.

Water professionals need to strive for good governance structures and also be able to recognise and manage accordingly when working in an environment of poor governance. In all situations, it is vital to understand who the key players are and whether their influence comes from legislative backing or some other basis. The effect of government, regulators, water utilities, commercial interests and the population in decision making exerts incredible influence over the success or failure of water management projects.

Key points to take away from this discussion are the differences between the enforceable aspects of legislation and the influential nature of policy. Depending on the role they take and the levels to which they climb, water professionals should expect to be frustrated by policy and legislation (or lack of it) that can often rail against logic and credible scientific evidence. It will be inevitable, whichever country or countries are involved. Such frustrations are however often what trigger new initiatives, new research and new partnerships to challenge the *status quo* and create change. In many cases these frustrations sow the seed for purposeful, inspiring and challenging careers.

Water management, and especially its high-level policy, can be very fickle, with topics coming in and out of fashion despite the underlying issues remaining the same. Focus on policy to improve water supply systems and engineering solutions can quickly shift to focus on ecosystem services and less engineered systems. Similarly, deep-routed traditional management systems orientated around centralised water utilities are being turned on their heads with interest returning to more decentralised approaches. The robustness of water supply systems is now talked about increasingly in terms of the 'resilience' of the cities dependent on supply and at risk of flooding. Focus in one area may be on strengthening domestic water supply policy, while there may be much higher levels of water demand and pressure being exerted through agricultural and industrial use. Water professionals need to be able to retain clarity on the core issues that need to be managed or resolved, while also being able to operate within and influence the dialogue of whatever topic is in fashion at the time.

References

AMCOW. 2008. Can Africa Afford to Miss the Sanitation MDG Target? A contribution to AfricaSan 2008, African Development Bank, World Bank and WSP p. 41.

Biswas, A.K. 2008. Integrated water resource management: is it working? *International Journal of Water Resources Development*, 24(1), 5–22.

Biswas, A.K. and Tortajada, C. 2010a. Future water governance: problems and perspectives. *International Journal of Water Resources Development*, 26(2), 129–139.

Biswas, A.K. and Tortajada, C. 2010b. Water supply of Phnom Penh: an example of good governance. *International Journal of Water Resources Development*, 26(2), 157–172.

Black & Veatch Corporation. 2014. Strategic Directions: U.S. Water Industry – a Black & Veatch Report. Black & Veatch, Kansas, USA.

Briscoe, J. 2011. Invited Opinion Interview: Two decades at the centre of world water policy. *Water Policy*, 13, 147–160.

Butterworth, J., Warner, J., Moriarty, P., Smits, S. and Batchelor, C. 2010. Finding practical approaches to integrated water resources management. *Water Alternatives*, 3(1), 68–81.

CEO Corporate Europe Observatory. 2005. Reclaiming Public Water. Achievements, struggles, and visions from around the world. Transnational Institute TNI.

Charbit, C. 2011. Governance of Public Policies in Decentralised Contexts: The Multi-level Approach. OECD Regional Development Working Papers, 2011/04, OECD Publishing. Available at: http://dx.doi.org/10.1787/5kg883pkxkhc-en (accessed January 2017).

CIWEM. 2011. Integrated Water Management. A CIWEM briefing report. Available at: http://www.ciwem.org/wp-content/uploads/2016/02/Integrated-Water-Managment-Report.pdf (accessed February 2017).

Clark, P. 2014. A World Without Water. Financial Times. Available AT: http://www.ft.com/intl/cms/s/2/8e42bdc8-0838-11e4-9afc-00144feab7de.html#slide0 (accessed January 2017).

Commonwealth of Australia. 2010. The Impacts of Water Trading in the Southern Murray-Darling Basin: an economic, social and environmental assessment. Commonwealth of Australia, Canberra, Australia.

Defra. 2015. Policy paper: 2010 to 2015 government policy: water industry. Available at: https://www.gov.uk/government/publications/2010-to-2015-government-policy-water-industry/2010-to-2015-government-policy-water-industry (accessed January 2017).

Dietz, T., Ostrom, E. and Stern, P.C. 2003. The struggle to govern the commons. *Science*, 302, 1907–1912.

European Commission. 2013. Assessing the Costs and Benefits of Regulation. Available: http://ec.europa.eu/smart-regulation/impact/commission_guidelines/docs/131210_cba_study_sg_final.pdf (accessed January 2017).

Food and Agriculture Organisation of the United Nations. 2002. Crops and Drops: making the best use of water for agriculture. FAO, Rome.

Galbraith, K. 2014. Ireland Sets Water Fees, Angering Thousands. The New York Times. Available at: https://www.nytimes.com/2014/11/13/business/international/ireland-sets-water-fees-angering-thousands.html (accessed February 2017).

Giordano, M. and Shah, T. 2014. From IWRM back to integrated water resources management. *International Journal of Water Resources Development*, 30(3), 364–376.

Global Water Forum. 2015. Current and future challenges in water resources management in Spain. Available at: http://www.globalwaterforum.org/2015/04/20/current-and-future-challenges-in-water-resources-management-in-spain (accessed January 2017).

Global Water Partnership. 2000. Integrated Water Resources Management. Technical Advisory Committee background paper no. 4. Global Water Partnership, Stockholm, Sweden.

Hoekstra, A.Y. and Chapagain, A.K. 2006. Water footprints of nations: water use by people as a function of their consumption pattern. *Water Resources Management*, 21, 35.

Hoekstra, A.Y. and Mekonnen, M.M. 2011. National Water Footprint Accounts: The green, blue and grey water footprint of production and consumption. UNESCO-IHE, Delft, Netherlands.

Horne, J. 2013. Economic approaches to water management in Australia. *International Journal of Water Resources Development*, 29(4), 526–543.

International Water Management Institute (IWMI). 2012. Non-integrated water resources management. Mark Giodarno: Presentation at 6th Botin Foundation Water Workshop, 13–14 November. Madrid. Available at: http://www.slideshare.net/IWMI_Media/nonintegrated-water-resources-management (accessed January 2017).

Jønch-Clausen, T. 2004 IWRM and Water Efficiency Plans by 2005: Why, What and How? TEC Background Papers No. 10, Global Water Partnership, Annex 4.5, p. 43.

KBH Energy Centre. 2015. New legislation seeks to curtail regulation of groundwater withdrawals for oil and gas projects. Available at: https://kbhenergycenter.utexas.edu/2015/03/11/new-legislation-seeks-to-curtail-regulation-of-groundwater-withdrawals-for-oil-and-gas-projects/ (accessed January 2017).

Kiem, A.S. 2013. Drought and water policy in Australia: Challenges for the future illustrated by the issues associated with water trading and climate change adaptation in the Murray-Darling Basin. *Global Environmental Change*, 23, 1615–1626.

Muradian, R. and Rival, L. 2012. Between Markets and Hierarchies: the challenge of governing ecosystem services. *Ecosystem Services*, 1, 93–100.

National Water Commission. 2012. Water Policy and Climate Change in Australia. Available at: http://apo.org.au/files/Resource/water-policy-and-climate-change-in-australia-full.pdf (accessed February 2017).

Noel, S. 2010. The Economics of Climate Change in Tanzania - Water Resources. Stockholm Environment Institute, University of Dar es Salaam, Tanzania.

Ofwat. 2006. Emergency Meeting to discuss the performance of Thames Water. Notes from the Ofwat Board meeting held 29 June 2006. Available at: http://www.ofwat.gov.uk/wp-content/uploads/2015/10/OFWAT-BOARD-Meeting-29-June-2006.pdf (accessed February 2017).

Page, D., Gonzalez, D., Torkzaban, S., Toze, S., Sidhu, J., Miotliński, K., Barry, K. and Dillon, P. 2014. Microbiological risks of recycling urban stormwater via aquifers for various uses in Adelaide, Australia. *Environmental Earth Sciences*, July, 1–5.

Piure, A. 2014. Markets, water shares and drought: Lessons from Australia. What can the water industry in England and Wales learn from Australia's water reform story? Anglian Water. Available at: http://www.anglianwater.co.uk/_assets/media/Markets_watershares_and_drought_FINAL_REPORT.pdf (accessed February 2017).

Plummer, J. and Slaymaker, T. 2007. Rethinking Governance in Water Services. Overseas Development Institute, London, UK.

Porter, E. 2014. The Risks of Cheap Water. The New York Times. Available at: http://www.nytimes.com/2014/10/15/business/economy/the-price-of-water-is-too-low.html?_r=0 (accessed January 2017).

Quiggin, J. 2007. Economic and political Overview: Issues in Australian Water Policy. Available at: http://www.ceda.com.au/media/29714/quiggin_water_ace_200702.pdf (accessed January 2017).

Rahaman, M.M. and Varis, O. 2005. Integrated water resources management: evolution, prospects and future challenges. *Sustainability: Science, Practice & Policy*, 1(1), available at: https://sspp.proquest.com/integrated-water-resources-management-evolution-prospects-and-future-challenges-c43839577953#.9u7ex2jhr (accessed February 2017).

Reed, J.M., Leng, M., Ryan, S., Black, S., Altinsacli, S. and Griffiths, H.I. 2008. Recent habitat degradation in karstic Lake Uluabat, western Turkey: a coupled limnological-palaeolimnological approach. *Biological Conservation*, 141(11), 2765–2783.

Rodriguez, D.J., Delgado, A., DeLaquil, P. and Sohns, A. 2013. Water Papers: Thirsty Energy. World Bank, Washington DC, USA.

Ryan, S. and Griffiths, H. 2001. The decline and probable extinction of Graecoanatolica macedonica (Gastropoda: Orientallnidae) in Balkan Lake Dojran. *Journal of Conchology*, 37(3), 261–265.

Rygaard, M., Binning, P.J. and Albrechtsen, H.-J. 2011. Increasing urban water self-sufficiency: new era, new challenges. *Journal of Environmental Management*, 92, 185–194.

Schuetze, T., Lee, J-W. and Lee, T-G. 2013. Sustainable urban (re-) development with building integrated energy, water and waste systems. *Sustainability*, 5, 1114–1127.

Shen, D. and Varis, O. (2000) World water vision: balancing thoughts after The Hague. *Ambio*, 34(8), 523–525.

Sukhdev, P., Wittmer, H. and Miller, D. 2014. The economics of ecosystems and biodiversity (TEEB): challenges and responses. In Helm, D. and Hepburn, C. (eds) Nature in the Balance: The Economics of Biodiversity. Oxford University Press, Oxford, UK.

Texas A&M University. 2014. Texas Water. Available at: http://texaswater.tamu.edu/water-law (accessed January 2017).

Texas Commission on Environmental Quality. 2015. Water Rights: Permits. Available at: https://www.tceq.texas.gov/permitting/water_rights/wr-permitting/water_rights.html (accessed January 2017).

Tiemann, M. 2011. Fluoride in Drinking Water: A Review of Fluoridation and Regulation Issues. Congressional Research Service. Available at: http://www.dentalwatch.org/fl/crs.pdf (accessed January 2017).

Tortajada, C. 2010. Water Governance: some critical issues. *International Journal of Water Resources Development*, 26(2), 297–307.

UK Water Industry Research Ltd. 2015. Assessing the performance of the existing regulatory framework to regulate and control municipal water reuse schemes in England and Wales. Establishing a Robust Case for Final Effluent Reuse - An Evidence Base – Phase 2. Available at: https://www.ukwir.org/reports/15-WR-29-4/129558/Establishing-a-Robust-Case-for-Final-Effluent-Reuse--Phase-2-Testing-the-UK-Regulatory-Framework (accessed February 2017).

UNESCO. 2014. Water in the post-2015 development agenda and sustainable development goals. Discussion paper. Available at http://unesdoc.unesco.org/images/0022/002281/228120e.pdf (accessed January 2017).

United Nations Environment Programme. 2013. TEEB – Economic of Ecosystems and Biodiversity: Guidance Manual for TEEB Country Studies. UNEP, New York, USA.

United States Environmental Protection Agency (EPA). 2013. Regulating Public Water Systems and Contaminants Under the Safe Drinking Water Act. Available at: http://water.epa.gov/lawsregs/rulesregs/regulatingcontaminants/basicinformation.cfm (accessed January 2017).

United States Geological Survey. 2002. Wetland Management and Research Wetland Protection Legislation. Available at: http://water.usgs.gov/nwsum/WSP2425/legislation.html (accessed January 2017).

US AID. 2010. Mali. Water and Sanitation Profile. United States Agency for International Development US AID). Available at: http://www.washplus.org/sites/default/files/mali.pdf (accessed February 2017).

UN-Water. 2008. Status Report on IWRM and Water Efficiency Plans for CSD16. Available at: http://www.unwater.org/downloads/UNW_Status_Report_IWRM.pdf (accessed January 2017).

UN-Water. 2013. United Republic of Tanzania: UN-Water Country Brief. Un-Water, New York, USA.

Viscusi, W.K. 1994. Risk-risk analysis. *Journal of Risk and Uncertainty*, 8, 5–17.

Water Partnership Program (WPP). 2010. Water Sector Governance in Africa. Volume 1: Theory and Practice. African Development Bank. Available at: https://www.afdb.org/fileadmin/uploads/afdb/Documents/Project-and-Operations/Vol_1_WATER_SECTOR_GOVERNANCE.pdf (accessed February 2017).

World Bank. 2006. Handbook for Evaluating Infrastructure Regulatory Systems. Washington, DC.

World Bank Group. 2007. Review of the Dutch Administrative Burden Reduction Programme. Available at: http://www.doingbusiness.org/reports/thematic-reports/review-of-the-dutch-administrative-burden-reduction-programme (accessed February 2017).

World Economic Forum. 2014. Global Risks 2014. Insight Report. Ninth edition. Geneva. Available at: http://www3.weforum.org/docs/WEF_GlobalRisks_Report_2014.pdf (accessed January 2017).

World Energy Council. 2010. Water for Energy. World Energy Council, London, UK.

World Health Organisation. 2006. Guidelines for the Safe Use Of Wastewater, Excreta and Greywater: Volume 1: Policy and Regulatory Aspects. WHO, Geneva.

World Health Organisation. 2011. Legislative Frameworks. Guidelines for drinking water quality. Available at: http://apps.who.int/iris/bitstream/10665/44584/1/9789241548151_eng.pdf (accessed February 2017)

World Meteorological Organization. 1992. The Dublin Statement on Water and Sustainable Development. Available at: http://www.wmo.int/pages/prog/hwrp/documents/english/icwedece.html (accessed January 2017).

World Summit on Sustainable Development (WSSD). 2002. *Report of the World Summit on Sustainable Development.* A/Conf. 199/20. Available at: http://www.unmillenniumproject.org/documents/131302_wssd_report_reissued.pdf (accessed February 2017).

WWF. 2003. Water transfers are increasing the water crisis. The case of the Tagus-Segura Transfer in Spain. Lessons from the Past. Available at: http://wwf.panda.org/about_our_earth/all_publications/?7047/water-transfers-are-increasing-the-water-crisis-the-tagus-segura-water-transfer (accessed January 2017).

World Wide Fund for Nature (WWF). 2006. Rich Countries, Poor Water. Available at: http://wwf.panda.org/?80000/Rich-countries-Poor-water (accessed February 2017).

8

Ownership and Investment

Whole books have been written on the ownership of water resources, on the perceived 'rights' for their use, and on the issue of whether public or private approaches to the delivery of water supply and sanitation services are the most appropriate. While these themes are important facets of water management, they are not a fundamental influence on the concepts of future water management we describe in this book. Nevertheless, in recognition of the reliance of a significant and growing proportion of the global population on the performance of water service providers, in this chapter we explore how water resource systems and their component elements of infrastructure can be funded and operated.

8.1 Public versus Private Ownership Models

Historically, as populations and communities first began to grow, the responsibility to supply water was taken by the state. In the 2000 years since Rome's water system was built, this model of the state supplying water to its citizens has been replicated across the world and still dominates to this day.

In privately owned and operated water service systems, the state has passed over some or all of this responsibility to a private company. The private company typically provides the investment required to build, enhance and operate the water system and receives fees from its customers. During the 1990s and early 2000s, over 90 countries introduced privatisation into the water sector with the result that, at that time, some 6% of the global population was provided with water services via private operators (Memon and Butler 2003). Opponents of private involvement in water service provision argue that as a business's overriding goal is to secure profit, water services (a component of which, it should not be forgotten, is a basic human right) may well be squeezed in order to extract maximum financial return. Proponents of private sector involvement argue that a business's focus on economic return is likely to drive innovation, productivity and maximum efficiency in the system. In modern water service networks, private sector involvement is typically regulated to support the interests of the customer and to guide system enhancement. Various forms of public-private partnership (PPP) have also been used; however, it is not within the scope of this book to describe the many variants.

Water Resources: A New Water Architecture, First Edition.
Alexander Lane, Michael Norton and Sandra Ryan.

Some of the first privately owned water service systems emerged in the late eighteenth and early nineteenth centuries, as the following examples illustrate.

- In New York in 1799, and following the refusal of the city council to raise loans and taxes to expand its water works, the Manhattan Company was tasked with supplying water to the city. By 1802, the company had installed over 30 km of log pipe; however its primary focus was elsewhere, to become part of the banking industry of New York. The Manhattan Company system proved to be wholly inadequate and, some 30 years later, responsibility for water supply was ceded back to the city government.
- In England, the rapid expansion of urban communities associated with the industrial revolution of the late eighteenth and early nineteenth century created threats to public health associated with water contamination. For a period of time, much of the responsibility for addressing this issue was taken up by the private sector. By 1860, private systems served around 60% of large towns and cities; however, this commercial approach to water supply was increasingly seen to be unsatisfactory. The role of private entities in the English water service system slowly declined and continued this trend until the latter half of the twentieth century. By the early 1970s water supply in England and Wales was provided by 198 undertakings, 33 of which were private.
- In France, water supply was seen as a private sector service from as far back as the late eighteenth century. The private company General des Eaux was awarded its first contract in 1853 to supply water to the city of Lyon, followed shortly after by another to supply Paris. General des Eaux is now known as Veolia, a global environmental service provider.
- In Berlin, a contract to develop the city's first water supply system was awarded to the private sector in 1852. However, reflecting poor progress in delivering on its responsibility, the private entity's contract was terminated by the city government in 1873.

8.1.1 A New Era of Privatisation

When the Conservative government of the United Kingdom launched its privatisation programme in the 1980s, water was one of the last utilities to be fully privatised. The tight fiscal controls applied by central government in the 1970s and 1980s and the high levels of debt in the incumbent public water authorities led to insufficient expenditure to meet the capital maintenance and investment requirements. The resulting inevitable service quality problems became particularly evident in the 1980s, regarded as increasingly unacceptable by a public with growing environmental awareness and influenced by increasingly stringent European environmental legislation.

In an attempt to address these concerns, the water industry of England and Wales was privatised in 1989, with the physical and human assets of the then ten water authorities transferred into limited companies. Capital was raised by floating the companies on the London Stock Exchange, via a one-off injection of public capital and through the write-off of significant government debt. To ensure the interests of customers and the environment were secured, three separate, independent regulatory bodies were also established:

1) National Rivers Authority (the environmental regulator, with responsibility for protecting the water environment);
2) Drinking Water Inspectorate (responsible for regulating drinking water quality); and
3) Office of Water Services (the economic regulator).

Exposure to the model of privatisation implemented in England and Wales and the perceived success of the French approach supported a growing opinion that privatisation could represent the best means of addressing those problems that were commonly and increasingly afflicting public water service models of the time, namely, inefficiency and a lack of access to capital. The international funding institutions, particularly the World Bank, adopted this ethos and embarked on strategies that required borrower governments to privatise their water utilities, particularly those serving large urban communities.

A spate of privatisations ensued, including Buenos Aires in Argentina (1993), Manila in the Philippines (1997), Cochabamba (1999) and La Paz (1997) in Bolivia, Chile's entire urban water supply and sanitation sector (1998–2005), Jakarta in Indonesia (1993), Tallinn in Estonia (2001), Guayaquil in Ecuador (1995) and Bucharest in Romania (2000). According to the World Bank, a total of 55 countries introduced privatisation models in 338 separate cities over the period 1990–2005.

8.1.2 A Backlash Against Privatisation

One of the often-cited negative examples of water privatisation is that of Cochabamba, Bolivia's third largest city. Located in a water-scarce region, Cochabamba has a long history of water-related disputes. Government funding of the public utility's attempts to secure more sustainable water supply had, according to a World Bank report of 1999, cost over one-quarter of a billion US dollars (Zenteno undated). Despite this investment, water service provision remained poor and new investments were needed to secure the necessary water supplies for an expanding population.

A first auction for a private water concession in Cochabamba (driven by strong support from the major international funding institutions) came in 1997. It was however declared void by the mayor who wanted the construction of a large dam, and a pipeline from the dam to the city, to be included in the concession. After a delay the government proceeded with the auction, including the dam project, but received only one bid which was from the Aguas del Tunari consortium. The bid was accepted with a contract guaranteeing Aguas del Tunari a minimum 15% return on investment that would be achieved by raising water tariffs by 35% (Zenteno undated).

Protests erupted with the imposition of the tariff hike in January 2000, with demonstrations and a general strike in the city. Although the leader of the protest groups was arrested, dissent spread across the country and the government declared a state of emergency. The employees of Aguas del Tunari fled Cochabamba.

To quell the public backlash and on releasing the protest leader, the government signed an agreement to end the concession. It then informed Aguas del Tunari that by leaving Cochabamba, the consortium had abandoned the concession. The company insisted that it had been forced out and filed a lawsuit in the International Centre for Settlement of Investment Disputes, a claim that was eventually dropped.

Back under a public sector delivery model, and despite expanding the water supply system with funding assistance from international institutions, water supply to many of Cochabamba's citizens remains intermittent. Despite the complexities of this case study, Cochabamba's experience of privatisation became a world news story and a symbol of a newly emerging view that water is not a market commodity and should not therefore be entrusted to the private sector.

By 2003, the World Bank was beginning to question whether privatisation was a viable solution to the challenges of urban water supply and sanitation. The problems experienced in Cochabamba and other cities had shown that private service models could often be unaffordable to the majority of the urban poor. As a result, the World Bank began to encourage new models of mixed ownership or public-private partnership (PPP) such as those pioneered in Cartagena and Barranquilla in Colombia (Breakout Box 8.1).

8.1.3 Reflections on the Public versus Private Debate

In a 2005 research paper entitled 'Thirst: a short history of drinking water', Professor James Salzman of Duke University concluded that 'while making for powerful rhetoric, treating drinking water management as a binary conflict of rights versus markets, of public versus private management, forces a false choice'. Salzman (2005) believes that the rights and markets approaches do not have to be mutually exclusive, pointing to ancient Rome where water by right and water by transaction coexisted and were dependent on each other.

What Salzman's study concluded, and what many of us who have worked in and for public and private ownership models have recognised, is that water is a natural resource like no other. The UN has declared that we all have a right to drinking water and to sufficient water to meet the needs of basic hygiene. Institutional, regulatory and business models to fund and deliver this right are no better or worse if they rely on private ownership and/or operation of the infrastructure. Perhaps what is clear is that the true and full cost recovery of water and sanitation is unaffordable to the poorest elements of society. This in turn means that successful models must incorporate cross-subsidy in some form or another and this requires that the wealthy elements of society accept that, in the special case of the natural resource of water, they must contribute to the cost of providing water and sanitation to the poor. Such acceptance prevailed in Rome 2000 years ago and many current tariff and regulatory regimes recognise that cross-subsidy is key (the case of Cartagena highlighted in Breakout Box 8.1 and the case of the England and Wales water sector are two examples).

Breakout Box 8.1 PPP in Cartagena, Colombia

Prior to 1995, water and wastewater service provision in Cartagena, a city of around 900,000 people, was extremely unreliable (less than 70% of households had a water connection and less than 55% had a sewage service) and the system operated at a financial loss. Social calls for action led to the development of a PPP between the public works department and a Spanish firm. The PPP (called AGUACAR) immediately commenced a program of water truck deliveries to service unconnected homes and also significantly reduced the percentage of non-revenue water. The financing for these investments came, in part, from a restructuring of the tariff system to incorporate cross-subsidies that ensured that more affluent customers helped support lower-income families. By 2005, water supply coverage had increased to 99% of the population and sewage coverage had risen to 75%. Source: UNDP (2012).

Section 2.3.1 refers to past civilisations (e.g. Rome) where many stakeholders were involved in making decisions on the allocation of water. That everyone needed and had a right to be involved was not questioned. Fast forward to the present day and there are many places where, because water is 'provided' by large private or public entities, decisions are generally made within those organisations and without widespread consultation. As a result, the general population takes the supply of clean and abundant volumes of water for granted. In this book we have highlighted instances where this lack of appreciation of the true value of water has led to profligate water use and the degradation of the environment. But perhaps now, with the proliferation of more holistic, collaborative and integrated approaches to water management, more active participation in the water debate and the growing power of consumer groups, the role of stakeholder consultation and collective buy-in to the management of our water resources can flourish.

Both public and private approaches to the supply of water can succeed, in isolation or in combination. However, any system must consider the multitude of interacting factors necessary to ensure that the exploitation of water resources remain within safe bounds.

8.2 Investment Models and the Economics of Water Management

In this section we discuss the investment needed to maintain and expand water infrastructure, where the funding is sourced from now, and where funding is likely to come from in the future.

8.2.1 Current and Future Forecast Levels of Investment

Despite efforts made over the past 20 years to increase investment in water, sanitation and hygiene (WASH) at a global scale, only modest progress has been made in reducing the number of people without access to clean water and in increasing the number of those who have access to safe sanitation. While the World Health Organization and the UN (2014) highlight a cost to benefit ratio of anything between US$ 4 and 8 for every US$ 1 invested in WASH, they also report that 80% of countries have insufficient finances for WASH. Interestingly, more than 70% of countries also report that cost recovery in water supply is less than 80% of operations and maintenance costs. Requirements for capital and ongoing expenditure in water supply and sanitation systems therefore cannot be met. Deficiencies in financing for WASH are highlighted in Figure 8.1.

The World Bank estimates that, based on current trends, universal access to sanitation and improved water supply is more than 50 years away for most African countries (McKinsey 2013). In order to make a real dent in the proportion of the global population currently unserved by safe water supply and sanitation systems, various organisations and advocacy groups have attempted to estimate the future investment required.

- In 2006, the OECD (OECD 2006) estimated that investment in the water sector would double from then current levels of US$ 570 billion a year to over US$ 1 trillion per year by 2025.

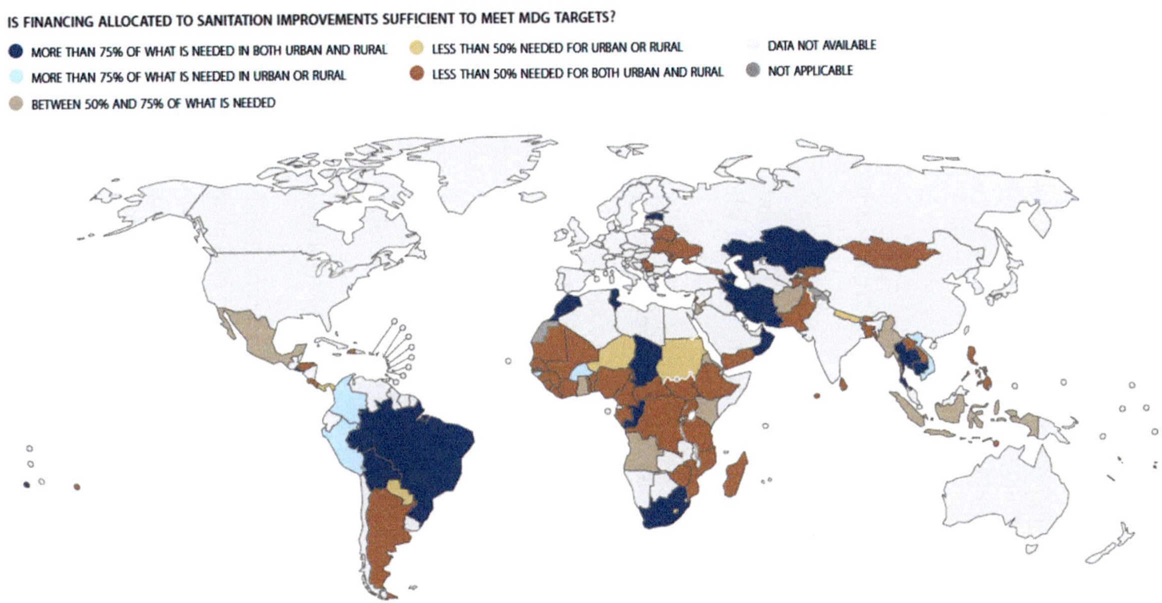

Figure 8.1 Deficiencies in financing for WASH.
Source: Reproduced with permission of WHO and UN (2014).

Table 8.1 Estimated average annual global infrastructure expenditure for selected sectors, US$ billions/year.

Type of infrastructure	2000–10	Approx. % of world GDP	2010–20	Approx. % of world GDP	2020–30	Approx. % of world GDP
Road	220	0.38	245	0.32	292	0.29
Rail	49	0.09	54	0.07	58	0.06
Telecoms	654	1.14	646	0.85	171	0.17
Electricity	127	0.22	180	0.24	241	0.24
Water	576	1.01	772	1.01	1037	1.03

Source: Adapted from OECD (2006).

- Merrill Lynch Bank of America (2014) forecast that investment in the water sector would be US$ 1 trillion by 2020, a figure which included forecasts for the municipal, industrial and agricultural sectors (see Table 8.1).
- McKinsey (2013) estimated that some US$ 11 trillion of investment in water systems would be needed between 2013 and 2030, simply to keep up with projected global GDP growth.

8.2.2 Meeting Investment Needs

Providing a water service requires continual investment, initially to construct new infrastructure and then to operate and maintain assets over their lifetimes. Figure 8.2 illustrates how these capital expenditure (CAPEX) and operating expenditure (OPEX) requirements are addressed through revenues and a variety of other funds that can be used (where available) to fill a funding gap.

In most water supply systems, revenues are generated through one of '3 Ts' (OECD and WWC 2015):

1) tariffs: direct charges to customers;
2) taxes: general or property taxes; and
3) transfers: from other revenue sources.

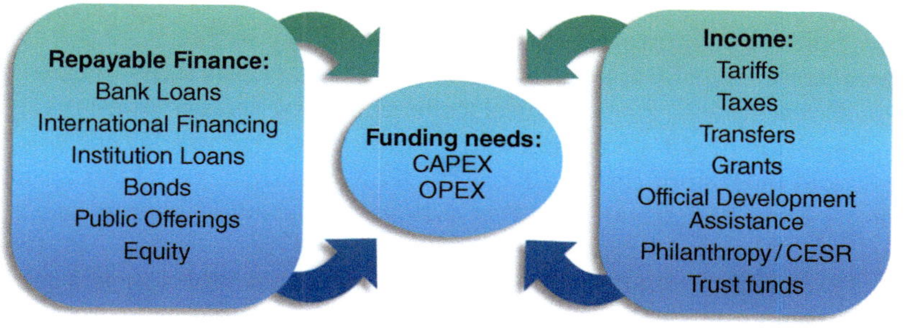

Figure 8.2 Sources of finance for water and sanitation.
Source: Adapted from OECD and WWC (2015).

These revenues are then used to pay operational costs, provide the investment for improvements in the network and, for private service providers, to recognise a profit. Approaches to meeting the financing gap from the '3 Ts' vary widely.

Where water and sanitation systems require significant capital expenditure (e.g. upgrading a water treatment plant), any annual financing gap has to be met through the raising of repayable finance. That finance could be sourced from a number of facilities, most commonly commercial loans and non-repayable grants. Recent years have seen a growth in the use of bonds (where the general public lend money) and equity (where entities such as pension funds lend money) on the promise or expectation of a return on their investment.

When the water industry in England and Wales was privatised in 1989, one of the principal drivers was said to be that private owners would be in a position to raise capital that the over-indebted public sector could not. In the USA on the other hand, one of the reasons why most water utilities remain in public hands is that they are readily able to raise funding to close the financing gap.

While it has been argued that privatisation of water utilities paves the way for access to cheap capital and self-sufficiency, hard evidence for this assertion is limited and often weak. Over the last 10 years, private investment activity in the water sector in developing countries has averaged only US$ 2.5 billion annually, or about 3% of the investments needed for water supply and sanitation (World Bank 2012). This is a significant decline from the decade leading up to 2000 which saw several large privatisations which contributed up to US$ 14 billion per year when new and existing private activity was considered together (World Bank 2012).

Realising that the provision of a water service needs to be financially sustainable if it is to be maintained and therefore add value to society, many water utilities have sought to achieve this position of security by raising tariffs. As a result, over the period from 2000 to 2008, the global average revenue raised by utilities for every cubic metre of water sold nearly doubled from US$ 0.37 to 0.71 (World Bank 2012). Given the critical importance of water to the most basic of human needs, right through to many of our most advanced technological processes, fierce debates over the affordability of water are no surprise, especially in countries and regions where water has historically been provided for free or a nominal fee. Even in a developed nation such as the UK, there is a political view that household water fees, equating to approximately US$ 1.9 per cubic metre (NUS Consultants 2006), are too high for many people on low incomes. In the USA, the average cost of water is some US$ 0.7 per cubic metre (NUS Consultants 2006).

Increased household expenditure on basic needs (especially if this results from tariff increases unaccompanied by appropriate public awareness campaigns) is clearly going to breed social resentment; many political leaders therefore refrain from actively pushing for the imposition of self-sustaining water tariffs, those that would otherwise ensure revenues could meet both the CAPEX and OPEX requirements of the water and wastewater network. The consequence is that too little investment is made, assets deteriorate and the water service provided ultimately suffers. In the USA, the American Water Works Association points to a ticking time bomb of dereliction of underground pipe systems which require US$ 1 trillion of investment over the next 25 years (AWWA 2011).

8.2.2.1 Investment to Achieve Basic Human Needs

What is the basic human need for water? Up to 50 litres per person per day? In 2015 there were almost 1 billion people, 1 in 7 of the global population, whose basic water needs were not being met. That there is insufficient funding to provide the infrastructure needed to deliver this basic human right may seem shocking; however, it reflects a combination of governance and economic constraints in the countries where those 1 billion people live. To support these (typically) developing countries, governments and citizens of other nations often provide assistance via charities and bilateral grants; however, progress towards universal improved water and safe sanitation remains slow. A reliance on aid can also hinder long-term progress in water service provision. For example, by becoming dependent on aid to meet short-term water needs, a country's focus will naturally shift away from what must be a long-term goal of developing financially sustainable water service providers, be they public or private.

Put into perspective, it should be possible to provide the basic human needs for water and sanitation for 1 billion people with less than US$ 200 per person, or less than US$ 200 billion, which is just over one-tenth of global defence spending of US$ 1676 billon (Stockholm International Peace Research Institute 2016). While this comparison may seem simplistic, it illustrates that the global financing gap to meet basic human needs is very modest. Notwithstanding this fact, the very poorest of nations will struggle to close that gap from tax revenue, or to raise commercial finance due to affordability constraints. In these cases, there seems little alternative to receiving charitable and bilateral support; however, these need to ensure the development of a long-term sustainable water sector.

8.2.2.2 Investment to Achieve Discretionary Domestic and Industrial Needs

In developed and most developing countries, per capita domestic water use far exceeds the basic human need of, say, 50 litres per day. In the UK the average is 150 litres per person per day; in the USA it is over 500 litres per person per day. The use of water in agriculture and manufacturing is also technically discretionary, but a vital means of adding value to the products produced.

In a number of countries, the response to this differentiation between essential and discretionary water use is to adopt tiered tariff regimes where the price of that volume of water required for basic human needs is lower than the price of water used for activities above and beyond the basic volume (i.e. for discretionary purposes). In this way, additional revenue is raised to meet system CAPEX and OPEX and therefore to close any infrastructure financing gap. While this approach appears logical, there remains in some developed nations a reluctance to embark on the systematic installation of water meters required to enable tiered tariffs to be applied (stemming from social and political concerns over subsequent water affordability).

Taking into account the relatively modest investment necessary to deliver basic drinking water needs to 1 billion people, and the associated sanitation infrastructure, the major part of the US$ 1 trillion of annual investment will be needed to meet discretionary use, including that in agriculture and industry, and including the associated wastewater infrastructure. How this will be financed has been studied by the Organisation for Economic Cooperation and Development and the World Water Council (OECD and WWC 2015). Traditional "3 Ts" models will continue to provide the baseload of investment, but increasingly we are likely to see overall needs being met by supplementary sources such as sovereign wealth funds, infrastructure bonds, specialised water funds, payment for ecosystem services and habitat banking, in addition to international financing institutions and philanthropic sources from the business sector.

8.3 Summary

In this chapter we have described the mechanisms through which water infrastructure capital and operating needs are financed. We have given a brief history of the evolution of ownership models of water resources and water infrastructure, and of the debate that continues, needlessly in our opinion, on the topic of public versus private ownership and operation.

We have discussed the role of tariffs as a means of raising revenue, and the need in most circumstances for those tariffs to incorporate cross-subsidies so that those with higher income are able to help meet the costs associated with supplying the poorest water users. We have noted the general reluctance of governments to adopt stepped or progressive tariff structures that would charge the basic needs of water at prices which could be affordable by the poorest, but would charge more per unit for discretionary use.

We have pointed to the shocking and shameful statistic that 1 in 7 of us still does not receive a safe and secure supply of drinking water, the cost of which to resolve would be of the order of one-tenth of one year of global defence spending of more than US$ 1.5 trillion. This in contrast to the high-level estimates of investment needs for water infrastructure provision which are thought to be of the order of US$ 1 trillion annually and which, if met, will be through the emergence of innovate financing models to complement traditional '3Ts' sources.

Having presented the physical challenges around water security in Part II (Chapters 3–6), and what we call the 'current water architecture' of management, regulation and financing here in Part III (Chapters 7 and 8), we now move into Part IV (Chapters 9–13) to describe a way forward, which we call **'New Water Architecture'**.

References

American Water Works Association. 2011. Buried No Longer: Confronting America's Water. American Water Works Association.

McKinsey, 2013. Infrastructure Productivity: how to save $1 trillion a year. McKinsey, New York.

Memon, F.A. and Butler, D. 2003. The role of privatisation in the water sector. *Water Perspectives*, 1, March, 99, 28–36.

Merrill Lynch Bank of America. 2014. Blue Revolution – Global Water Primer. Merrill Lynch Bank of America.

NUS Consultants. 2006. 2005–2006 International Water Report & Cost Survey. NUS Consultants.

OECD. 2006. Infrastructure to 2030: Telecom, Land Transport, Water and Electricity. OECD.

OECD and World Water Council. 2015. Water: Fit to Finance? OECD and World Water Council.

Salzman, J. 2005. Thirst: a short history of drinking water. Duke Law School, Legal Studies Research Paper Series, no. 92.

Stockholm International Peace Research Institute. 2016. Fact Sheet. Stockholm International Peace Research Institute.

United Nations Development Programme. 2012. Cartegena, Colombia: Case Study – Water. UNDP, New York, USA.

World Bank Water Partnership Programme. 2012. Investing In Water Infrastructure: Capital, Operations and Maintenance. World Bank Water Working Paper no. 74051 (Diego J. Rodriguez, Caroline van den Berg and Amanda McMahon). World Bank.

World Health Organisation and United Nations. 2014. UN-water global analysis and assessment of sanitation and drinking-water (GLAAS) 2014 – report: Investing in water and sanitation: increasing access, reducing inequalities. WHO, Geneva.

Zenteno. Undated. The water war to resist privatisation of water in Cochabamba, Bolivia. Global Water Partnership. Available at: http://www.gwp.org/en/ToolBox/CASE-STUDIES/Americas--Caribbean/Bolivia-The-water-war-to-resist-privatisation-of-water-in-Cochabamba-157/ (accessed January 2017).

Part IV

Moving to a New Water Architecture

9

Challenges and Opportunities

9.1 A New Water Architecture: An Introduction

In the preceding Parts I to III of this book, we described the role of water as the 'bloodstream of the biosphere'; how people use water; how ecosystems and biodiversity depend on water; how water is managed and regulated to allow us to live, eat and consume; and how water stress and its management is becoming a crucial factor in sustainability of the biosphere.

The remaining chapters describe a new approach to water management that we term **'New Water Architecture'**. In conceptualising this architecture we seek to address the underlying challenges associated with the practical implementation of IWRM, the failure of water governance leadership, and the lack of societal and governmental appreciation of the true value of water. We consider this not simply as a reactive problem-solving exercise, but also as a means to generate opportunities in areas such as water quality improvement, wastewater reuse and careful decentralisation. Our **New Water Architecture is a systems-based framework of conceptual, institutional and physical integration** that builds on the principles of IWRM.

New Water Architecture is not a one-size-fits-all formula. Our experience of working in a range of political and cultural environments makes us wary of the pitfalls of assuming that approaches that are successful in one country can and should be transferred to another. As argued by Carter (1998) and IWMI (2006), 'off-the-shelf policy proposals not tailored to local contexts but championed by donors and some international organisations should be examined critically, as they may be unsuitable governance tools in many countries despite their effectiveness in others.' Similarly, 'policies that are developed without sufficient finance for implementation complicate sector governance by adding to existing collections of unenforceable or unrealistic legislative or policy initiatives. Likewise, the pursuit of targets or objectives set out in policy without sufficient attention being given to the processes and resources needed to attain them also inhibits effective sector governance' (Water Partnership Program 2010).

While what we are describing as New Water Architecture can be framed in systems-thinking language, we have chosen instead to use 'connectivity' and 'integration' as words to underpin the description of the approach. We envisage three levels of integration: physical (focused on optimising water infrastructure); institutional (focused on improving water governance); and conceptual (focused on our collective mindset and attitude towards water).

Water Resources: A New Water Architecture, First Edition.
Alexander Lane, Michael Norton and Sandra Ryan.

- **Physical integration** is the development of water infrastructure as one integrated infrastructure that recognises that the management and allocation of water transcends the more usual approach of sector-siloed solutions. Physical integration includes all natural and engineered assets that capture, transfer, treat, distribute and collect water and wastewater.
- **Institutional integration** seeks to dissolve the barriers that exist between water management and regulation entities. Institutional integration should cascade down through an overarching national policy (superstructure), through interlocking supporting institutions (substructure), and incorporate business and local community participation (foundations).
- **Conceptual integration** refers to mindset. Physical and institutional integration cannot evolve without first establishing a new societal mindset of the role of water in the biosphere.

Before we describe New Water Architecture, we summarise the headline challenges facing water management that have emerged from the stresses and strains that are explored in Part II of this book and of the current architecture of water management described in Part III. There are significant opportunities to improve, refresh or develop new water policy and management in terms of economic, environmental and social benefits, and so we examine how water professionals can realise those opportunities. We believe that it is not just possible to create a New Water Architecture, but that this is critical if we are serious about addressing the problems of water stress and water inequality.

9.2 Challenges

9.2.1 Stresses and Strains

Part II of this book paints a simple picture of the global water balance; each year accessible water resources are renewed by around 15,000 km^3, and overall we only withdraw about a quarter of that (4,500 km^3). Even allowing for environmental flow needs, there should be sufficient water to support our Live, Eat and Consume demands for the foreseeable future. However, we cannot ignore the evidence of the deteriorating quality of waterbodies and of struggles to access water. The challenge is matching local demands in space and time with local and distant renewable resources. Our ways of doing this are currently combinations of physically transferring water, storing water, and making use of water from elsewhere by importing virtual water within goods and products.

Seven billion people exert multiple pressures on the water environment simply by where and how we live, the food we eat and the products we surround ourselves with. That number is expected to increase to 8.5 billion by 2030 and upto 10 billion by 2050.

We use water at home. Most of us are lucky enough to have it supplied directly to us; many others do not have that luxury. The access statistics are shocking: one in seven of us does not have access to safe drinking water, that most basic of requirements; and one in three does not have access to basic sanitation. For those of us who do not need to collect and carry it, only a small part of the water we use is actually vital (i.e. used for drinking, cooking and essential sanitation). The majority of our use is completely discretionary.

Our demand for energy stresses our water resources via its requirement for mineral extraction, power plant cooling and for the water required to grow biofuels. In turn, our demand for water also stresses our energy resources, as we need to use energy to move it and treat it. As more and more of us live in cities and our environments and lifestyles become more urbanised, our water and energy demands will continue to intensify. While this could be considered beneficial in terms of an ability to harness economies of scale, we will need to take care to avoid unwanted impacts from intensification, such as overreliance on particular water or energy sources.

There are a number of challenges to be met if we are to provide sufficient water to live for some 10 billion people well before the end of this century. Inadequate funding, under-pricing of water, misaligned institutional structures and irrational end-use allocation of water are at the root of these physical challenges.

Our demand for food drives the largest component of our personal water footprints yet arguably receives the least attention as a water-related challenge. While rain-fed agriculture still provides well over 50% of our food at the global scale, the increased agricultural productivity of irrigation sees this proportion falling year by year. Changes in diet are exerting more and more pressure on water as more people enter the middle class and can afford to choose to eat more meat, a much more water-intensive food than fruit and vegetables. Our demand for food would also be much less of a pressure if one-third of it was not thrown away or perished before it reached a consumer. Climate change is also influencing, both positively and negatively, crop growth and water availability in different regions of the world.

The challenges of achieving 'more crop per drop' are being addressed in some rain-fed and irrigated systems, while other countries are reducing pressure on their own water resources by importing food from others. However, the true cost of the virtual water trade in these circumstances is not reflected in the cost of the goods due to under-developed concepts of ecosystem services. Feeding people creates the toughest of the physical water challenges due to the immense volumes and fluxes of rainfall, green water, evapotranspiration and blue water withdrawals involved. Misaligned domestic food and trade policies, intentional and unintentional food price subsidies, health and diet, and food waste are at the root of these challenges.

Let's consider the pressure generated by our thirst to consume. The unrecognised role of water in national economic policy and the way in which big businesses view water are key concerns. There is wide variation between countries in the volumes of water abstracted for industry. Industrial and agricultural components have risen exponentially over the past 30 years as businesses have attempted to address the challenge of meeting demand for food and products. There are major variations in sectoral demand for water and the cotton industry is highlighted as particularly water intensive. Agricultural demand for water must not be simplified into 'water for food', and cotton is a prime example of an agricultural product that does not feed people. The shrinkage and disappearance of the Aral Sea caused by demand to irrigate cotton crops is just one example of the catastrophic environmental impacts that unchecked and under-regulated demand can have. It is not just agriculture making devastating impacts. Demand for minerals to produce the vast array of products we use exerts complex impacts on water resources. Businesses are beginning to assess water risks – physical, regulatory and reputational – in their supply chains but, incongruously, the same framework is not used by nations to assess their supply chains.

The water-related challenges facing nations and businesses to generate more economic growth are essentially the same challenges for securing water to Live and Eat. In the case of Consume, the challenges are heightened by the slow uptake of real corporate environmental and social responsibility, as opposed to the so-called 'greenwash' of politically correct but insincere corporate responsibility. Perhaps even more critically, there seems a lack of insight within government as to the multiple roles of water for public health, driving economic growth, and sustaining ecosystems.

9.2.2 Current Architecture of Water Management

In Part III (Chapters 7 and 8) we review how our water resources are currently managed, consider how that management is regulated and funded, and explor the interminable private versus public ownership debate. We identify many challenges, which we here characterise into three areas: **management and allocation**; **regulation and governance**; and **pricing and funding**.

The **management and allocation** of water resources has so many variations in approach across the world that it is difficult to summarise. The diametrically opposed approaches are:

- major publically owned and managed integrated water capture and transmission systems such as those in California; and
- traded water rights and privately owned discrete capture and transmission systems such as those in Chile and Australia, and emerging decentralised approaches to water infrastructure.

Issues arise around transboundary rivers, unsustainable projects, first-come-first-served water rights, competition for water resources, impacts of climate change, financial mismanagement, lack of technological development, and impacts of municipal and industrial pollution on the environment and people. IWRM was promoted in the late twentieth century as the solution to these challenges but, while the intent of IWRM is easily grasped and understood, experience has shown that putting it into effective practice has been patchy.

The **regulation and governance** of water resources and water use also has many variations from the well-regulated regimes in the OECD countries to less well-regulated frameworks in many developing and least-developed nations. Issues arise around excessive bureaucracy, corrupt decision making, inadequate monitoring, lack of funding and capacity in approach and reporting, over-regulation leading to stagnation, politicised agendas, excessively risk-averse policy, and lack of insight into the water-energy-food system. It could be argued that most of these issues stem from a failure of leadership, or more simply as a failure to take the issues seriously.

The **pricing and funding** of water resources and drinking water receive much media attention. Whether it is drought or flood, the stories are visually striking and can elicit commonly held views that water is free because it falls from the sky, or that keeping it away from property is the job of the government. There are many issues which arise including keeping up with population growth, the cost of capital for investment funding, dereliction of underground pipe networks, lack of full economic pricing (including of ecosystem services), provision of irrigation water at zero cost, profits of privately owned utilities, and affordability for the poorest sections of society. Most of these issues arise because there is a poor societal and political appreciation of the true value of water.

9.3 Opportunities

9.3.1 Emergence of Virtual Water Concepts in Water Policy

In Part II we describe the concepts of virtual water and water footprint. We show that the virtual water trade network has over 6,500 connections between countries and that the virtual water flow through the network is over 600 km^3 per year. But to what extent do national water policies recognise virtual water and water footprint? Breakout Boxes 9.1 and 9.2 discuss the cases of the UK and USA.

If national policies don't recognise virtual water, is there an opportunity for a better understanding of these principles to encourage the more rational use of water, locally, regionally and globally? This brings into play the concept of ethical water management, the fair trade of water that would otherwise consider the water context (such as the level

Breakout Box 9.1 UK case study: policy on virtual water?

The UK's water footprint of consumption is 75 km^3 per year of which 55 km^3 is imported, 20 km^3 from water-scarce nations (Water Footprint Network data). The UK has reached this position of heavy reliance on food and clothing imports over decades of population growth, rising standards of living and progressive standing down of national agricultural arable land.

Does the UK national water policy recognise that over one-quarter of the UK's water needs are aggravating environmental stress in other nations? Almost certainly it does not but, if it could, how could its policy respond to this situation? Could the UK grow more of its own food, thereby reducing the external component of the UK water footprint, reducing imported virtual water, and relieving pressure on water-scarce nations? In our view, this is feasible through a connected water and agriculture policy that seeks to reuse set-aside arable land and to stimulate an agricultural renaissance with associated economic and social benefit. We believe it is both feasible and environmentally responsible because most of the UK has lower water stress than many of the nations we currently import food from. The UK still has reliable winter rainfall and great potential to store more of that rainfall (from runoff) for irrigation use during summer. In an average year the UK sees 250 km^3 of rainfall, with 150 km^3 of water reaching the oceans after less than 10 km^3 of net blue water consumption. Put another way, it still has substantial green and blue water resources available, and currently only stores 90 m^3 per capita in reservoirs compared to over 2000 m^3 per capita in the USA (Aquastat data).

Would the UK government modify its water and agriculture policies to stimulate this change? Politically this would be difficult because it might make it incumbent on the government to invest in strategic water storage and transfer to facilitate additional irrigation. However, if water is accurately valued, the economic and social case for such investment is strong.

Breakout Box 9.2 USA case study: policy on virtual water?

The USA has an annual per capita water footprint of consumption (2800 m^3) double that of the UK (1300 m^3) (Water Footprint Network data) and yet has been able to both meet most of that from its own water resources and at the same time drive a healthy export industry based on virtual water. This is in part a consequence of a much higher available water resource per capita than UK, but is also a result of national water and export industry policies that have encouraged optimal use of the nation's water resources. It can be argued that the USA's export strength has been driven by water and that national policy has recognised the impact of virtual water.

In a letter to the President of the Senate and Speaker of the House in 1965, Lyndon B Johnson said:

'A nation that fails to plan intelligently for the development and protection of its precious waters will be condemned to wither because of its shortsightedness. The hard lessons of history are clear, written on the deserted sands and ruins of once proud civilizations.'

What Johnson was driving at was the role of water in sustaining ecosystems and driving the economy, and his vision helped the USA to achieve those twin goals. The interesting aspect of his statement was the use of the word 'development'. While the UK national water policy has focused on 'protection', the US government has actively encouraged development. Examples can be seen all over USA such as the Hoover Dam and the California North to South water transfer schemes.

While we can be sure that explicit use of the words 'virtual water' and 'water footprint' have never featured in US national or state water policy literature, they have been implicit in policies to stimulate water storage and transfer to expand agriculture and provide hydroelectricity.

of water scarcity) of those countries from which another nation imports virtual water. We see no evidence that this highly important issue is afforded consideration in trade negotiations.

We are not aware that any national government at this time makes explicit reference to virtual water in national policy. That said, Fulton *et al.* (2014) describe the emergence of virtual water concepts into state policy in California: 'In California, the Department of Water Resources has taken the step of integrating virtual water and water footprint concepts into a framework of sustainability indicators being developed for long-term state water resource planning.'

Most national water policies are concerned primarily with management and use of national water resources, with objectives for drinking water and sanitation coverage and standards, and with prevention and control of pollution of the aquatic environment. However, national policies relating to industry, agriculture, energy, and export and import will all impact virtual water flow, albeit unintentionally in most if not all cases. For example, a national policy of increasing self-sufficiency and food security through expanded agriculture might have significant implications for national abstraction of blue water for irrigation. The impact on virtual water flows would be to decrease imported virtual water. If instead the policy were to expand agriculture to drive food exports, then the impact would be to increase exported virtual water.

By not explicitly considering virtual water, national water policy may overlook the impact on water stress in another country. In Part II we illustrate how a number of developed nations are importing virtual water from water-scarce nations. But who cares? The importing nation receiving food and goods that are not being priced to include ecosystem services? The exporting nation who is driving economic growth though competitive advantage against other nations? The supermarket shopper in the importing nation who is pleased to be able to purchase food and goods at lower prices than home-produced items? In fact, those that care are those who suffer the local impact of increased water stress through lower water availability and environmental water quality, people who may not have a voice in their own nation and even less so in the importing nation.

As water scarcity deepens with population growth and climate change impacts, our view is that virtual water considerations will enter into national water policies. Already we are seeing water-scarce wealthy nations employing what have been called 'neo-colonial' agreements with land and water-rich developing nations to improve food security. There are many examples of this phenomenon in Africa. In fact, it can be argued that most of these agreements are based on a need to improve water security by guaranteeing that the virtual water flow will be available to them.

A debate continues to unfold about the usefulness of the concepts of virtual water and water footprint. Antonelli and Sartori (2014) conclude:

> Despite not being a policy tool itself, the virtual water concept can reveal aspects related to production, consumption and trade in goods which monetary indicators do not capture. Because of the ambiguity associated with the meaning of the virtual water concept, generated by its trans-disciplinary nature, its potential as an indicator for informing decision-making in water management and policy, as well as in commodity trade policy, still has to be fully appreciated.

Our view is that the potential referred to by Antonelli and Sartori (2014) is being realised and that we will see policies such as that reported by Fulton *et al.* (2014) emerge in other sub-national and national contexts.

9.3.2 Emergence of Multi-Stakeholder Approaches to Water Policy

Several global-scale initiatives have emerged in the last decades that have sought to facilitate more collaborative approaches to the development of water policy. For example, since 2006 the WEF and its members have been bringing the interrelated global risks of water supply and food shortages to the attention of policy makers. The WEF Water Initiative embarked upon its Water Partnership Project workstream with the objective of creating collaboration between government, development agencies, NGOs and WEF industry partners in regions of special interest to WEF members. The partnerships would seek to develop a flow of water-related projects with economic benefit that would be attractive to private sources of finance. In January 2010, WEF published a paper (WEF 2010) on partnership case studies in which the following principles were advocated:

- advocating the benefits of increased cooperation to water management;
- connecting stakeholders through diverse multi-stakeholder platforms;

- catalysing cooperative water interventions; and
- anchoring the process through coordination, facilitation, and incentivisation.

These principles provided the basis for the drive for more collaborative policy development, as evidenced by the coming together of the World Bank and a number of multinational businesses (some of them members of WEF) to launch the 2030 Water Resources Group (2030WRG). The group sought to develop a new fact base of potential levers and associated costs for addressing water scarcity; the ultimate objective was to provide tools that could be used in multi-stakeholder settings coming from the WEF partnership workstream. In November 2009, 2030WRG published its groundbreaking report 'Charting our Water Future: Economic Frameworks to Inform Decision-Making'. 2030WRG works as a public-private-civil society partnership, using a three-step approach (analyse, convene, transform) to stimulate multi-stakeholder partnerships based on the following guiding principle: 'The 2030 WRG brings transformative change to water resources planning by convening national multi-stakeholders platforms and structured processes – including key public decision-makers, concerned private sector champions and civil society representatives – who catalyse sustainable, rational, economics-based solutions to close the water supply demand gap.'

At the heart of the 2030WRG approach is the development of multi-stakeholder platforms from which to drive the 'transform' stage of water sector interventions. It is an approach that seeks to mobilise private sector and civic society effort and funds towards solving water sector challenges that traditionally would fall to the public sector. But this is not traditional 'private sector participation'; it is the search for innovative ways in which private-public-civic sectors can work together.

Since 2006, the approach has been applied in South Africa, Jordan, China and India, and is in the process of being applied in Tanzania, Peru, Bangladesh, Kenya, Mongolia and Mexico. In 2013 and 2015, 2030WRG published catalogues of case studies that illustrate how water scarcity is being addressed around the world, a number of which came from the 2030WRG multi-stakeholder approach.

Our view is that these multi-stakeholder approaches are already demonstrating speedier and more effective decisions about management of water than traditional, sequential approaches that start with public sector assessment and end with consultation. That is not to say that they are simpler to apply; in fact, they require a deeper understanding of stakeholder interests, engagement and coordination.

9.3.3 Reform of Water Policy as Opportunity

As we discuss in Chapter 7, reforming water resources and supply policy to ensure that the management of water quality becomes a fully integrated rather than subordinate component is a particularly important opportunity to bring about environmental and economic gains. This requires much more than simply acknowledging the need to supply safe drinking water. By maintaining the quality of water throughout the whole hydrological cycle (e.g. through the application of catchment land management practices and sustainable drainage systems such as green roofs) detrimental impacts on ecosystem services are reduced and physical infrastructure (such as water treatment plants) has to work less hard to provide the services we demand.

Specific examples of what we mean by water policy as opportunity include:

- In Orange County, southern California, aligned thinking helped the authorities responsible for water supply and wastewater management to integrate their respective strategies and invest in an aquifer storage and recovery scheme that stores treated wastewater underground for times of scarcity. In the 1990s, the water supply provider was becoming increasingly aware that traditional water sources would be insufficient to sustain future supplies. The sanitation authority, meanwhile, was faced with a US$ 200 million bill to build a pipeline to discharge treated wastewater to the ocean. Rather than focusing on individual solutions, the two authorities combined their strategies to develop the Groundwater Replenishment Scheme in 2003 (Circle of Blue 2014a).
- As an example of the benefit of a portfolio of more localised initiatives, Philadelphia is investing US$ 800 million over 25 years on green roofs, street side buffers and wetlands that provide clean water benefits at a cheaper cost than traditional systems (Circle of Blue 2014b).
- In the UK, years of water resources assessment are now being used to support a new abstraction licensing regime which seeks to allocate water in a rational and time-limited manner, replacing the first-come-first-served paradigm which has prevailed for decades (see Breakout Box 9.3).

In a more general sense, a new and modern approach to decentralisation is receiving much attention as a policy reform that recognises increased needs for resilience and local stakeholder engagement. It refers to transferring political, financial and administrative authority, including decision making and management, from central government to more local levels. While the process of decentralisation is widely advocated and is now common in many developing nations, the devolution of water-sector decision making is occurring with varying degrees of success, with many less-successful cases reflecting excessive central government control.

A lack of human resources can also be a constraint to successful decentralisation. While there is wide consensus on the importance of decentralising water and sewerage services delivery and expenditure management, this must be accompanied by recognition of the need to first improve the managerial and technical capacities of local authorities in question.

Breakout Box 9.3 Abstraction reform in the UK

In the UK there has been over a decade of water resources availability assessment that has matured the argument surrounding management of water abstractions. Realisation of the extent of the pressure that limited water resources are under has led the UK government to consider reforming the abstraction licensing regime. A 2014 Water Act includes a 5-year timeframe for the Secretary of State to lay a report in front of parliament on the progress of abstraction reform. UK regulators now have a legal basis on which to reform the abstraction licensing regime.

A cultural shift in attitudes towards abstracting water is continuing with a change in terminology. Licences may now be rebranded 'abstraction permissions', invoking a sense of temporary rather than permanent right.

But in centralised or decentralised policy frameworks, what criteria could or should be used to prioritise who gets to use the water? In the UK, reaction from the business community of water users is quiet; online forums and media indicate a few concerns that licences could be limited or even revoked. It may be that until the affected business community is actually confronted with new allocation recommendations, they will continue to be less than fully engaged.

Existing and future water policy practitioners have major challenges ahead of them to craft new and improved governance systems and regulatory models that will be appropriate for the twenty-first century and longer-term perspectives. Water resources planning and management today inevitably involve multiple goals or objectives, many of which may be conflicting. It is difficult, if not impossible, to please all stakeholders all the time.

While the traditional authorities may be sluggish in some cases, NGOs on the other hand have the same 'light feet' as business. They essentially operate without borders and are able to conceptualise the links between globalisation of water and the very real specific impacts on waterbodies, habitats and livelihoods. The future of local water management will need to be responsive to regional and global processes and pressures, and so the future structure of water management is likely to have to change to keep up with the 'new' dominant players (who will increasingly include NGOs) in decision making and implementation.

9.4 A Systems Approach to Water Management

9.4.1 Principles of Systems Thinking

One of the best-known works on systems thinking is *The Fifth Discipline: The Art and Practice of the Learning Organization* published by Peter Senge (2005), professor at MIT Sloan School of Management. It isn't the role of our book to describe systems-thinking concepts, but the following quotes from Peter Senge give an insight into how systems thinking can throw light onto complex interrelationships between apparently simple components:

- 'The smartness we need is collective. We need cities that work differently. We need industrial sectors that work differently. We need value change and supply changes that are managed from the beginning until the end to purely produce social, ecological and economic well-being. That is the concept of intelligence we need, and it will never be achieved by a handful of smart individuals.'
- 'Business and human endeavors are systems...we tend to focus on snapshots of isolated parts of the system. And wonder why our deepest problems never get solved.'

A way to understand systems thinking is to consider how a farm functions. A farm is a collection of apparently simple components such as crops, livestock, chickens, earthworms, soil, etc. Although each can be managed independently of the other, the experienced farmer understands that the farm is a system and that changing any single component may have unintended impacts on others. Systems thinking considers the looped, not linear, links between actions and consequences. As shown in Figure 9.1, these loops may be reinforcing (R) or balancing (B).

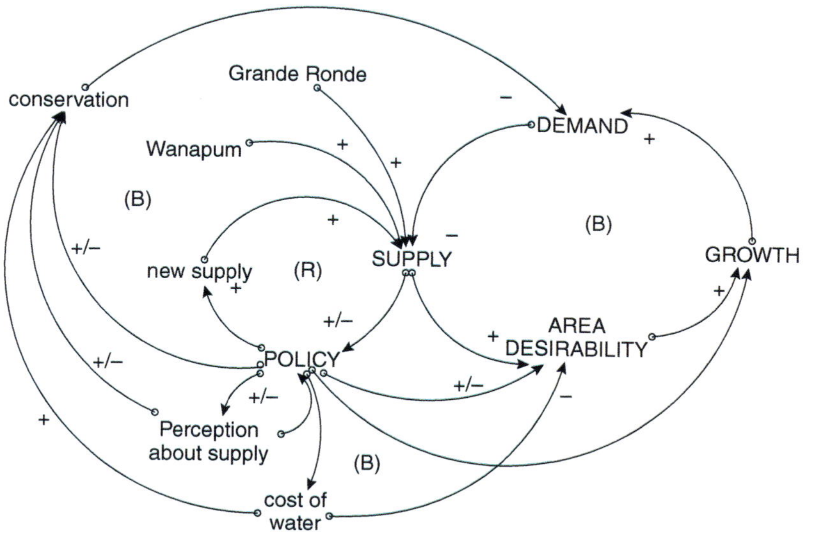

Figure 9.1 Example of systems thinking and feedback loops in water resources planning. *Source:* Beall *et al.* (2011). Reproduced with permission of MDPI.

It seems to us that unless a systems approach is taken to water management, we can never fully understand its role in the biosphere nor, more importantly, how we can manage water in the manner needed for future societal, economic and environmental benefit. To understand why a systems approach is needed, we need to appreciate the complexity of water fluxes and man's interactions with them at scale.

9.4.2 Integrated Management of Water at a Catchment Scale

The systems concepts of physically managing water at a catchment scale are illustrated in Figure 9.2. The figure represents a catchment, its surface and subsurface hydrology and hydrogeology, and fluxes and uses of water. It was conceptual representations such as this that underpinned river basin water management models such as those introduced in England and Wales in 1973. Unfortunately, the full benefits of this model were not realised by the time the model was re-cast in 1989 (Breakout Box 9.4).

The Global Water Partnership (GWP) was founded in 1996 to foster IWRM, which it defines as the coordinated development and management of water, land and related resources in order to maximise economic and social welfare without compromising the sustainability of vital environmental systems (Figure 9.3). The GWP definition of IWRM therefore introduces system components to the physical elements of Figure 9.3.

The GWP has pursued a strategy of bringing together partnerships all round the world based on these IWRM principles. But almost 20 years on from the bold step taken by the founders of GWP, we still see too few examples of IWRM in practice. For example, within the European Union, each of the Member States is charged by the Water Framework Directive with developing and implementing River Basin Management Plans. Progress towards this objective is extremely slow however, with national environmental regulators often struggling to identify the broad array of affected water

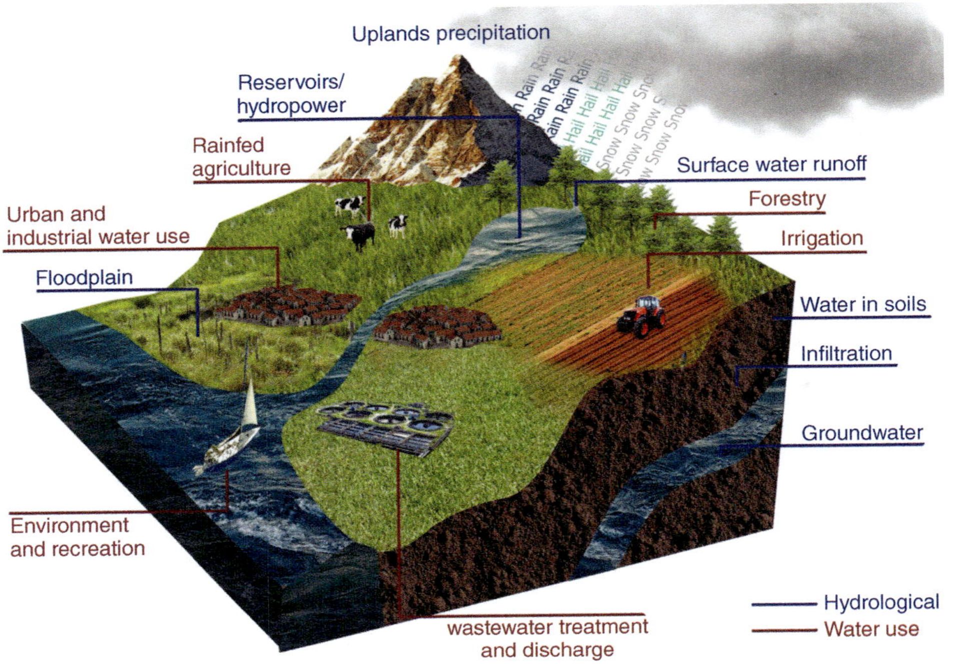

Figure 9.2 Catchment-scale management of river basins.

Breakout Box 9.4 Water management in England and Wales

England and Wales provides a good example of how the way water is managed can be changed. In 1973, water supply services were delivered by water boards, sewerage by municipal and urban councils, and water resources and flood control by river authorities before all these management functions were transitioned to water authorities (defined on the basis of river basins). It was another 15 years before the term Integrated Water Resources Management emerged, but in 1973 a transformation took place in England and Wales that was heralded world-wide as a new and visionary model for water management. A pre-existing water resources board supplemented the river basin water authorities, providing national oversight of water resources and being responsible for developing long-term initiatives that would serve the national good.

This model of IWRM was eventually broken up in 1989 when the UK government privatised drinking water supply and sewerage provision and, in doing so, separated responsibilities for water supply and sewerage from those of water resource and flood management. The latter were then passed to the National Rivers Authority, which later became part of the Environment Agency. It is arguable that the purity of the 1973 IWRM model was lost in 1989 and that, since that time, England and Wales have struggled to regain a river-basin-based approach to water management.

The current institutional landscape in England and Wales is one that does not easily lend itself to IWRM. While the subject area ministry Department of Environment, Food and Rural Affairs (Defra) and its daughter regulator the Environment Agency (EA) promote well-intentioned initiatives (e.g. Water for Life 2011), they have no government mandate to do anything other than encourage IWRM. The finance ministry (HM Treasury) wields the real power to allocate funds to nationally important IWRM interventions, but instead continues to allocate the majority of funds to transportation and, to a lesser extent, energy infrastructure. By encouraging the water supply regulator Ofwat to keep water prices as low as possible, the UK central government sends its citizens the signal that water has a low value. This is at odds with the expressions of intent of Defra in Water for Life (2011) around how society should have a better appreciation of the value of water.

Figure 9.3 The Global Water Partnership vision of IWRM.
Source: Reproduced with permission of Global Water Partnership.

stakeholders, to align their needs and to mobilise the collective initiative required to implement change.

In England and Wales the Department for Environment, Food and Rural Affairs (Defra) established a series of 'demonstration catchments' that employ some of the concepts behind IWRM, at the same time that several of the privatised water utilities set up their own 'catchment projects'. We are in favour of these initiatives, because they reflect an important step away from a siloed management mentality and support moves away from focusing purely on the role of 'end of pipe' solutions to the management of water quality. Demonstration projects are effective tools for reaching out and engaging with farmers and other landowners to take an active role in managing the catchment to support the water ecosystem.

While a voluntary and largely informal approach would make it easier to initiate this type of project, we believe that a more structured mandate could and should follow as wider stakeholders become more receptive. People don't like to be told what to do, but generally they appreciate structure and organised systems. We also believe that raising the status of this type of programme to a more formal level is needed to inject urgency into tackling the critical issues that are already upon us.

But why then, did England and Wales draw back from 'full blown' IWRM? Why is it taking so long for river basin management plans (RBMPs) to be adopted in Europe? Why have the efforts of the GWP been slow to realise the benefits that were hoped for?

We believe that there are two factors that have separately and collectively acted to impede the implementation of IWRM and catchment-scale management of water. Firstly, the spatial and temporal distribution of water means that it is unlike any other natural resource, making the type of management regimes that are used for other natural resources inappropriate. Secondly, it is much more complicated to allocate water to different users and for different purposes than it is for any other natural resource, partly because of its distribution fluxes but also the social, political and economic issues. People see water as a right, and that in itself is highly complex and emotive. Who should make the decisions? On what basis? Who is to say which stakeholder and which use of water is more important or valuable than another? Allocation of water rights is a highly contentious issue all around the world, and any attempt to reduce someone's access to water is likely to be met with strong opposition.

9.4.3 Cyclical Management and Allocation of Water Resources

The movement of water through the water cycle, both locally and globally, and through time makes it a unique natural resource. It also makes it the most complex natural resource to manage. We have established sophisticated land and satellite monitoring systems able to measure and transmit flux data from clouds, of rainfall, surface water flows, evapotranspiration, soil moisture and groundwater. However, despite the monitoring systems available to us, observations and reporting still tend to be *ad hoc*, focusing on a single or small number of fluxes in isolation. By failing to monitor the water cycle as a connected system we remain largely unaware of the specific nature of flux relationships, and there remain reinforcing and balancing loops that we are largely unaware of.

Temporal variations in the distribution of the fluxes are the result of even broader systemic effects. For example, global weather patterns drive daily and seasonal cloud and rainfall distribution; heat waves drive evapotranspiration spikes; El Niño and La Niña oscillations result in extreme drought and flooding events; and land-use changes drive responses in runoff regimes.

Given this complexity, it is hardly any wonder that rational allocation of access to the fluxes is missing or incomplete. While bold attempts have been made to allocate blue water in rivers, lakes and aquifers in a rational and integrated manner, even in this small part of the larger system there are few successful examples. The Murray-Darling basin case study described in Part II has in many ways shown a path forward; however, at the same time, it illustrates the systemic complexity of the challenges.

If we could imagine water as the complex system of fluxes that it is, whose distribution varies in time and space, then our New Water Architecture proposition is to improve understanding of the system to enable more rational allocation decisions. The concept here is that by using systems thinking, the economic models that drive allocation decisions will be more representative of externalities and that, in turn, management and regulatory frameworks will more effectively reflect the structure of the systems.

In the following three Chapters (10, 11 and 12) we describe each of the three areas of integration which comprise New Water Architecture: conceptual, institutional and physical.

References

Antonelli, M. and Sartori, M. 2014. Unfolding the Potential of the Virtual Water Concept. What is still under debate? Working Paper no. 74, The Center for Research on Energy and Environmental Economics and Policy, Bocconi University.

Beall, A., Fiedler, F., Boll, J. and Cosens, B. 2011. Sustainable Water Resource Management and Participatory System Dynamics. Case Study: Developing the Palouse Basin Participatory Model, Sustainability 2011. MDPI.

Carter, J. 1998. Postmodernity and welfare: when worlds collide. *Social Policy and Administration*, 32(2), 101–115.

Circle of Blue. 2014a. Orange County Recycled Water Systems Shows Importance of Collaboration. 21 November 2014. Available at: http://www.circleofblue.org/2014/world/orange-county-recycled-water-system-shows-importance-collaboration/ (accessed February 2017).

Circle of Blue. 2014b. Report: US Water Systems, Deteriorated and Slow to Change, Need New Policy – and Money. 15 October 2014. Available at: http://www.circleofblue.org/2014/world/u-s-water-systems-deteriorated-slow-change-need-new-strategy-money/ (accessed February 2017).

Fulton, J., Cooley, H. and Gleick, P.H. 2014. Water Footprint Outcomes and Policy Relevance Change with Scale Considered: Evidence from California. Springer Science and Business Media.

IWMI. 2006. Water governance in the Mekong region: the need for more informed policy-making. Colombo, Sri Lanka: International Water Management Institute (IWMI). 6p. (IWMI Water Policy Briefing 022), doi: https://doi.org/10.3910/2009.339.

Senge, P. 2006. The Fifth Discipline: The Art and Practice of the Learning Organization. MIT Sloan School of Management.

Water for Life. 2011. Water for Life. White Paper. HM Government, Department for Environment Food and Rural Affairs. Available at: https://www.gov.uk/government/uploads/system/uploads/attachment_data/file/228861/8230.pdf (accessed February 2017).

World Economic Forum. 2010. Innovative Water Partnerships: Experiences, Lessons Learned and Proposed Way Forward. Available at: https://www.weforum.org/reports/innovative-water-partnerships-experiences-lessons-learned-and-proposed-way-forward (accessed February 2017).

World Partnership Program (WPP). 2010. World Partnership Program (WPP) annual report 2010: driving change in water. Available at: http://documents.worldbank.org/curated/en/279221468157790439/Water-Partnership-Program-WPP-annual-report-2010-driving-change-in-water (accessed February 2017).

10

Conceptual Integration

Chapter 9 introduces how water, people and the biosphere interact with each other as a system. It describes how the spatial and temporal distributions of water complicate the system, and it explores some of our various attempts at managing the system.

For most of human history the system was simple; all people really had to worry about was dealing with the temporal nature of water availability. Then, as now, water became available and unavailable cyclically, with occasional unexpected droughts or storms to contend with. As population has increased, moved into new areas and concentrated in others, we have placed this system under stress and introduced new spatial relationships to contend with, moving water from source to demand in ever more elaborate water transfer projects for example. In our globalised world, we now have the ability to exploit water resources far from the point of demand and transfer water virtually through national, regional and global trade networks.

In a New Water Architecture, water management embraces a systems-based approach that combines a suite of integrated initiatives that:

- derive greater overall benefit than would be the case if initiatives were pursued in isolation;
- are in themselves low-regret actions (actions that derive benefits even if factors change); and
- build resilience into those water systems that support our ecosystems and biodiversity, as well as those that serve society directly.

While integrating physical and regulatory infrastructure is critical to achieving these goals, more fundamentally, society must recognise the value and role of water. A New Water Architecture also demands, perhaps more controversially, that politicians and decision makers urgently change how they view the role of water in the lives and livelihoods of the citizens they serve.

This chapter presents the first of the integration themes of New Water Architecture, that of conceptual integration. It does so by exploring the value of water to mankind and to the biosphere. In so doing, it makes the case for society to re-evaluate how it views and appreciates water and, in turn, the way that it should be managed.

Water Resources: A New Water Architecture, First Edition.
Alexander Lane, Michael Norton and Sandra Ryan.

10.1 Societal View of the Value of Water

10.1.1 The 'Free' Resource

How often when talking to friends and family do we hear 'water should be free because it just falls from the sky'? This perception drives a societal view that water has a low *value* and therefore that its *price* should also be low or even zero. For centuries politicians have understood that there are "few votes in water"; to suggest that investment is needed in water is tantamount to proposing to spend tax revenue on something that is 'free'. Consequently, even when we are forced to face up to the reality of water scarcity, politicians urge their finance ministries and regulators to ensure that the price of drinking water is kept to a minimum.

10.1.2 Price Signals in Drinking Water Supply

The terms 'value' and 'price' tend to be used interchangeably by society. In a perfectly functioning market-driven economic world, the value and price would be the same number. In reality however, the two are often far apart. This is especially the case for water.

Let's consider the average price for water and sanitation paid by UK citizens served by privatised utilities: around £3.50 per cubic metre. Does this mean that one cubic metre of water has a value of £3.50? Sadly not. The price paid in this case is a number generated by assessing the costs (as defined by an economic regulator) associated with transporting water from its source to households, treating it to a standard stipulated by other regulators, and then collecting and treating the resultant wastewater to stipulated standards (by another regulator). The price does not reflect the intrinsic value of water as a scarce resource in terms of what it may be worth in an open market, or of its worth to society as a critical enabler of ecosystem services.

Furthermore, as of 2016 only about half of all households in England and Wales have a water meter installed, and therefore many have no financial incentive to use less water. The price that they pay is linked to the theoretical value of their property, an approach that encourages people to think of the 'water rate' as just another tax, rather than a payment for a service or good received. This perception reinforces political desires to keep water prices as low as possible.

In an Australian study (TruCost and Yarra Valley Water 2013), researchers demonstrated that the price of drinking water greatly underestimates the value of water. In that study, the value of water was estimated by the benefit it provides to society through ecosystem services, and this was shown to vary temporally according to the degree of water scarcity. Even during times when water was relatively abundant, the researchers demonstrated that its value was considerably more than the price paid by drinking water customers. However, when water was scarce, its value was determined to be more than four times the price paid. The report concluded that:

> ... water prices are typically related to the capital required to supply water and do not reflect the 'true' value of the resource to society. Decisions around water use usually only consider the value of water in a monetary sense through its direct uses but the non-monetary value, such as what it is worth to society or the environment as a result of indirect uses, can be considerable compared to the price paid to consume water resources.

One of the major challenges for water professionals and stakeholders is choosing between different options to improve water resources, secure supplies or to enhance environmental conditions. All options have a long list of pros and cons (benefits and dis-benefits, or costs) and, inevitably, there are always tradeoffs. There are many different decision-making models to choose from, but an assessment of the net cost/benefit usually has a major influence on the final decision reached. The problem is how to take account of the things that people value but which are not easily quantifiable? For example, people value the natural character of a landscape and therefore to them there is a cost impact of flooding it to build a reservoir. Conversely, other people value the recreational facilities that a new reservoir can offer. These are different issues to valuing the security to supply that a reservoir can provide, or the reduced need to abstract from an unsustainable river or groundwater source. Even defining these types of values can be difficult, but in recent decades this concept of ecosystem services has gained traction.

It is not just large engineering options that have unquantifiable or non-monetised 'values' associated with them. Making cost-based decisions on demand management options is also made difficult if factors such as the level of disruption that digging up highways to repair or replace leaking pipes will cause are taken into account. The main point here is that economics usually play a major role in decision making, and economists like dealing with quantifiable monetised values. Regulators and water utilities around the world have begun to develop tools to make it easier to quantify qualitative values; however, in recent years, acceptance of the need to adequately measure ecosystem services into account has driven international initiatives to incorporate non-monetary valuation techniques (Kelemen *et al.* 2016). These include the Millennium Ecosystem Assessment, The Economics of Ecosystems and Biodiversity, and the Intergovernmental Platform on Biodiversity and Ecosystem Services. These and other initiatives are moving decision makers towards better appreciation of the role of water in the natural environment, particularly in water-scarce areas. Such factors should be included in the price that people pay for water, to truly reflect its value (particularly for volumes that go beyond essential human needs). Unfortunately, despite the obvious need to take account of unquantified issues, non-monetary valuation techniques remain fairly arbitrary and subjective (Seppelt *et al.* 2011), thereby diminishing the benefit of their application.

10.1.3 Price Signals Related to Water in Food and Other Goods

Above and beyond the price we pay for our direct water supplies, it is just as important to consider whether the true value of water to society is reflected in the price we pay for things that we eat or otherwise consume. Are people aware of the proportion of the price they pay for food that could be attributed to water? A study published by the Business Coalition of The Economics of Ecosystems and Biodiversity (TEEB 2013) observed a failure to account for the value of water in the pricing of food grown in southern Asia. Taking wheat as an example, while the water consumed in growing the crop was estimated to have an annual natural capital value of US$ 214 billion, the revenue from the wheat sales was estimated at only US$ 32 billion.

Take the UK as another example. It's one of the largest importers of virtual water embedded within food and other products (75 Gm^3 per year) and almost the largest importer of virtual water from water-stressed countries (20 Gm^3 per year). The unpalatable truth is that UK citizens (among many others) enjoy imported food at low prices at the expense of environmental damage in other countries. Put another way, the value to the environment of the water used in the growing and processing of that food has not been fully accounted for in its price.

As awareness of the natural capital value of water and its role in supporting the potential revenues of so many goods begins to grow, so too does an awareness of the risk that poor water management poses to the availability of cheap food and other goods. As a result, more and more companies are seeking to make the water supplies that support their production processes more resilient. This urge, for example, might encourage micro-chip industries to relocate away from traditional production areas to areas where water is more abundant.

As water becomes more commoditised, those responsible for making decisions on how to allocate water (e.g. a decision over whether to allocate water to grow food for overseas supermarkets or to manufacturing) will feel pressure to support the user that is prepared to pay the most for the resource. Ethical, economic and political issues affecting how water is allocated are complex and vary considerably around the world. In our globalised society, this issue will continue to create difficulties for how we manage our shared resources.

10.2 Water as an Under-Valued Resource: The Consequences

10.2.1 Profligacy

The low value that society currently ascribes to water encourages politicians to either deliver drinking water for 'free' or at prices which do not reflect its true value. It also means that the role of water in producing goods is ascribed a very small value or ignored altogether, thereby maximising the interests of business and macro-economic productivity. As a consequence, water is often not used carefully but is used with profligacy. Using more than is needed carries little or no additional cost, and wasteful behaviour carries with it little or no concern or conscience. In many countries this practice has led to environmental degradation, for example by depriving ecosystems of water or delivering polluted wastewater into them.

10.2.2 Poor Water Management and Decision Making

The profligacy exhibited by end-users of water is a consequence of society having an incomplete view of the value of water to society. There is however a more insidious consequence of the perceived low value of water, and this is observed in the way in which water is managed by national and provincial governments and by regulators intended to be independent.

We have described how the ideals of IWRM in England and Wales have been hampered by government approaches to the provision of water supply services. It is interesting to contrast this approach to that adopted in Scotland where responsibility for drinking water and sewerage services has remained in public hands. In 2010, the devolved Scottish Parliament published a forward-thinking 'Hydro Nation' concept that considered water as a potential driving force for the economy, as long as it could be managed in an integrated manner. The report set in motion means to attract to Scotland those industries that could use water sustainably.

The Scottish Hydro Nation concept has few, if any, counterparts around the world. Most governments and their ministries struggle with the conundrum of whether water is: a public health service (as per UN Resolution 64/292 on water as a human right) which should be provided free or at cost; an economic good which should be allocated through unhindered market forces; or a natural capital resource for the maintenance of biodiversity and ecosystems. It is all of these things of course, and this is what makes the management and regulation of water so challenging. The awareness of governments of the fundamental role of water is, on the whole, limited, and this means that water is often seen as a much less important resource, service or infrastructure than, say transportation or energy.

Because of these governance constraints, we see that water is all too often managed and regulated in the silos of other sectors. Water management takes place from a platform which undervalues water and which fails to appreciate its wider roles. The power of IWRM and effective decision making is therefore compromised.

10.3 Moving to Conceptual Integration

10.3.1 A New Appreciation of the Role and Value of Water

In a modern, consumer society it could be argued that appreciation of the true value of water is unlikely unless price signals are sent and received. In the case of drinking water this sounds simple; however, there are some barriers that often get in the way.

- Water metering is not universal. Many people do not have water meters, have no idea how much water they use and care even less. Changing this situation has emotive and political ramifications, and it also requires investment to install, maintain and monitor the equipment.
- We actually have very little information describing how much of the water that people consume at home is used for drinking and basic hygiene. This is a major problem because it is only that component of the total water use that relates to the basic human right and that, as such, could be argued should be delivered free of charge or 'at cost'. Opinions on what constitutes essential versus discretionary water use are subjective, and this makes it exceptionally difficult to define the dividing line. Quantifying that threshold is even more challenging as this will vary from person to person and will change over time in response to external factors such as weather and lifestyle.
- At the current time there is very little discussion that the price people pay for water could be determined by applying a 'real time' scarcity factor. This would be akin to the ever-fluctuating price people pay for petrol or diesel in response to changing oil prices. The lack of any real discussion on this precludes any moves towards its acceptance.

Notwithstanding their size, these barriers can be overcome with political will and leadership and with appropriate water pricing structures. Volumetric (sometimes known as progressive, incremental or rising-block) tariffs provide one such mechanism and these can be supplemented by measures that ensure affordability for the most vulnerable sections of society, often termed 'social tariffs'.

Economic theory tells us that if market-based principles are applied to all users of a given good, in this case water, market practice should ensure their effective allocation. In turn, this should mean that decision makers are better equipped to develop water policy that is optimised for all those resource sectors that require water.

Unfortunately we don't live in a world that fits this idealised economic model. We have real people and organisations determining how much water different sectors need, the quantity and quality of water required to sustain biodiversity and ecosystems, and the value to society that those ecosystem services provide. It's incredibly easy to make flawed or suboptimal decisions either because of incomplete information or knowledge, a predetermined agenda or simply poor judgement. Unfortunately, poor or suboptimal water management decisions are commonplace and we have to deal with the consequences.

A reinvigorated societal appreciation of the entrenched problems affecting water management would put in place the foundation necessary for a conceptual integration of the role and true value of water.

10.3.2 The Role of Water Professionals

Back in 2009, the UN's third triennial report on Water Development painted a stark picture of increasing water scarcity and environmental degradation. Although much of the report was factual, its single most powerful message was an emotive one, directed at water professionals – engineers, scientists, economists and lawyers – to 'get out of the water box' if messages about water scarcity are to be truly understood by society and politicians. The 'water box' referred to (see Figure 10.1) is a comfort zone of intellectual debate and discourse that all water professionals are familiar with. This domain includes our day-to-day work of understanding and solving water management issues in terms of both quantity and quality using standard practices. It also includes what may controversially be described as a 'circus' of national and international conferences populated by a high proportion of the same faces essentially debating and recycling well-worn topics.

To 'get out of the box' means to enter into the less comfortable world of politicians, policy makers, corporate business and influencers. In such an environment we face a much more challenging water management problem, one that includes more complex and messy relationships than those we apply our traditional practices and procedures to. Pure scientific and even economic logic is confronted by the realpolitik of decision making at the highest levels of government and industry.

For most water professionals, the space 'outside the box' is one seldom seen, heard or experienced. It is a world that we may not understand or respect; how often do we say or hear phrases such as 'the decision making is out of my hands'? If however we are to foster a new appreciation by society and politicians of the fundamental value of water, if we are to see conceptual integration emerge, then water professionals have a pivotal role to play. It is only when conceptual integration is established that a compelling case for improvements to physical and regulatory water management frameworks can be made.

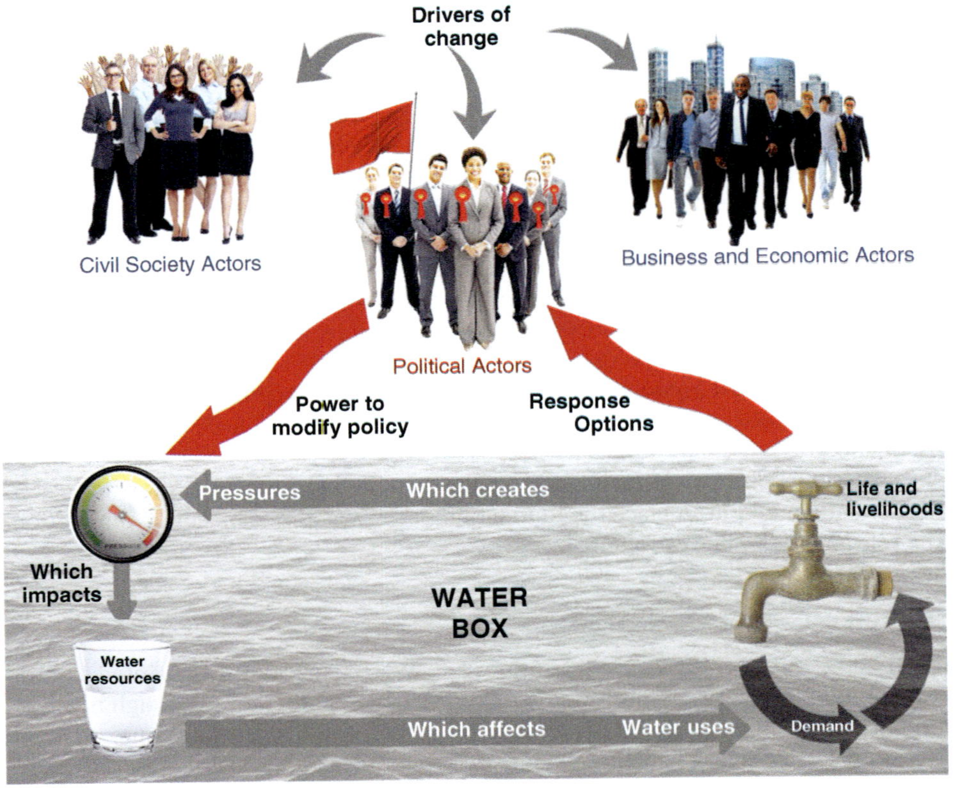

Figure 10.1 The 'Water Box'.
Amended from UN (2009).

References

Kelemen, E., García-Llorente, M., Pataki, G., Martín-López, B. and Gómez-Baggethun, E. 2016. Non-monetary techniques for the valuation of ecosystem service. In Potschin, M. and K. Jax (eds) OpenNESS Ecosystem Services Reference Book. Available at: http://www.openness-project.eu/library/reference-book/sp-non-monetary-valuation (accessed February 2017).

Seppelt, R., Fath, B., Burkhard, B. et al. 2011. Form follows function? Proposing a blueprint for ecosystem service assessments based on reviews and case studies. *Ecological Indicators*, 21, 145–154.

The Economics of Ecosystems and Biodiversity (TEEB) and TruCost. 2013. Natural Capital at Risk: the Top 1000 Externalities of Business. TEEB, London, UK.

TruCost and Yarra Valley Water. 2013. Valuing water to make drive effective decisions. Yarra Valley Water, Australia.

United Nations (UN). 2009. Water in a changing world. World Water Development Report No. 3 (WWDR3). Available at: http://www.unesco.org/new/en/natural-sciences/environment/water/wwap/wwdr/wwdr3-2009/ (accessed February 2017).

11

Institutional Integration

Chapter 10 highlights the urgent need for society to embrace a more holistic and integrated understanding of the importance of water. Some individuals, social groups, businesses and governments already subscribe to this belief, and there are encouraging signs that recognition of the true value of water is growing. The much-needed step change in our perception remains elusive however and, despite the rational arguments for a New Water Architecture presented in this book, it should be remembered that humans are highly emotive, prone to acting in self-interested ways that might not be to the benefit of the collective community. This tendency can be seen in the overconsumption of natural resource pools such as forests, mineral resources, soils and waterbodies, and it is because of this 'tragedy of the commons' that the success of a New Water Architecture relies on institutions and organisations to help guide our collective actions.

Institutions are defined by North (1990) as the 'rules of the game', the formal requirements and informal expectations of behaviour that guide human action by providing political, economic and legal governance. Like institutions, organisations such as governments and businesses also influence human action. However, because the aim of organisations is typically to 'win the game', the relationships between organisations and institutions are often blurred and complex.

This chapter considers the roles and responsibilities that institutions and organisations can play in achieving a New Water Architecture. In doing so, it hopes to demonstrate that by focusing on five key enablers (presented in Figure 11.1), organisations will be more likely to adopt outward-looking and long-term water management strategies that deliver a competitive advantage, while also helping to decouple economic growth from the damaging overconsumption of natural resources.

11.1 Requirements for Delivering Integrated Solutions

Solving sustainability issues that transcend resource sectors presents a fundamental challenge to the conventional structure of most governance structures and organisations for two primary reasons:

1) it requires coordination between different levels of governance (vertical integration); and
2) it requires collaboration between different stakeholders (horizontal integration).

Water Resources: A New Water Architecture, First Edition.
Alexander Lane, Michael Norton and Sandra Ryan.

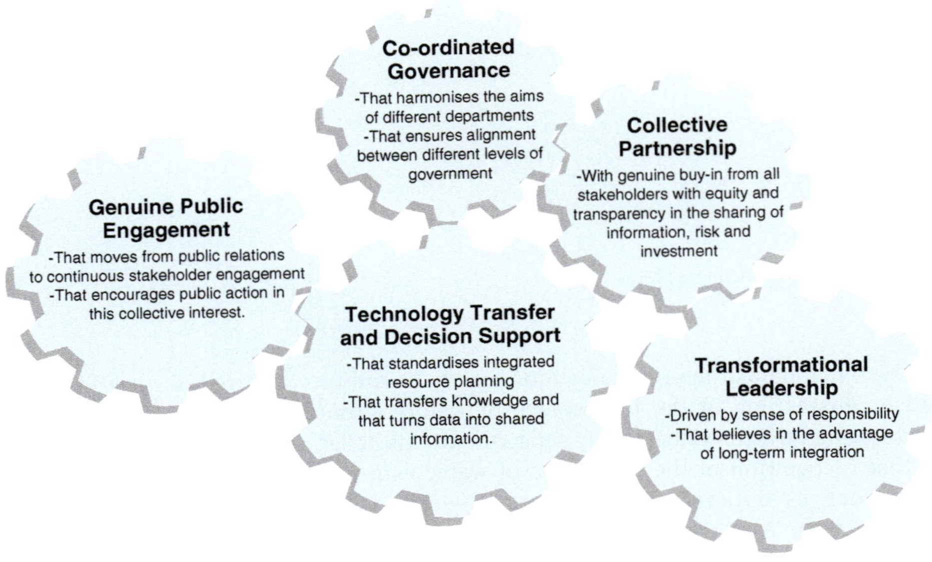

Figure 11.1 Enablers for a new water architecture.

The need to consider the various cultural values and perceptions of water held by different stakeholders is also crucial. For example, how should we define the distinction between water as a social good and as an economic good? Through which paradigm should water be managed: the traditional hydraulic approach; the water-centric ideals of IWRM; the 'nexus' approach that stresses the interconnections between all natural resources and human activities; or a New Water Architecture which stresses the critical role of water across all sectors by focusing on conceptual, institutional and physical integration?

11.1.1 Vertical Integration

Figure 11.2 illustrates the vertical levels at which water is governed. While the subsidiary principle supported in much water management literature advocates managing water at the lowest appropriate level of governance, recent recognition of the importance of globalised systems highlights the need for water to be managed at multiple levels simultaneously. In turn, there is a critical need for the actions taken at each governance level to be synchronised and based on common goals.

Global organisations such as the UN and UN-Water (its coordination body on water resource matters) can help to set these overarching agendas; however, their limited authority and weak legal backing limits their impact. For example, ratification of the 1997 UN Convention on the Law of Non-Navigational Uses of International Watercourses (which addresses transboundary water governance) by the requisite 35 UN members took 17 years, finally entering into force in August 2014. Obstacles like these mean that water governance often currently lacks a master plan, or set of guiding principles, that all levels of governance buy in to.

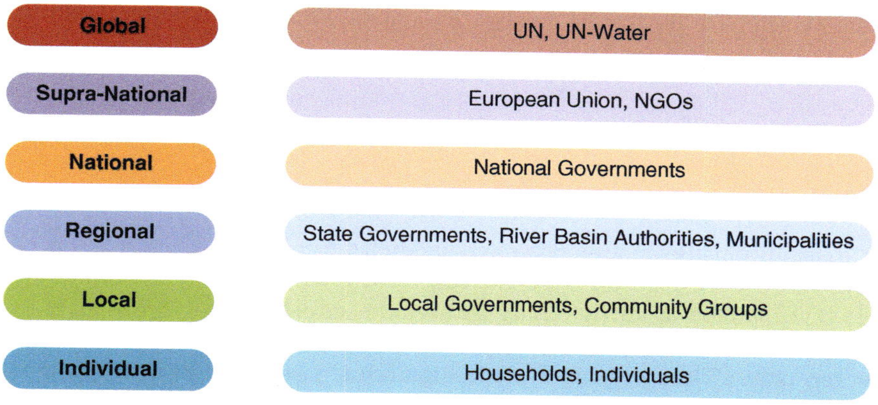

Figure 11.2 Multi-level governance.

In theory, the Sustainable Development Goals and the preceding Millennium Development Goals provide an existing avenue through which to achieve this alignment. In reality however, blurred lines of responsibility and/or duplication of responsibility between different tiers of jurisdiction and administration act to impair the adoption of common approaches to water resource management, in turn reducing the efficiency and effectiveness of their outcomes. This issue is especially magnified in cities where accountability for water management is often shared between local, regional, national and even supra-national government bodies, as well as with private utilities and service providers.

Other natural resources that interact with water may also be managed at different vertical governance levels. For example, while water is often regulated at a range of levels, predominant responsibility for energy policy almost always rests with the national government.

11.1.2 Horizontal Integration

The physical nature of natural water resources takes no account of political or administrative boundaries and so successful water management relies on neighbouring authorities adopting complementary strategies and policies. Intra- and international cooperation is therefore vital to sustainable water resource management (as it is to the management of any natural resource) and, where it is lacking, environmental impacts and conflict can result. The depletion of the Aral Sea, of the aquifer underlying the North China Plain and of the Amazon rainforest, as well as geopolitical tensions along the Nile, the Mekong and the Indus are all well-known examples.

Different economic sectors also exhibit different patterns of horizontal integration. These sectors, energy or agriculture for example, each have their own vested interest and perceive water in a different way, for example as an economic or social good, as a source of profit, as an ecosystem service or as a political lever. Like the goal of achieving vertical integration of governance and administration, horizontal integration requires an overarching agenda that acknowledges these different values and perceptions, but that nevertheless manages to identify common goals around which collective actions can be coordinated. The question of what level of governance this agenda should be set at binds the concepts of horizontal and vertical integration into a complex web of influence.

11.2 The Challenges of Delivering Integrated Solutions

11.2.1 The State of Play

The need to institutionalise a more integrated and holistic approach to water resource management has been recognised for some time. The UN's 2002 World Summit for Sustainable Development in Johannesburg secured the agreement of 193 countries to develop 'integrated water resource management and water efficiency plans by 2005'. By 2012 however, only 67 countries had made 'significant progress' towards this goal (UNEP 2012).

Historically, private organisations have been equally slow to evolve. The 2013 CDP Global Water Report (CDP 2013) found that while 70% of approximately 600 global businesses identified water as a substantive business risk, the vast majority responded by searching for inward-facing solutions only, setting targets for direct operations (63%) focused on water recycling and reuse. Only 6% of companies set goals related to community engagement and just 4% set targets for their supply chains, despite the fact that more than one-third acknowledged water-related supply chain risks. This reluctance to break away from 'business as usual' is symptomatic of a general failure to recognise that the collective management of water and other natural resources is increasingly the best and only means of establishing a competitive business advantage.

11.2.2 Causes and Barriers

Organisations typically function by creating departments with specific remits, for example education, energy or environment. These departments then develop and attempt to deliver on isolated targets; their focus is turned inwards with the result that cross-department data collection, sharing and strategising is rare. Similarly, budgets are developed and allocated on a departmental basis, encouraging the promotion and defence of self-interest and further discouraging cooperation.

Traditional or engrained political standpoints can also hold back otherwise sensible ideas. The reluctance of many governments to adopt metering of domestic water supply is one such example, despite widespread support for their use among many water managers. The historic strength of certain stakeholders in water management also frequently acts as an impediment to change. As important economic engines, the vested interests of large corporations are often preferentially addressed in water management planning.

Planning timescales create an additional and critical constraint. Working to deliver on commitments over short-term business and political cycles means that longer-term initiatives, although having the potential to realise a greater and more sustained benefit, are far less likely to receive investment. Because of factors such as these, the tendency of organisations to depend on entrenched practices and pathways often limits the range of options considered when deciding on future water resource management. It often takes a severe external shock, such as an acute drought or flood, to break this inertia and path dependence.

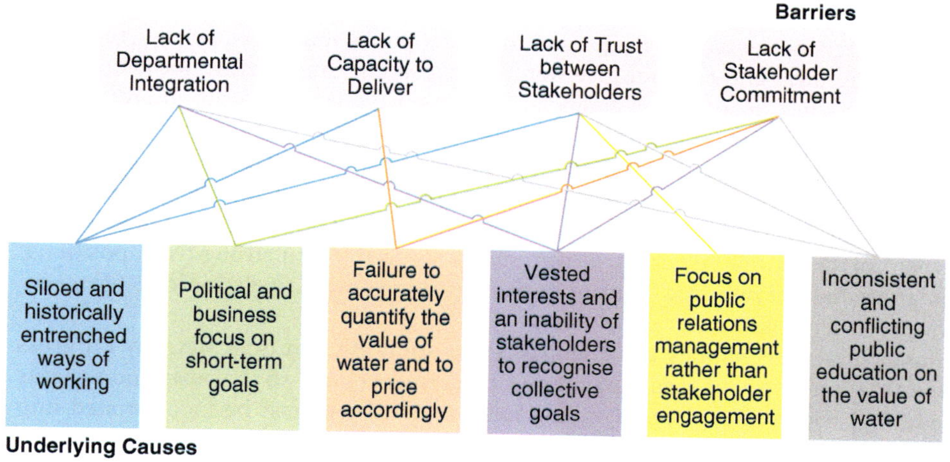

Figure 11.3 Institutional and organisational barriers.

As well as increasing the likelihood of inefficient natural resource management, this approach also prevents the transfer of knowledge and skills between professions, as well as between organisations in different regions and countries. It is a disappointing reality that the solutions to many of the water management challenges being faced by developing countries have already been investigated, designed and successfully implemented in developed nations.

Genuine buy-in to collaborative natural resource management is also held back by the lack of public trust in authority, a situation that often reflects public misunderstanding and exposure to conflicting messages. For example, while 90% of scientists agree that anthropogenic greenhouse gases are the dominant cause of recent global warming (Verheggen *et al.* 2014), the opposing public communication strategies of different organisations means that only 11% of the UK public is aware of the strength of this consensus (ComRes 2014). There is also a crucial need to acknowledge and understand the differing values and perceptions of water held by different stakeholders. Without an understanding of these values, partnerships have no reference point. Social scientists are hugely underrepresented in natural resource management and should be recruited in order to narrow these gaps in understanding.

Figure 11.3 summarises the barriers to the implementation of a New Water Architecture and presents some of the factors contributing to their influence.

11.3 The Role of Governments

Governments impart a particularly strong influence on the actions of their citizens. They have a responsibility to set the vision and to create the enabling environment that allows sustainable initiatives to flourish. To deliver on this responsibility, governments must carefully devise policies that inspire their citizens to feel part of a collective movement while also encouraging businesses to invest, innovate and succeed, and thereby provide the resources (both financial and human) to improve public services. At the same time, businesses, social groups and the wider public must acknowledge the benefits that long-term and collaborative water management can provide, not just to their own interests (e.g. a return on financial investment), but also to the interests of all stakeholders.

As identified in Chapter 10, the first challenge for governments is to devise and implement a means of accurately valuing and, in turn, pricing water. This task is vital not just to the success or otherwise of New Water Architecture, but also to the sustainable management of all natural resources. Governments (even those in strong fiscal positions) that ignore natural resource limits in decision making are putting their long-term economic security at risk.

Market-based economies are a global reality (only Bhutan does not rely on gross domestic product as a measure of country performance) and so the costs and benefits of altering, depleting and/or degrading natural resources must be incorporated into decision making. This is an exceptionally complex task that requires the pooling of expertise from a broad range of professions. Economists must engage with civil servants, social scientists, engineers and water resource planners in order to identify the most effective means of accurately monetising the value of water.

Assuming that this task can be achieved, market forces should optimise the efficiency with which water is allocated between its competing users. Unfortunately, for a variety of factors, it is virtually impossible to establish a perfectly functioning market. Common distortions such as unequal access to information and monopolies mean that government intervention is often required to encourage sustainable outcomes. Compounding the complexity of this issue, many government responses to market problems (e.g. subsidies for particular industries) often inadvertently act to broaden rather than narrow market failures.

Table 11.1 lists a variety of mechanisms that can be used by governments to encourage a more integrated approach to water resource management.

It is vital that governments select the portfolio of available water resource management measures (some of which are listed in Table 11.1) that is most suited to their situation. Governments should develop an overarching water resource strategy (potentially even a National Resource Policy) that acknowledges the water-related linkages between all natural resources. This would allow the better-informed resolution of tradeoffs between the management of water, food, energy and land, securing better social outcomes, environmental performance and ultimately economic health and geopolitical security as a result. Factors that such a National Resource Policy should address include:

- recognising the true value of water and committing to developing means to accurately monetise it;
- acknowledging that water use is both essential and discretionary, and thereby encouraging management of the latter component through market mechanisms;
- ensuring that policies which address the management of one natural resource effectively consider the implications for and interactions with all others.

In nations where an overarching resource governance framework is lacking, natural resource policies are more likely to have unexpected, unintended and potentially detrimental consequences. For example, the natural resource management challenges of India are discussed in a series of reports by Circle of Blue (2013).

Table 11.1 Government mechanisms to encourage improved water management.

Policy and regulation	
Water entitlements	Allocating entitlements or licenses for surface and groundwater resources is an effective means of distributing water between competing demands. In the UK, the water rights system functions by establishing the total sustainable abstraction level for a given catchment (giving priority to environmental needs) and then accepting applications for licenses up to the remaining amount. In California and Texas, 'senior' rights holders are preferentially allocated water over their 'junior' counterparts, even though many of these rights date to the early twentieth century. Allocations are also complicated in some places by rules which dictate that water can only be used once before it must be returned to the water owner. Such an approach inhibits opportunities for sensible water management practices such as water reuse. Other countries and regions have attempted to allocate water on a financial basis, the Murray-Darling Basin in Australia for example. While successful in several respects, the challenge of incorporating an allowance for the environment while also balancing agricultural demands has proved challenging (see Chapter 5 for further details).
Trade in virtual water	Global trade presents governments with choices over how self-sufficient to be in particular commodities (and therefore how much domestic water resources are used in their production) or, alternatively, how reliant to be on imported goods. For example, Egypt imports more than half its food requirements, saving more than a Nile's worth of water each year in the process (Roth and Warner 2008). Relying on virtual water might also expose countries to risks such as food price volatility and the impacts of natural disasters in foreign countries.
Land-use policy	Land zoning policies relevant to water and natural resource management include: • encouraging industries with similar resource needs to co-locate; • avoiding locating critical infrastructure in flood risk areas; and • maintaining blue corridors to minimise the disruption caused by flooding.
Waste policy	Regulating how waste is managed can encourage a shift in the perception of 'waste' to one of a 'resource'. The European Commission (2014) is considering a range of mechanisms to increase resource efficiency. This includes setting targets, such as a 70% reduction in municipal waste landfill by 2030, and mandating design standards to phase out inefficient products.
Product rating/ labelling	Product rating or labelling is fraught with complexity, but can be highly successful in raising consumer awareness of sustainability issues. Given the complexities of water resource management and of virtual water flows, developing accurate impact labelling related to water is extremely challenging. Simple labels that indicate the amount of water used to make a given product and the relative water scarcity of the source country would help to raise public awareness.
Financial mechanisms	
Taxes	Taxes help to generate the finance for public projects and can help to optimise sustainable outcomes. For example, a focus on the 'polluter pays' principle and the taxing of landfill has been highly effective in reducing landfill rates in Europe (European Commission 2014). Tax breaks can be effective when encouraging sustainable practices. These could include lower taxes on products that have been recycled, or reducing the taxes businesses pay on research and development activities. Care must however be taken to avoid creating false economies that may otherwise collapse when the tax break is removed.

(Continued)

Table 11.1 (Continued)

Financial Mechanisms	
Subsidies	Given the basic human need for and right to safe water, subsidising water supply, particularly in developing countries, can provide very high social returns. All subsidies have the potential to distort or prevent the emergence of effectively functioning markets, however. For urban domestic water supply, the vast majority of subsidies in developing countries are based on subsidising consumption (le Blanc 2007). These subsidies are sometimes ineffective however, because the subsidised block of water is often too large. Subsidising water connection fees could be more effective. Simple subsidies that could benefit water management include incentives for drought-tolerant landscaping or for the installation of water meters, rainwater tanks or water-efficient appliances.
Fines	Punitive measures are effective in signalling bad practice, but need to be carefully implemented to prevent public and/or business resentment.

India is home to more than one billion people and is projected to become the world's most populous country by 2028. Its demographic, like that of many countries at a similar stage of economic development, is largely rural (around 70%) but urbanising fast. Rural livelihoods rely on agriculture, so appeasing this industry is often the primary driver of government policy. In Gujarat state and many others, attempts to reduce hunger and improve crop yields have been supported by state energy subsidies, very low fees for groundwater use and a generous guaranteed price for crops. India spends US$ 6 billion a year on energy subsidies and its farmers pay just 13% of the true cost of electricity (WorldWatch Institute 2014). The result is that in some Indian states, more than 50 new groundwater tubewells are drilled every day and, as a consequence, aquifers are depleting fast. This water scarcity has promoted interstate tension. In 2004, Punjab repealed existing agreements on interstate water sharing and refused to build the last section of a canal that would have linked the Satluj and Yamuna rivers to the benefit of Haryana state (Roth and Warner 2008).

The Indian government's policies have also had important knock-on effects for the energy sector. Indian demand for electricity is growing at approximately 7% a year, with a power generation sector projected to be 70% dependent on coal by 2030 and inefficient transmission infrastructure struggling to keep up (Circle of Blue 2013). Greenhouse gas emissions have risen rapidly (by more than 50% between 1994 and 2007), while at the same time blackouts are common (transmission losses from India's inefficient grid are estimated at 25%). In turn, these power cuts stunt business innovation and economic growth.

In rural regions, large stockpiles of grain pile up in local government depots unable to be efficiently delivered to India's teeming cities due to poor supply chain infrastructure (roads and appropriately refrigerated supply chains in particular). The IME (2013) estimates that, every year, 21 million tonnes of wheat from these stockpiles perishes due to poor storage and inefficient distribution, a greater volume than the entire wheat production of Australia. For similar reasons, it is also estimated that as much as 40% of India's fruit and vegetables rot before they can be eaten (IME 2013), wasting not just a huge food resource but also the water and other inputs that went into their production. Future climate change could also have a major impact on Indian agriculture. The Asian Development Bank (ADB) suggests that more frequent severe weather events may result in GDP losses of up to 8.7% by 2100 (ADB 2014).

Via the web of interconnected feedbacks presented in this case study, it is clear that a government energy policy intended to reduce hunger and support rural livelihoods has unintentionally promoted a number of knock-on effects. The case study also highlights the balance that must be struck between managing natural resources as social welfare programs on one hand, and as economic goods on the other. Perhaps with a more holistic and integrated resource policy from the outset, many of these detrimental impacts could have been avoided.

Before moving on, it should be noted that several Indian states are now taking steps to reform their resource policies. In a test case in Gujarat state, by limiting electricity subsidies and encouraging groundwater metering, the local government was able to decrease public spending while at the same time reducing the frequency of debilitating power outages (WorldWatch Institute 2014). Importantly, because farmers were exposed to the price signals of energy scarcity, the voluntary adoption of on-farm water conservation measures increased significantly. Flood irrigation methods that were widespread when the energy subsidies were in place were replaced with drip irrigation systems that can achieve efficiencies as high as 90%. Thanks to the change in government policy, spending on energy subsidies fell by 50% in five years, farm power use by tubewells fell by 35%, more consistent power supply improved rural quality of life and the groundwater overdraft was reduced (Shah *et al.* 2008).

11.4 The Importance of Education

While horizontally and vertically integrated institutions and organisations are crucial enablers for effectively managed natural resources, a well-informed public is equally essential if policies are to deliver successful outcomes in the long term. Many existing examples of sustainable and innovative water management approaches (such as direct potable reuse) relied on severe water source constraints such as drought or flood to secure their initial public support. In such situations, public acceptance of the need for radical change, rather than genuine support perhaps, allowed the projects to be implemented. In such scenarios however, traditional public perceptions are likely to return once the immediate pressures ease. Water managers have to overcome this reliance on 'shock' to stimulate action, instead building consensus and understanding through sustained education that focuses on the following aims (Ross *et al.* 2014):

- *fairness*: people are likely to be more accepting of unfavourable outcomes when they are determined by fair institutional procedures;
- *identity*: people are more willing to trust leaders with whom they share a social connection;
- *shared values*: building on a strong identity leads to genuine shared values, objectives and goals; and
- *technical competency*: the public must perceive their water managers as credible and scientifically and technically astute.

If the public are genuinely brought on board, government policy has to work less hard to facilitate the sound management of natural resources. As well as building this foundation, public communication strategies must make effective use of contemporary communication pathways, particularly social media. Examples of successful education programs are described in Table 11.2.

Table 11.2 Government education programs.

Example	Description
Integrated water resource management in Singapore	Singapore is regularly cited as a best-practice example of how to establish an integrated and holistic approach to water management. Its government has made sustained efforts to link policies related to economic development, social capital and natural resources and has strengthened its management of water through investment in what it terms its Four National Taps: local catchment water; imported water; recycled water; and desalinated water.
	Investments in physical infrastructure have been accompanied by a sustained and coherent communication strategy that has evolved with the broader strategy. A visitor centre is the focal point for public education and is supported by a WaterHub where business and academia are encouraged to share research and best practice (Rygaard *et al.* 2011).
Potable reuse schemes in the USA	When planning to implement an indirect potable reuse scheme in Georgia, USA the local authority recognised that motivating the public to engage in the scheme would be critical to its success. They developed a Citizens Advisory Board which had oversight responsibility for the performance of the wastewater treatment plant. It controlled its own annual budget, imparting responsibility on to its members, and acted as the point of contact between the public and the government (Hartley 2006).
Water conservation in Australia	Australia is exposed to regular cycles of drought and flood that typically occur on decadal timescales. During the Millennial Drought (1997–2010), public education was highly effective at reducing municipal demands for water. In Melbourne, the utility and state government broadcast water storage levels on television, radio and print news services, and installed billboards that summarised the latest water storage data, weekly rainfall and inflows, and provided advice on how to save water. Commercial and industrial users were also encouraged to develop water conservation plans (Grant *et al.* 2013). These investments helped the public and businesses to understand the severity of the situation and meant that compliance with mandatory water restrictions, and a voluntary target, was high. Per capita municipal water demand declined by 46% over 12 years (Grant *et al.* 2013).
	In southeast Queensland, a similar public communication campaign was successful in cutting domestic water consumption by half (to less than 140 L/day/person). Furthermore, and somewhat unusually, this behavioural change largely survived the drought, with average per capita consumption in 2011–12 of around 160 L/day (Head 2014).
	Other government responses to the Millennial Drought have been less effective, however. Several states invested in expensive desalination plants that in a number of cases (e.g. Adelaide, Melbourne and the Gold Coast) were completed after the drought broke. With customers unwilling to pay the high prices associated with desalinated water when more traditional sources were again available, several of these plants were mothballed. While still providing important, climate-independent contingencies to future droughts, some in the community perceive that public funds have been wasted.

(Continued)

Table 11.2 (Continued)

Example	Description
Social media	In 2012, approximately 1.5 billion people used social networks. They were also used by 66% of US government agencies, 80% of US businesses and 89% of US NGOs (InSites Consulting 2012). For example, DC Water uses Twitter as a channel for customers to report problems, resulting in fewer e-mails sent to the general purpose inbox. The utility has also been able to help customers understand where their water comes from and how it is priced. Many water utilities all around the world now use Twitter to inform their customers of problems such as reduced water pressure, discolouration and local infrastructure works. In the UK, several apps are available which allow users to monitor flood risks in near-real time. In one example, information from government-operated river level gauges is published via Twitter. Users of the platform can subscribe to any of 2,400 different accounts, each corresponding to a separate gauge. In Brisbane, Australia residents can subscribe to receive free severe weather alerts via email, SMS or phone from the municipal government. The service monitors a variety of data sources to send targeted alerts to specific areas of the city and broader surrounds likely to be affected by storms, flooding and other severe weather events as they develop.

11.5 The Role of Private Organisations

Private business has finally woken up to the risks of poor water management. In 2011, 53% of companies reporting through the CDP Global Water Report thought that water represented a substantive risk to their operations. By 2015, this proportion had increased to 65% (CDP 2015). Water abundance or shortage increasingly causes the cessation of business operations; it also promotes regulatory change that impacts business performance or fosters community resentment that, in turn, impairs the social license of a business to operate. For example, in 2011 the global agri-business firm Bunge lost US$ 56 million as a result of severe droughts in its main growing regions (Oxfam 2012). The UK retailer ASDA has mapped its entire global fresh produce supply chain and found that 95% is under threat from future climate change (ASDA 2014). As companies increasingly recognise risks such as these, and learn of the failure of others to do so, more and more are attempting to control their potential impact. Few take a truly holistic approach to this process however, choosing to focus on measures over which they have strong control. For example, a business may locate infrastructure outside drought-prone areas rather than consider improving the management of watersheds or the product supply chains from which a significant proportion of their risks may have their root cause (see Figure 11.4).

Some collaborative organisations are beginning to help businesses address water risks beyond their immediate fence-line. The Alliance for Water Stewardship is one such organisation. Its membership includes the likes of Nestle and General Mills as well as government bodies and NGOS, and these partners have worked together to develop a standard and framework for responsible water stewardship.

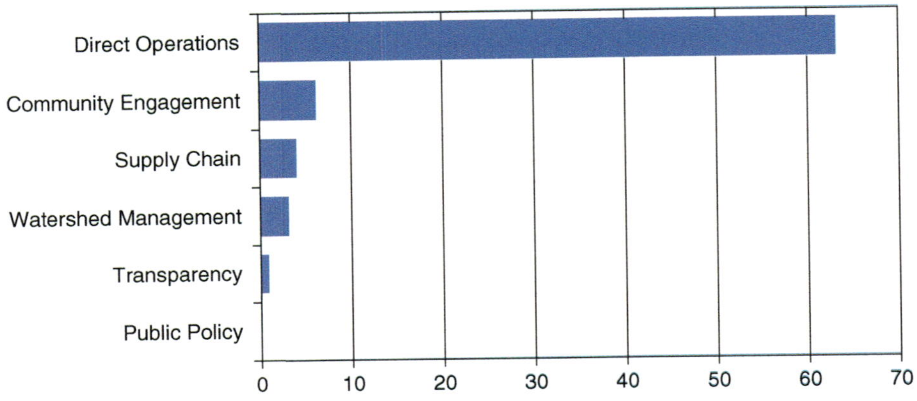

Figure 11.4 Percentage of respondents setting targets or goals by type.
Source: Adapted from Carbon Disclosure Project (2013).

For many businesses, long and complex supply chains mean that the ultimate source of a product and the immediate responsibility for its management may lie with a far-removed partner many thousands of kilometres away. In such a scenario, it is easy to see how the exploitation of natural resources can become disorganised and haphazard. The risks of this lack of oversight to the company at the head of the supply chain are potentially huge, and may only become apparent when it is too late. The sustainable management of water and other natural resources requires transparent supply chains within which each business partner adopts the same set of principles and resource management procedures. Rather than just signing a contract for a volume of a given product at a given quality and on a given date, suppliers should also be required to sign a code of ethics and a commitment to report on their resource consumption in a transparent and consistent manner.

The extent to which business and even whole sectors of economies rely on certain resources and inputs is also often underappreciated. For example, it is estimated that 16 separate sectors of the US economy, from seed suppliers to retailers, depend on corn as a key ingredient in their products. In 2013, the top 45 companies in the US corn value chain earned US$ 1.7 trillion in revenue, more than Australia's entire annual GDP (Ceres 2014). Now consider that US corn receives the most irrigation water of any US crop and that 87% of US irrigated corn is grown in regions with high or extremely high water stress (Ceres 2014). In other words, US$ 1.7 trillion of US business revenue is reliant on a single product that is in turn reliant on a highly and increasingly vulnerable resource.

The entrenched focus of businesses on securing quick returns on investment also represents an important cause of poor natural resource management. Overcoming this tendency requires business leadership founded on a sense of long-term responsibility as well as a recognition that sustainability can be the route to competitive advantage.

Where businesses can adopt long-term metrics of performance, it is more likely that strategies involving collaboration between stakeholders will emerge as the most attractive. The concept of industrial symbiosis (of co-locating different businesses and industries with complimentary resource requirements) is a good example in this respect. Although requiring a commitment to upfront capital expenditure, cyclical resource

Table 11.3 Private sector initiatives.

Example	Description
Collaborative initiatives	The number of businesses collaborating in sustainability initiatives is growing rapidly. The World Business Council for Sustainable Development (WBCSD) is a CEO-led organisation that includes membership by dozens of industry-leading companies in all regions of the world. The WBCSD's initiatives related to water include a series of tools that help to standardise the ways in which companies understand and report on their water-related risks. The WBCSD has also prepared business guides on industrial water reuse and water valuation.
Innovative business models	A variety of new business models are emerging that focus on the concepts of zero-waste, resource efficiency and circular economy. They include: • using renewable, recyclable or biodegradable inputs to products in order to maximise opportunities for reuse and to minimise final wastes; • extending the lifetime of products through the provision of recycling, repair and aftersales services; and • encouraging the sharing of assets and products through pay-per-use and rental schemes that focus on delivering performance rather than products. In many cases, these ideas are enabled or supported by the revolution in digital products and services, such as social media, analytics and cloud computing.
Individual business responses	While often less effective than collaborative initiatives, individual business responses can be successful when they recognise that integrated and holistic water resource management provides a competitive advantage. Companies in the food and beverage market have often been quickest to adopt these agendas. For example, Coca-Cola supports water conservation efforts in the Amazon, Mekong and Zambezi. It also achieved a 20% improvement in the water efficiency of its manufacturing operations between 2004 and 2012 and hopes to achieve an additional 25% gain by 2020 (Clancy 2014).

management systems such as that now operating in Kalundborg, Denmark can realise large financial benefit. In that example, the required US$ 78.5 million of investment yielded annual savings of US$ 15 million that had totalled US$ 310 million by 2006 (Domenech and Davies 2011). Table 11.3 presents a selection of private sector sustainability initiatives that highlight the benefits that can be realised when transformational leaders invest in approaches that synchronise strategies for water stewardship and business growth.

11.6 The Importance of Knowledge Transfer and the Benefits of the Digital Revolution

The last decade has witnessed a truly meteoric rise in the proliferation of products, services and systems founded on flows of data. The capture of billions of pieces of data about all facets of human activity means that our ability to monitor what must be managed has never been more advanced. The challenge is making sure we actually monitor the data that are most beneficial and that we transform those data into useful information that can be shared with those that need it, when they need it. Private industry increasingly recognises the opportunities that this digital revolution presents and governments and public bodies should embrace the potential partnerships and opportunities that result, particularly for water management in cities. By integrating sensors (the foundation of any so-called 'smart' network) with analytical software and adaptive response systems, the operating efficiency of urban water systems can be significantly improved. For example, leaks in pipelines can be more quickly identified, pinpointed to within a few metres and alerts sent to engineers to respond. Sensors at the point of supply can also relay information on water demand profiles to operators who can, in turn, more effectively balance their reliance on different water sources. Taking the concepts further even allows for the inclusion of energy considerations into water management decision making, such as variations in the energy cost of supplying water from different sources.

Smart networks also have huge potential in agriculture where they can be used to integrate soil moisture and nutrient sensors with climate records and drip irrigation systems to ensure that just the volume of water and nutrients required by the crop is supplied at just the right time in order to maximise yields for the lowest input of resources. Cloud-based computing systems even allow farmers to tend to their crops from hundreds of kilometres away, a useful tool for those managing huge farming estates in countries such as Australia.

Notwithstanding these opportunities, smart networks also present obstacles to effective water management. For example, a highly integrated network is more vulnerable to localised failures or sabotage that can quickly propagate to affect the whole system. Incorporating decentralised supply sources, such as stormwater or grey water reuse, could help to reduce this vulnerability and increase resilience. Also, crucially, to be most effective, smart sensors and tools require widespread (ideally universal) adoption across all components of the network. This is a significant challenge in already-established cities where new infrastructure usually has to be retrofitted incrementally as existing pipelines and pumps undergo maintenance. Rapidly urbanising regions therefore have the most to gain from these new technologies, giving them the chance to achieve highly efficient urban water management from the outset.

While the proliferation of new water management technologies is occurring rapidly in the developed world, the transfer and sharing of technology and human capacity with the developing world remains highly inefficient. In a 2014 report, McKinsey Global Institute concluded that while the volume of flows of knowledge are now rivalling the labour and commodity flows that supported the industrial revolutions of the twentieth century, most of these knowledge transfers occur only between developed country cities. This means that technologies that may otherwise significantly improve the livelihoods of developing country citizens exist, but are hidden from those that need them most. Take agricultural productivity for example. Yields of wheat, rice and maize are 40%, 75% and 200% less in developing countries than they are in developed nations (FAO 2011). The IME (2014) argues that up to 25% of developing country food waste could be eliminated through the adoption of improved refrigeration technologies that are already proven in developed countries.

All institutions need to do more to improve the equity with which information on effective natural resource management is shared. For example, in 2013, the municipal governments of Amsterdam, San Francisco and Barcelona established a multi-city platform to share urban resource management data and to pool and discuss ideas both between their respective governments and with the public (Andrews 2013). To support this process, decision support frameworks that encourage and remind practitioners to consider the multitude of factors that may influence a given water management problem can be invaluable (e.g. those developed by the WBCSD and referred to in Table 11.3). They can help to standardise integrated decision making and ensure that it becomes an everyday process. In turn, the tendency of organisations to maintain the *status quo* of siloed working that otherwise stifles innovation and solution sharing can be broken.

11.7 The Role of Non-Governmental Organisations

Non-governmental organisations (NGOs) have an important but often underappreciated role to play in the institutional integration of water resource management. They typically represent that organisation most trusted by the public and therefore possess a unique ability to guide social action. NGOs can also play a valuable role in holding both public and private sector organisations to account. When NGOs collaborate with public bodies and private corporations, trust in the overall partnership is increased and broad public support is more likely to be realised.

However, notwithstanding their social standing, NGOs (like public and private organisations) are also subject to pressures from stakeholders with vested interests. They must secure donations to operate and therefore have the potential to be coerced into supporting the aims of large funders. NGO strategies most prone to being influenced in this way include those regarding contentious topics such as support for or opposition to genetically modified crops, as well as decisions to collaborate with particular businesses. In these instances, the stance of NGOs has the potential to go above and beyond that which is communicated through the scientific community and therefore worsen rather than improve public understanding of important issues.

To overcome these pitfalls and to solidify public trust, the internal dynamics, goals and drivers of NGOs must be transparent.

11.8 How to Finance Change

Water use is both essential and discretionary, and therefore approaches to its valuation must account for this distinction. This book repeatedly presses the point that the value that society ascribes to water is significantly less than the benefits it provides. Assuming this social oversight can be rectified (see Chapter 10), flexible means of charging for water based on scarcity and demand would allow this resource to be managed in a far more effective manner than is often the case today. In several nations, many households are charged a flat fee for water irrespective of the volume they use. In a far larger number of countries, the charges imposed on domestic, industrial and agricultural water users fail to recover the full costs of supply, thereby jeopardising the long-term financial stability and viability of the utility provider.

To help achieve full cost recovery and to promote water conservation, incremental block tariffs on water consumption can be used to charge progressively higher fees as the profligacy of water consumption rises. Allowing for a social tariff to ensure that the essential water use component of domestic consumption can be met, such a pricing structure can be effective in both ensuring the financial viability of the service provider and in improving public awareness of water scarcity. By incorporating incremental block tariffs with smart data systems, rate structures can be optimised to appropriately reflect local conditions as they change (e.g. seasonally or even periods of peak daily demand). Examples of alternative water tariff structures are presented in Table 11.4.

Obviously, universal water metering is a precondition for volumetric tariff structures. Implemented effectively, metering clearly informs the user of their consumption patterns and enables the supplier to establish pricing mechanisms that can adapt to variations in water demand and supply scarcity.

An often-cited argument against the widespread metering of water supply is the perceived difficulty and cost associated with fee collection. While true in some remote

Table 11.4 Water pricing mechanisms.

Mechanism	Description
Fixed charge	The user is charged a fixed fee irrespective of the volume of water used. As a result, the consumer experiences no incentive to reduce their use. Most countries are moving away from the use of fixed water fees. In France for example, a 1992 Water Law prohibits the use of flat-fee tariffs (le Blanc 2007). They still persist in a surprisingly large number of countries, however. In the UK, for example, for all unmetered homes, consumers are charged a set annual rate based on the rateable value of their home. Where it exists, an absence of universal metering acts as a severe constraint on a transition away from fixed-charge tariffs.
Volumetric charge	Tariff structures based entirely on volumetric charging are rare, as they can be perceived to disregard the essential component of water use. Some countries such as Hungary, Poland and the Czech Republic do employ these tariff structures.
Two-part tariff	A simple two-part tariff structure could comprise a fixed component to cover supply costs and a variable charge based on the volume consumed. Several OECD countries, for example Australia, Austria, Denmark and Finland, use a two-part tariff structure (Rogers *et al.* 2002). Tariff structures based on this principle are also used for the supply of electricity, for example in France (le Blanc 2007).
Incremental block tariffs	IBTs are a refinement on the two-part tariff that incorporates extra tariff bands that progressively kick in as consumption rises. The first volumetric block can be charged at cost or even below cost (to ensure the essential component of water use is met) while additional blocks can increasingly penalise overconsumption. Profligate uses can therefore subsidise essential needs. Spain, Italy, Greece and some regions in Belgium and the US all make use of IBTs (Rogers *et al.* 2002). During the Australian Millennial Drought, the tariff structure for domestic water supply in Melbourne was altered from two to three tiers. The higher charge on the third volumetric tier helped to signal the scarcity of water and also helped to pay for major investments in supply augmentation (Grant *et al.* 2013).

geographies and common in past developing country examples, modern advances in smart data systems will continue to reduce the extent of this barrier. In developing countries, there exist numerous examples of how mobile phones can be used to improve access to finance and improve fee collection. Three-quarters of the world's population already has access to mobile phone communication, and around 80% of mobile phone subscriptions are in developing countries (UN International Telecommunication Union 2012). Mobile devices allow water users to monitor their own consumption, report faults and pay bills, thereby enabling significant improvements in system efficiency. With improved information on their consumption patterns, customers could be encouraged to reduce their water use during peak demand periods. Because water suppliers must deliver on this peak demand, they have historically had to invest in large, centralised infrastructure. If more balanced demand profiles can be achieved, decentralised water infrastructure that is often less energy intensive becomes more viable.

To maintain and progressively enhance municipal water supply networks as urban populations increase and sprawl, the reinvestment of water fees often has to be supplemented by third-party finance. Securing the right type of funding is an often-overlooked influence on the long-term success and viability of urban water management. For example, because general maintenance activities generate a low but steady return on investment, they are often unattractive to traditional financiers that demand a quick return on investment. The steady paybacks do however closely match the requirements of other potential funding partners such as pension and sovereign wealth funds. Identifying patient capital is important for ensuring long-term successful schemes and municipal governments could help to connect appropriate parties.

11.9 Conclusions: Institutional Enablers

The case studies and concepts described in this book show that the relationships between water and other natural resources are too complex to be managed successfully in isolation. Similarly, this chapter highlights how the actions of different institutions and organisations result from and influence the actions of others. It is because of these influences that the most sustainable outcomes can only be achieved through collaboration, not just between the various institutions and organisations responsible for water resource management, but between those organisations with a stake in the management of all natural resources. Unfortunately, the internal functioning of most governments, public bodies and a large number of private organisations currently encourage an entirely opposite way of working, one characterised by inertia, siloed strategising and single-minded project delivery. To transform this approach, this chapter can be condensed into five key institutional enablers originally introduced in Figure 11.1:

1) *Transformational leaders*: that act out of a sense of responsibility, not just out of a need to deliver a return on investment or to secure re-election. Such leaders realise that the objectives of successful and popular governance, competitive business advantage and effective resource management, so long thought of as conflicting goals, can go hand-in-hand when the costs and benefits of the consumption of natural resources are accurately valued at appropriate geographic and temporal scales. Leadership of this kind helps to ensure that innovative ideas are not stifled by inertia.

2) *Collective partnerships*: within which all stakeholders are driven by recognition that their own interests are best served through the achievement of collective goals. This requires that the values and perceptions of water held by all partners are well-known and understood. Equitable sharing of information, risks and investments are also crucial characteristics of a successful and committed natural resource management partnership.
3) *Coordinated governance*: that harmonises the aims of different departments and organisations in different sectors (horizontal integration) and that balance bottom-up and top-down governance (vertical integration). Implemented effectively, polycentric governance of this kind would support the emergence of a coherent National Resources Policy supported by a common national or even global agenda.
4) *Genuine stakeholder engagement*: that moves away from 'public relations' to embrace continuous and transparent stakeholder engagement, ensuring that it becomes an everyday process, not just forced by regulation. In turn, public understanding, buy-in and a sense of accountability for the collective good all grow.
5) *Technology transfer and decision support*: to ensure that integrated resource management becomes standardised as an everyday process and to ensure that success stories and solutions are shared openly and globally. Decision support frameworks can help to standardise means of accounting for natural capital, allowing the fair comparison of different management alternatives. They can also provide the framework for combining tools, finance and organisations in multi-stakeholder platforms.

By focusing on strengthening these enablers, governments, private businesses, NGOs and the public will be able to develop and endorse natural resource management strategies that embed the conceptual ideals of a New Water Architecture presented in Chapter 10. They also secure the preconditions for the third component of a New Water Architecture, the enhanced physical integration of water resource infrastructure.

References

Andrews, J. 2013. Amsterdam, Barcelona and San Francisco sign up to new data platform. Available at: http://cities-today.com/2013/11/amsterdam-barcelona-and-san-francisco-sign-up-to-new-data-platform/ (accessed February 2017).

ASDA. 2014. The challenge of a changing climate. ASDA, UK.

Asian Development Bank. 2014. Assessing the Costs of Climate Change and Adaptation in South Asia. ADB, Philippines.

CDP. 2013. Moving Beyond Business as Usual: a need for a step change in water risk management. CDP Global Water Report 2013. Carbon Disclosure Project, UK.

CDP. 2015. Accelerating Action: CDP Global Water Report 2015. Carbon Disclosure Project, UK.

Ceres. 2014. Water and Climate Risks Facing US Corn Production: how companies and investors can cultivate sustainability. Ceres, Boston, USA.

Circle of Blue. 2013. Scarcity in a Time of Surplus: Free Water and Energy Cause Food Waste and Power Shortage in India. Available at: http://www.circleofblue.org/waternews/2013/world/scarcity-in-a-time-of-surplus-free-water-and-energy-cause-food-waste-and-power-shortage-in-india/ (accessed February 2017).

Clancy, H. 2014. 10 companies making waves in water innovation. Available at: http://www.greenbiz.com/blog/2014/08/04/10-companies-innovating-water-making-waves-water-innovation (accessed February 2017).

ComRes. 2014. Headline Findings: EXIU Launch Survey. Available at: http://eciu.net/assets/ECIU_Launch-Survey_Headline-Findings_4-August-2014-1.pdf (accessed February 2017).

Domenech, T. and Davies, M. 2011. Structure and morphology of industrial symbiosis networks: The case of Kalundborg. *Procedia – Social and Behavioural Sciences*, 10, 79–89.

European Commission. 2014. Towards a circular economy: a zero waste programme for Europe. Available at: http://eur-lex.europa.eu/resource.html?uri=cellar:50edd1fd-01ec-11e4-831f-01aa75ed71a1.0001.01/DOC_1&format=PDF (accessed February 2017).

Food and Agriculture Organization of the United Nations. 2011. Looking Ahead in World Food and Agriculture: Perspectives to 2050. FAO, Rome, Italy.

Grant, S.B., Fletcher, T.D., Feldman, D., Saphores, J.-D., Perran, L., Cook, M., Stewardson, M., Low, K., Burry, K. and Hamilton, A.J. 2013. Adapting urban water systems to a changing climate: lessons from the millennial drought in southeast Australia. *Environmental Science and Technology*, 47, 10727–10734.

Hartley, T.W. 2006. Public perception and participation in water reuse. *Desalination*, 187, 115–126.

Head, 2014. Managing urban water crises: adaptive policy response to drought and flood in Southeast Queensland, Australia. *Ecology and Society*, 19, 2.

InSites Consulting. 2012. Social Media around the World 2012. InSites Consulting, New York, USA.

Institution of Mechanical Engineers. 2013. Global Food: Waste Not, Want Not. IME, UK.

Institution of Mechanical Engineers. 2014. A Tank of Cold: Cleantech leapfrog to a more food secure world. IME, UK.

le Blanc, D. 2007. A Framework for Analysing Tariffs and Subsidies in Water Provision to Urban Households in Developing Countries. UN, New York.

McKinsey Global Institute. 2014. Global flows in a digital age: How trade, finance, people, and data connect the world economy. McKinsey, London, UK.

North, D. 1990. Institutions, Institutional Change and Economic Performance. Cambridge University Press, Cambridge, UK.

Oxfam. 2012. Physical Risks from Climate Change: a guide for companies and investors on disclosure and management of climate impacts. Oxfam, USA.

Rodgers, P., de Silva, R. and Bhatia, R. 2002. Water is an economic good: How to use prices to promote equity, efficiency, and sustainability. *Water Policy*, 4, 1–17.

Ross, V.L., Fielding, K.S. and Louis, W.R. 2014. Public acceptance of recycled water: the impact of trust. *Journal of the Australian Water Association*, 41(1), 89–92.

Roth, D. and Warner, J. 2008. Virtual water: Virtuous impact? The unsteady state of virtual water. *Agriculture and Human Values*, 25, 257–270.

Rygaard, M., Binning, P.J. and Albrechtsen, H.-J. 2011. Increasing urban water self-sufficiency: new era, new challenges. *Journal of Environmental Management*, 92, 185–194.

Shah, T., Bhatt, S., Shah, R.K. and Talati, J. 2008. Groundwater governance through electrical supply management: assessing an innovative intervention in Gujarat, western India. *Agricultural Water Management*, 95, 1233–1242.

UNEP. 2012. The UN-Water Status Report on the Application of Integrated Approaches to Water Resources Management. United Nations Environment Programme, New York, USA.

United Nations International Telecommunication Union. 2012. Measuring the Information Society. UN, New York.

Verheggen, B., Strengers, B., Cook, J., van Dorland, R., Vringer, K., Peters, J. Visser, H. and Meyer, L. 2014. Scientists' views about attribution of global warming. *Environmental Science Technology*, 48, 8963–8971.

WorldWatch Institute. 2014. Reforming Energy Subsidies Could Curb India's Water Stress. Available at: http://blogs.worldwatch.org/nourishingtheplanet/reforming-energy-subsidies-could-curb-indias-water-stress/ (accessed February 2017).

12

Physical Integration

Physical water infrastructure is essential for sustaining social, economic and environmental health in a modern society. Expanding populations, their increasingly urbanised nature and their exacting demands require a complex web of water supply services consisting of pieces of built (often termed grey) infrastructure that harness natural water resources, treat them for use and transport them to where they are required. More grey infrastructure is subsequently required to convey, treat and return wastewater to the environment.

Over human history, the evolution of water resource management has been driven by changes in the dominant social and institutional concerns of the time. When western countries first industrialised, the focus was on sanitising rapidly expanding urban environments, isolating water sources and wastes via a network of pipes, and treating them in order to maintain public health. This approach was incredibly successful and, as a result, waterborne diseases were practically eliminated in the developed world. However, as the pressures of population growth, urbanisation and climate change have continued to grow, some elements of rigid infrastructure networks are increasingly seen as inefficient and challenging to operate in a manner that effectively addresses changing water availabilities and demands.

In recognition of these limitations, new theories of how water should be managed have emerged, particularly in the urban and peri-urban environments that will continue to be the focal points for population growth. These concepts include:

- incorporating decentralised components and satellite systems (e.g. rainwater harvesting) into existing centralised water infrastructure networks;
- providing water of a fit-for-purpose quality rather than treating all water to the standard required by the most demanding customer;
- harnessing natural ecosystems (referred to as green infrastructure here), such as upland watersheds and wetlands, to complement grey infrastructure in order to improve the overall efficacy and efficiency of water management; and
- reusing wastewater and its embedded resources (e.g. energy, nutrients and organic matter) through circular and regenerative cycles.

This chapter explores some of the alternative ways in which centralised water management networks could be enhanced to realise a much broader range of benefits than is currently the case. The focus is on urban environments, although many of the initiatives are readily applicable to rural settings. The benefits are also diverse, extending beyond the more effective and efficient management of water resources to deliver

Water Resources: A New Water Architecture, First Edition.
Alexander Lane, Michael Norton and Sandra Ryan.

mutually positive outcomes to the other essential resources of energy, food, land and biodiversity. Furthermore, thanks to their flexibility and collective resilience, the initiatives proposed are more likely to support long-term sustainability in spite of the uncertainties associated with future population growth and climate.

12.1 The Need for Change

12.1.1 Existing Limitations

A brief journey through the past two centuries of human development helps to explain the current configuration of our water management networks, in turn highlighting their successes as well as the root causes of their limitations.

Prior to industrialisation, the supply of water to populations was largely uncoordinated and discrete, focused on meeting the needs of sparsely distributed and small communities (Marlow and Tjandraatmadja 2014). This approach to water management persists to the present day in many developing countries, where it often struggles to improve access to safe water and sanitation to the level required to ensure public health and, in turn, to support livelihoods and economic growth.

Industrialisation and its associated urbanisation triggered the development of the centralised water management networks typical of most modern cities. By concentrating the exploitation of water sources using large-scale storage, pipeline networks and treatment plants, water could be supplied at economies of scale to rapidly expanding populations and industries. While this investment enabled economies to flourish and urban populations to boom, the wastes from factories and their burgeoning workforces were largely ignored, quickly resulting in public health crises. This spurred a second infrastructure response, focused on isolating flows of wastewater and conveying them away from populated areas as quickly as possible. Discrete water and wastewater systems therefore evolved, often managed by separate entities.

Further infrastructure transitions have followed in most cities, required in order to protect communities from the flooding caused by converting natural environments to meet large-scale agricultural, industrial and residential needs. Again, the process has relied heavily on grey infrastructure, this time to sever the links between the human and natural environments with walls and barriers in a bid to control overland flows of water.

Collectively, these three infrastructure paradigms are borne out in the water management networks that characterise almost all contemporary urban environments. The centralised approach can also be seen in road networks, solid waste management systems, energy grids and communication pathways. It truly is the defining feature of urban organisation.

Figure 12.1, based on work by Brown *et al.* (2008), conceptualises the evolution of urban water management. It presents the transitions explained above and also considers how future water management paradigms (explored in subsequent sections of this chapter) might evolve.

Notwithstanding the public health benefits of centralised water management networks, they encourage overreliance on a small number of water sources and tend to concentrate pollutants in receiving environments. In most cities, the degradation of

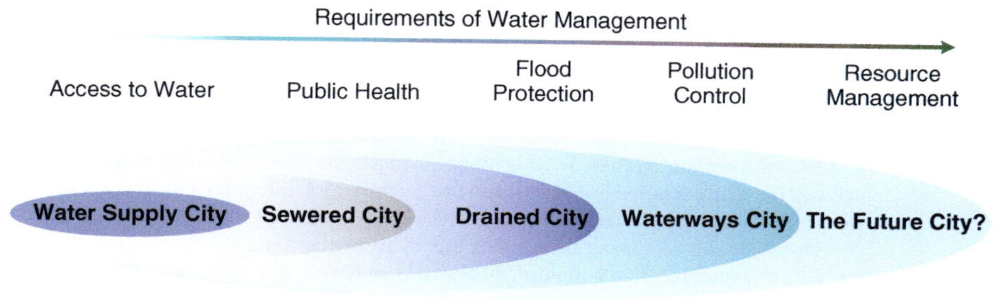

Figure 12.1 Water management infrastructure paradigms.

adjacent and downstream natural environments has prompted efforts to control point and diffuse sources of pollution. Broader shifts in the infrastructure paradigm, for example to those that place greater focus on the more holistic management of natural environmental assets, have been largely absent however, held back by a number of barriers to change that are discussed in Section 12.1.2. Urban communities are however increasingly experiencing unexpected side-effects that can be linked to the limitations of existing water and wastewater service systems. Table 12.1 considers each of these limitations in turn.

Table 12.1 Limitations of existing water management infrastructure.

Limitation	Impact
Inflexibility to changes in demand	Pipelines and centralised treatment facilities have lifetimes of 50 years or more and are therefore relatively inflexible to changes in the quantity of water demanded by customers. With urban populations expanding at rapid rates, and in patterns that may be difficult to accurately predict or control, infrastructure must either be oversized to allow for high-side predictions of population growth or undergo expensive retrofitting as required. Neither approach represents a cost-effective means of accounting for uncertainty. The use of large centralised water treatment facilities also requires that water is treated to the highest quality demanded by all users; no mechanism is provided for fit-for-purpose water supply.
Inflexibility to changes in supply	Most fixed infrastructure assets assume no change in climate over their operating lifetimes. Unless they are significantly oversized, this makes them vulnerable to shifts in patterns or extremes of rainfall or temperature that may in turn affect water supplies. For example, the Hoover Dam of the Colorado River was designed based on observations made during three of the wettest decades of the past millennium. Lake Mead, the dam's reservoir, now consistently stores only around 30% of its designed capacity (Matthews *et al.* 2011).
High energy consumption	Pumping water along pipes from water sources to demands (often uphill) is energy intensive. As a result, 50–75% of a utility's total capital and operating expenditure is spent operating pipeline networks (Marlow and Tjandraatmadja 2014). Alternative management options that reduce the dependency on pipeline networks can therefore be very cost effective.

(Continued)

Table 12.1 (Continued)

Limitation	Impact
High maintenance costs	Centralised networks reliant on grey infrastructure require large stocks of materials and labour for maintenance. Furthermore, once a centralised system is in place, utilities have little choice but to continue to maintain its individual components so that the performance of the overall network can be sustained. The pressure to keep costs low also means that utilities may defer routine maintenance and upgrades. As a result, unavoidable major maintenance works may end up costing more than would otherwise have been the case. The US EPA (2009) estimates that just maintaining the USA's current levels of water service provision requires more than US$ 16 billion of investment every year.
Failure propagation	Because centralised water management networks are typically reliant on a limited number of water sources and treatment facilities, when a component of the network fails the whole network is vulnerable to its impacts. These impacts could be chronic (such as the gradual degradation of a key water source) or acute (such as a pollution incident or cyber-threat). Systems that include decentralised elements are more likely to be able to limit the impact of such failures by switching to other water sources or infrastructure components.
Wasting resources	Traditional water networks are characterised by a linear, 'take-use-dispose' approach to water management. They also tend to focus on a single service such as water supply, therefore making them wasteful of a range of potential resources: • in some developed country cities, the amount of water lost to leaks in pipeline networks can be as high as 40% (Hering *et al.* 2013). In developing countries, illegal connections are also often a drain on the financial performance of utilities; • reliance on pumping makes centralised networks highly energy intensive; and • wastewater represents a bountiful and climate-independent source of a variety of resources that are typically foregone by traditional water management. Phosphorus for example is vital for food production and also present in useful quantities in human waste. Anaerobic digestion of sewage can also be used to produce biogas that can then be burnt to generate electricity.

In addition to the limitations presented in Table 12.1, centralised models of urban water management encourage a perception by customers that water is abundant and ubiquitous, a perception that is enhanced by generalised tariffs and regulation. This mindset promotes profligate water use and, in turn, encourages utilities to focus on supply-side approaches to water management (e.g. large dams or increased river abstractions).

There are also impacts and tradeoffs associated with other centralised grey infrastructure networks. For example, Shilling *et al.* (2007) postulate that, while the construction of roads has been a major driver of poverty reduction in developing countries, the new networks have also acted as avenues for resource depletion, clearly witnessed in aerial images of deforestation. The authors argue that a significant proportion of the environmental damage experienced in China and India (estimated to cost the respective economies between 4% and 8% of gross domestic product) can be traced to the impacts of built infrastructure. Breakout Box 12.1 considers some of the approaches that China has taken to manage its water resources.

Breakout Box 12.1 Water resource management in China

China's approaches to water resource management have focused heavily on supply-side measures dominated by large-scale infrastructure. Liu *et al.* (2013) estimates that the nation has more than 87,000 dams (equating to around 10% of the world's storage capacity and around 20% of its hydropower) as well as inter-basin water transfer projects with a combined length in excess of 7,000 km (this will increase to more than 10,000 km with the completion of the South–North Water Transfer Project). While these investments have sustained rapid growth in economic development and population, they have also degraded natural environments and displaced approximately 22 million people (Liu *et al.* 2013).

12.1.2 Barriers to Change

A number of physical, social and institutional barriers currently act to hold back or actively discourage alternatives to existing water management networks.

12.1.2.1 Path Dependency

Path dependency is a key issue. The performance of each component of a centralised network is reliant on the system as a whole, so transitioning to a new approach requires huge capital investment and severe disruption to service provision. Furthermore, because each pipe or component piece of infrastructure has a long lifetime (often 50 years or more) and because each degrades at a slightly different rate, repairs are required at different times. In order to maintain overall system performance, like-for-like replacement of individual components is therefore the only real option. As a result, pipelines quickly become sunk assets and their continued maintenance and operation typically form the greatest proportion of ongoing expenditure of a utility.

Newly developing countries have an important advantage in this respect, possessing the opportunity to learn from the limitations of existing schemes in other nations. For example, Hong Kong invested in the development of a dual water distribution system that supplies seawater to homes and business for toilet flushing. Now in operation for more than 50 years, it has cut municipal freshwater use by 20% (Grant *et al.* 2012).

12.1.2.2 Siloed Decision Making

Water supply and wastewater services may, in some cases be delivered by separate organisations that share little common understanding and which may therefore function without sufficiently considering their inherent relationship to the other. As a result, decisions made in one domain have the potential to cause unintended detrimental impacts in the other. In Australia for example, researchers investigating rapid corrosion of concrete sewerage networks were able to trace the cause back to the treatment of water supply. Cleaning additives raised sulphate levels in treated water which persisted into wastewater. In sewers, take up of the sulphate by microbes led to the formation of sulphuric acid and the subsequent corrosion of the concrete pipes at rates ten times faster than those which would otherwise have been expected (Pikaar *et al.* 2014). Examples such as this highlight the need for investment in improved decision support tools and frameworks to help planners make more informed investment choices.

12.1.2.3 Perceptions of Ecosystem Services

The concept of ecosystem services – the provisioning, regulating, supporting and cultural benefits that humans gain from nature – is increasingly becoming mainstream terminology. An appreciation of how nature helps mediate the complexities of the water-energy-food nexus by storing, moving, cleaning and buffering flows of water, making drought and flood less severe, replenishing soils and making food and energy production more reliable, is also increasingly acknowledged by decision makers (Krchnak *et al.* 2011). Despite this recognition however, ecosystem services often continue to be marginalised as a conservation issue and an afterthought with the focus being on mitigating specific environmental impacts. Working with and enhancing ecosystem services, rather than just aiming to prevent their degradation, will ensure that the benefits of ecosystem services can be both harnessed and sustained.

12.1.2.4 Business Models

Utilities typically operate by leveraging economies of scale in water supply, meaning they tend to be naturally attracted to large centralised schemes that support this approach. Many water utilities are also legally obliged to supply water of a sufficient quality and quantity to their customers (Savić *et al.* 2013). They are therefore discouraged away from innovative options (for which long-term performance data may be lacking) towards more risk-averse options that, while perhaps less efficient, can be easily benchmarked against existing schemes.

12.1.3 Overcoming the Barriers

Societies have often required the influence of a severe shock to trigger concerted action. For example, it took a sustained drought and the real threat of water supplies running dry for the town of Wichita Falls, Texas to invest in a water recycling facility that facilitates the direct reuse of treated wastewater for potable use. In many cases, the institutional responses to shocks such as these are reactionary and incompletely thought through. As a consequence, the resultant solution may fail to live up to initial expectations.

To avoid reactionary responses to shocks, a fundamental shift in our perception of water is required. Rather than a simple input to be taken for granted and exploited, the social and institutional changes advocated in Chapters 10 and 11 must lead to a recognition that water is a scarce, fragile and vital resource that plays a pivotal role in virtually all human activities. In turn, the benefits that these activities gain from water must be accurately costed so that water becomes an appropriately valued enabler of economic growth, environmental health and human development.

Where accurate valuation is used to derive flexible water fees, users can be more readily exposed to the scarcity component of water, making them more likely to respond to demand reduction initiatives. Importantly, the revenues will mean that utilities are also more likely to achieve financial stability, enabling them to better plan their long-term investments. Amazingly, even in a highly developed country such as the USA, it is estimated that only a third of utilities earn enough revenue to operate a financially sustainable business (Black & Veatch 2014). In low-income countries, 50% of utilities don't secure enough revenue to meet their operation and maintenance costs, a percentage that doubled between 2000 and 2010 (Danilenko *et al.* 2014).

The proliferation of more integrated approaches to water management will also rely on recent technological advances in data collection, analysis and application. Smart data systems and their associated decision support tools are vital components of infrastructure systems that are more flexible to uncertainty and responsive to changes in supply and demand. Technological innovations in energy supply and agriculture also promise much for the more efficient management of water resources. Examples include more water-efficient cooling facilities at power plants, the development of drought-resistant crop varieties, precise irrigation and nutrient application systems, and improved on-farm data collection and interpretation (e.g. of key parameters such as soil moisture).

In the time before these conceptual and technological transitions are fully embraced however, the path dependency of existing water networks will continue to weigh heavy on decision making. This means that the retrofitting of decentralised components and satellites within existing centralised networks, a 'hybridised' systems approach, is likely to represent the most feasible means of addressing the backlog of vital improvements required to our water management networks.

12.2 Integrating Green and Grey Infrastructure to Slow Down Water

Many of the limitations associated with traditional water management networks reflect their tendency to speed up the passage of water across the landscape. Rain falling onto intensively farmed land or impermeable urban landscapes runs off rather than moving through it, picking up pollutants or eroding soils and their entrained nutrients as it goes. These contaminants force downstream water treatment plants to work harder to produce the high-quality water we demand. The high volumes of runoff also lead to human impacts in the form of flooding, while the lack of natural impediments to flow means that floods also arrive more quickly, giving communities less time to prepare.

More efficient and effective management of water resources requires that the natural environment be regarded as an essential component of water infrastructure. Nature does not waste resources; it seeks to use the minimum amount of energy for a given task and, as such, delivers its services with incredible efficiency. Green infrastructure also acts to moderate the impacts of extreme precipitation and temperature, providing communities with a buffer against the projected impacts of climate change (Foster *et al.* 2011). Harnessing these characteristics has huge potential to significantly increase the performance and resilience of our water management networks.

Figure 12.2, published by UNEP (2014), lists examples of green infrastructure and the broad variety of ecosystem services that each can provide. Importantly, and in contrast to most grey infrastructure, each example of green infrastructure tends to deliver benefits to a number of services. For example, afforestation has the potential to benefit water supply regulation, flood control, water purification, carbon sequestration, enhanced biodiversity, improved air quality, recreation and tourism. Green infrastructure options therefore frequently represent low-regret or no-regret opportunities.

Where green infrastructure can be incorporated or retrofitted into existing water networks, it can often be used to replace or enhance the services traditionally delivered by grey infrastructure alternatives. At the same time, it directly supports the health of

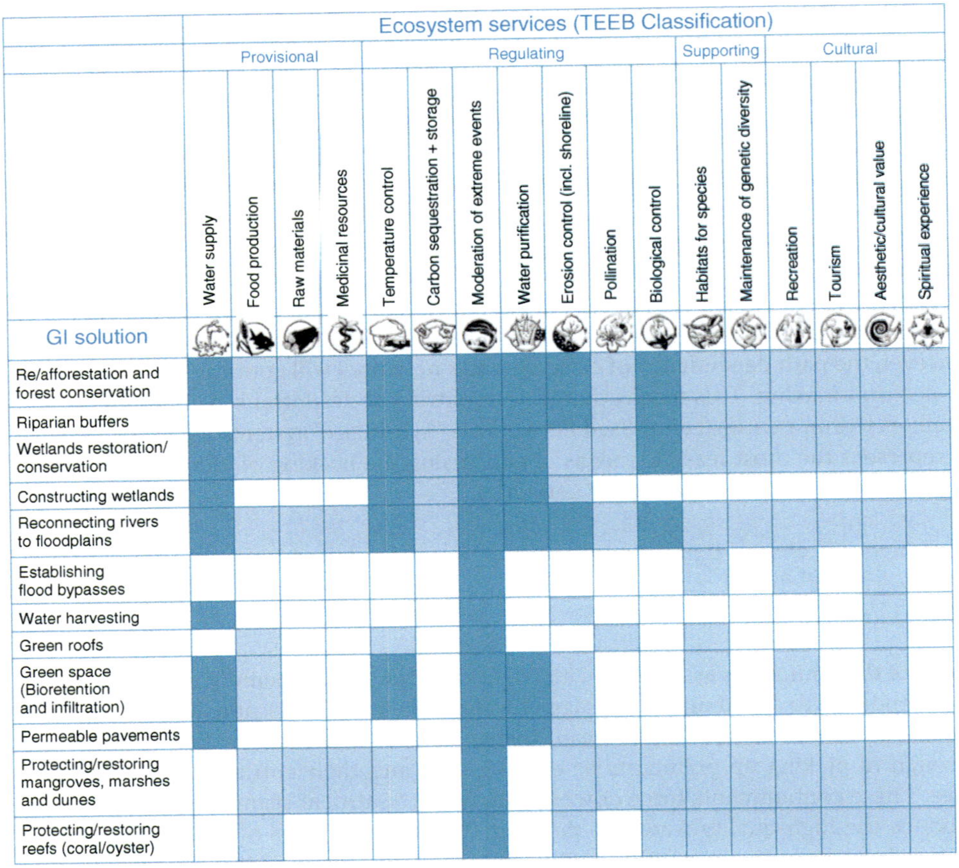

Figure 12.2 Ecosystem services provided by green infrastructure.
Dark shading indicates services directly provided; light shading indicates co-benefits.
Source: Reproduced with permission of UNEP (2014).

the environment, which is the foundation of all ecosystem services. Table 12.2 illustrates how almost all green infrastructure solutions have a grey counterpart in traditional centralised networks. Table 12.2 also shows how green infrastructure can be used throughout a given catchment, not just within urban environments or immediately adjacent to watercourses, but also in upland settings.

Crucially, green infrastructure solutions should be viewed as a means to enhance grey infrastructure networks and not as a reason to abolish them. Although singular adoption of green solutions may be feasible in local settings, it should be remembered that grey infrastructure continues to provide a vital and often irreplaceable public health and industrial support service. The focus should therefore be on integrating green infrastructure within existing grey networks or, in the case of new developments, formulating and implementing that mix of green and grey options that most efficiently provides the functions required or demanded of the system. As an example, treated wastewaters discharged to the Santa Ana River of southern California are managed so that they flow through and experience passive treatment from the Prado Wetlands

Table 12.2 Functions of green and grey infrastructure.

Function/ ecosystem service	Green infrastructure	Grey infrastructure
Water supply regulation	Water harvesting; permeable pavements; re/afforestation	Dams; groundwater abstraction; desalination plants; water recycling; water distribution systems
Water quality regulation	Riparian buffers; wetland restoration; re/afforestation	Water treatment plants
Moderation of extreme events	Reconnecting rivers to floodplains; re/afforestation; water harvesting; green roofs; permeable pavements; restoring mangroves and dunes	Dams; levees; flood walls; urban stormwater drains; coastal protection systems

Source: Adapted from UNEP (2014).

(a piece of natural green infrastructure). High river flows are then captured and pumped underground (grey infrastructure) for storage in an aquifer to be recovered and used by the community during times of scarcity (Hering *et al.* 2013).

Even in those instances where grey infrastructure is the only option, we can still learn from the efficiency of natural systems by mimicking their characteristics. This concept of 'biomimicry' is growing and its application can be seen in a number of products across a whole host of industries. Chapter 4 explains how a mixing impeller designed to replicate the patterns of flow observed in a whirlpool is now being used in more than 200 cities to prevent water stored in reservoirs from stagnating. In testing, the product achieved reductions in energy consumption compared to standard impellers of up to 90% (Harman 2013).

12.3 The Storage Continuum

It is useful to visualise the integration of green and grey infrastructure proposed in this chapter as a storage continuum. The concept, introduced by McCartney and Smakhtin (2010), refers to the simultaneous use of a variety of different types of green and grey infrastructure that each act to slow down the passage of water across the landscape. By applying the initiatives collectively, in a cascading system and at the catchment scale, the overall benefits are optimised in both their magnitude and influence across the water-food-energy nexus. Figure 12.3 schematises the approach, building on the work of Norton and Lane (2012) to show how effective and sustainable water storage can act as the catalyst for the development of a new, wholly integrated approach to water resource management. Its benefits include:

- high flows that may otherwise jeopardise downstream assets are captured for use in agricultural or industrial applications or to recharge aquifers for later use during times of scarcity. Energy can also be generated through this process;
- re/afforestation and farming practices such as the careful tillage of land and the maintenance of ground cover help to reduce flood peaks, while also supporting food production and the provision of raw materials;

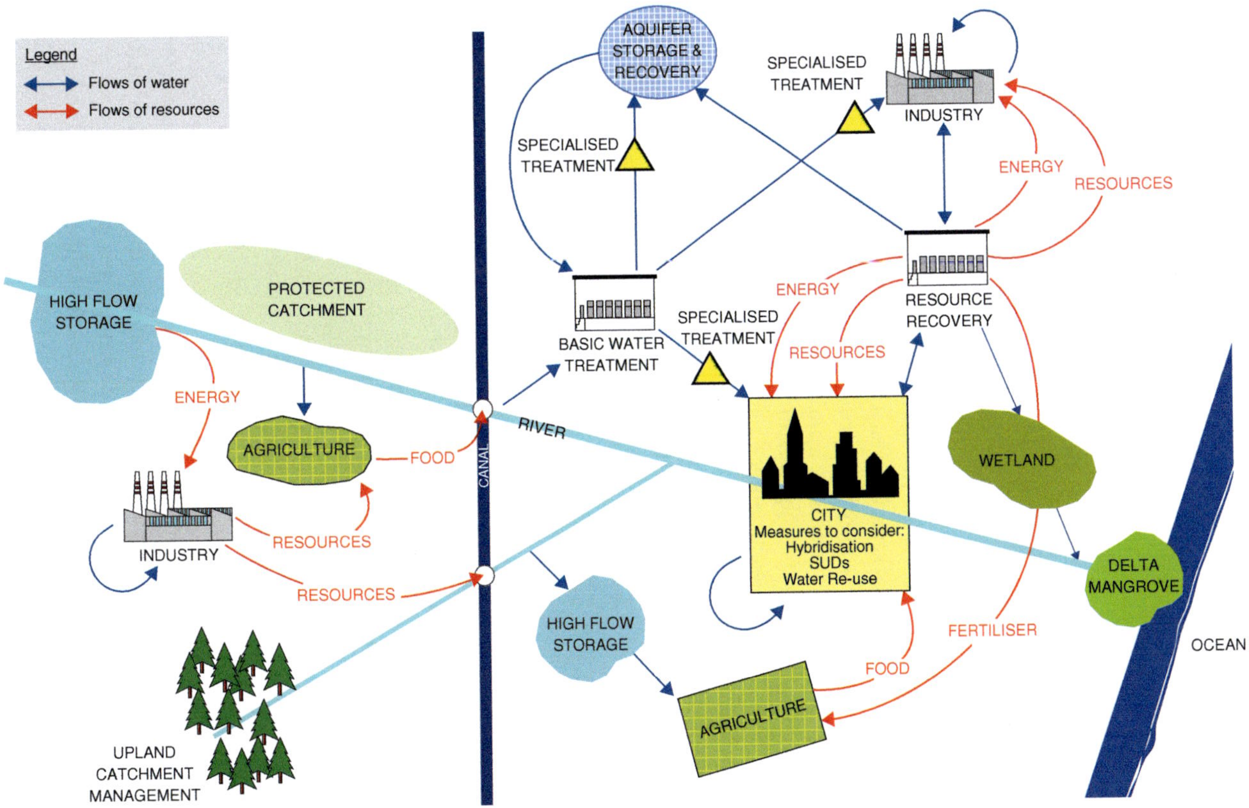

Figure 12.3 The New Water Architecture storage continuum.

- upland catchment management practices slow down water, ensuring that less soil is eroded and that fewer pollutants and nutrients are carried to downstream water treatment plants (hence requiring them to use less energy and chemicals in the treatment process);
- in urban areas, sustainable drainage systems such as permeable paving, rainwater harvesting and green roofs help to reduce instances of acute flooding that may otherwise arise from stormwater runoff. Captured water can be reused while green roofs can also help to reduce the urban heat island effect, limiting energy consumption and improving public wellbeing. Community schemes such as these are also likely to benefit from a local sense of ownership and therefore a proactive willingness to maintain them;
- approaches to wastewater treatment utilise green infrastructure and passive techniques (e.g. wetlands) to improve process efficiency. The embedded resources in 'wastewater' are also recovered and supplied to industries and municipalities;
- natural defences, such as mangroves, estuaries and deltas, are protected and restored so that they effectively buffer the impacts of intense weather events, thereby helping to ensure such events cause less disruption to communities.

Figure 12.3 also includes several, more systemic, enhancements to the current water management paradigm. For example, it highlights how basic water treatment functions could be delivered at a centralised facility with more tailored and specialised treatments applied closer to the end use, thereby facilitating a fit-for-purpose approach to water supply. This concept is discussed further in Section 12.5.

Aquifer storage and recovery (ASR), small water storages and large reservoirs each represent crucial components of the New Water Architecture presented in Figure 12.3. Projections of future climate change point to an increased frequency and magnitude of intense rainfall events, and the success of water management systems will therefore depend in a large part on our ability to capture and exploit high river flows that would otherwise jeopardise downstream infrastructure. For example, in India annual precipitation is concentrated in the four months of the monsoon and then only in a few hours of these months. In fact, most of the country only receives rain for around 100 hours a year (Keller *et al.* 2000). Better utilisation and control of this source of water is therefore vital both to the country's human development and to sustaining its economic growth.

Table 12.3 critiques the characteristics of ASR, small water storages, and large reservoirs. Although each type of grey infrastructure storage has advantages in its own right, it is important to remember that each is also characterised by a number of constraints. In existing water networks that rely on one or a limited number of water storage or supply options, these constraints limit network performance. By taking an alternative approach that incorporates all three storage types into an integrated and cascading system, the overall performance of the system can be significantly enhanced.

Green infrastructure often represents the most effective means of enhancing the performance of the grey infrastructure in a storage continuum. For example, catchment management practices such as the careful tillage of fields for cropping and the protection and restoration of peatlands help to reduce erosion and therefore limit the rate of sedimentation in downstream water storage infrastructure. The development of blue

Table 12.3 Advantages and disadvantages of major storage infrastructure.

Storage type	Advantages	Disadvantages
Small reservoirs	• Operationally efficient and capable of being operated to effectively respond to specific demands. • Relatively cheap and straightforward to construct.	• High proportion of evaporative losses. • The small storage volume is unlikely to allow for seasonal or annual carryover of surplus water. • Subject to sedimentation and the associated reduction in capacity.
Large reservoirs	• Reduced proportion of evaporative loss as compared to small reservoirs. • More likely to provide multi-year carryover of excess water. • Often provide multiple benefits for example through energy generation, flood control, water supply, tourism and navigation.	• Construction often necessitates relocation of communities. • Subject to sedimentation and the associated reduction in capacity. • Complex operative demands from multiple customers make tailoring operations to individual demands challenging.
Aquifer storage and recovery	• Very little evaporative loss although not all stored water can be subsequently recovered. • Movement of water through geological units acts as a form of passive water treatment.	• Rising or falling groundwater levels could mobilise pollutants. • Controlling access to the aquifer may be difficult in some jurisdictions (overuse of the stored groundwater could therefore be an issue).

Source: adapted from Keller *et al.* (2000).

corridors in urban areas represents another example of this approach, acting to safely control flows of water during flooding while otherwise providing valuable recreational and green open space.

An often-quoted case study of the water resource benefit of catchment management comes from New York. In the early 1990s, faced with the challenge of meeting water quality standards, the city was faced with a US$ 10 billion bill to design and construct a filtration facility. Instead, the municipality chose to invest in purchasing land and restoring habitat and buffer zones in its water supply catchments of the Catskill and Delaware Mountains (Hering *et al.* 2013). Annual operation and maintenance costs for these land management initiatives are estimated at around US$ 300 million (Hering *et al.* 2013), representing a significant financial saving to the city.

Examples of the efficiencies associated with enhancing water management networks by integrating green infrastructure can also be found in industry. For example, the cement manufacturer Lafarge found that by maintaining existing natural ecosystems, it avoided the need for erosion control and nutrient removal infrastructure at two of its quarries in the USA; these interventions would otherwise have cost the company more than US$ 2 million a year (Wong *et al.* 2014). In the Sarapiquí watershed of Costa Rica, a hydropower company pays upstream landholders to maintain and restore degraded forests in order to avoid the costs of reservoir dredging and to benefit from the resulting more reliable stream flow (Krchnak *et al.* 2011).

Although actively managed green-grey water systems are more likely to provide multiple benefits and therefore improve overall system performance, it should be remembered that green infrastructure can exhibit a number of constraints that require careful control. For example, in green treatment systems (such as engineered wetlands), the plants and microbes that do the work tend to be less active in winter as well as being strongly influenced by the local characteristics of a given site. This means that the treatment efficacy of green infrastructure tends to vary over time and space to a much greater degree than that of grey treatment alternatives. This geographic variability in performance also means it is often inappropriate to transfer the knowledge gained from one application of green infrastructure to another, a constraint that acts to hold back wider uptake of such systems. That said, the growing number of successful green infrastructure schemes is beginning to allow planners to more accurately quantify the potential financial benefits. For example, using case studies in Portland and Seattle, Emerton and Bos (2004) were able to conclude that for every US$ 1 invested in watershed protection, up to US$ 200 in opportunity costs for new water treatment facilities could be saved.

12.4 Creating Hybrid Water Management Systems

Hybridisation refers to the process by which, in developed country cities, enhancements to existing water management networks are most likely to be made. In modern urban environments, the effects of infrastructure lock-in and path dependency mean that broad overhaul of the existing water networks is usually infeasible. Retrofitting green infrastructure and integrating decentralised satellite networks do however represent viable and low-regret enhancements that have the potential to deliver benefits to a broad array of human activities. Table 12.4 summarises some of their key characteristics.

Table 12.4 Characteristics of hybrid water management systems.

Characteristic	Description	Benefit
Multiple water supply sources	Hybrid systems avoid reliance on a single or limited number of water sources and instead combine multiple sources within an integrated network. Localised water supply satellites are also often embedded within the broader system (such as through the capture and reuse of stormwater in an urban development).	Flexible water supply; improved resilience to supply impairment.
Staged and modular infrastructure	Future scenarios of water supply and demand are inherently uncertain and so hybrid systems prioritise infrastructure that can be constructed in stages or enhanced with additional modules. This allows infrastructure to be adapted as demographic responses to population growth and hydrological responses to climate change become clearer.	Allows for progressive system improvement in spite of future uncertainties.

(Continued)

Table 12.4 (Continued)

Characteristic	Description	Benefit
Fit-for-purpose water supply	Hybrid systems focus on supplying water of a quality that matches that demanded by its end-user. This aim could be achieved by treating all water to a basic standard at a centralised facility and then utilising specialised treatment infrastructure in closer proximity to the customer. As a result, energy and resource consumption for treatment would be reduced.	Efficient and integrated use of resources; minimal waste generation.
Cycles of resource reuse	By improving collaboration between different resource users, cycles of reuse can be established whereby the wastes from one activity provide beneficial inputs to another. The final and ultimate waste volumes generated by a hybrid water management system are therefore significantly reduced.	Efficient and integrated use of resources; minimal waste generation.
Improved data management	Advances in data monitoring, analytics, management databases and response systems enable integrated infrastructure networks to be operated precisely and in a manner that permits effective and quick response to variations in supply and demand.	Effective response to changes in supply and demand; improved resilience to system failures.

No single option or scheme, whether green or grey, is likely to simultaneously maximise water reuse, enhance system resilience and limit the generation of waste. Identifying the right mix of options is a significant challenge, so effective decision support frameworks and planning tools are an important requirement for the industry. Insight could be gained from the other centralised infrastructure networks that characterise our cities (e.g. electricity supply) and that, as is the case for water, are increasingly experiencing pressures that force decision makers to look to alternative service models.

A good example of a hybridised approach to water management can be seen in the evolution of water supply planning in eastern Spain. Regional water supply in Spain is complicated by a strong geographical imbalance between water availability and demand, with populations concentrated on the east coast and the majority of renewable water sources hundreds of kilometres to the northwest. Figure 12.4 shows how policies to address this constraint have evolved over the past 25 years. They began with calls for a complex network of inter-basin transfers, moved to a proposal for a single, large water transfer scheme and are now transitioning to a more holistic approach which prioritises the diversification of water sources (e.g. through wastewater reuse) and the derivation of greater value from existing assets. Muñoz *et al.* (2010) evaluated the lifecycle environmental impacts of the proposal for a single inter-basin transfer (the Ebro River Water Transfer Scheme) and the hybridised approach (the AGUA Programme), and found that the latter had a lower environmental impact in almost all categories thanks to its reduced energy demands and lower resource intensity.

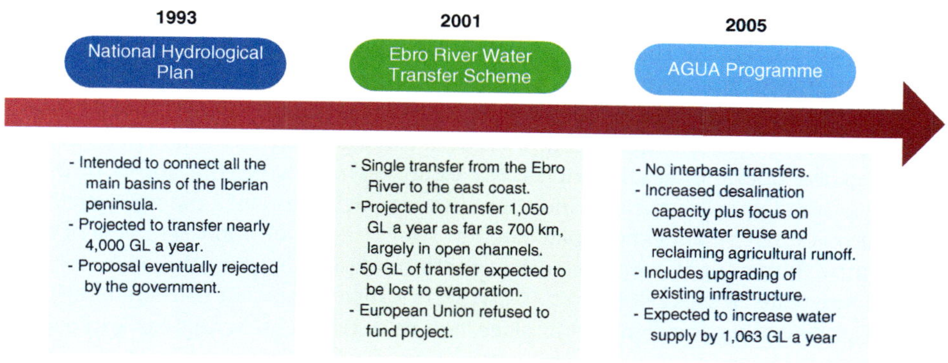

Figure 12.4 Water supply schemes for eastern Spain.
Source: Adapted from Muñoz *et al.* (2010).

12.4.1 The Challenge of Maintenance and Long-Term Responsibility

Operational responsibilities are key issues for green and decentralised infrastructure, with requirements for maintenance often being much more frequent, varied and complex than those typical of conventional networks. As a result, staff with specific specialised skills may be needed, a requirement that must be accounted for in long-term planning. Despite this complexity, the outsourcing of maintenance responsibilities for decentralised infrastructure to individuals or NGOs is common in developing countries and an increasing trend in developed nations, driven by municipalities that lack the human resource capacity or that are seeking to reduce public spending. Unfortunately, the newly responsible party is often no better placed to sustain effective maintenance in the long term. As a result, decentralised and green infrastructure systems often prematurely fail or suffer from rapidly deteriorating performance. In 2009 for example, a water recycling satellite scheme was constructed for a residential development in Gold Coast, Australia. Although the scheme initially allowed the development to achieve a high degree of water self-sufficiency, a lack of sustained investment in maintenance lead to the reintroduction of a fully centralised water management system after just five years of operation.

In cases where individuals or businesses are made responsible for the maintenance of decentralised equipment, there may be a strong reluctance to accept ongoing and active involvement. This is particularly true if the initiative was imposed without their engagement and especially if a less onerous alternative, such as that presented by a centralised water network, is available (Nanninga *et al.* 2012). As an example, McCartney *et al.* (2013) found that of the 4,000 rainwater harvesting ponds constructed between 2000 and 2008 for households and small communities in the Amhara region of Ethiopia, less than half remained functional in 2009.

Better accounting of the maintenance requirements of green and decentralised infrastructure will naturally improve (albeit slowly) as the number of working examples increases. In order to accelerate this process however, knowledge transfers need to be streamlined; this requires practitioners to more readily engage outside of their immediate professional spheres (see the 'water box' referred to in Chapter 10). Community perception and buy-in to decentralised and green infrastructure is a critical determinant of their performance, much more so than for grey alternatives. As a result, fully inclusive engagement from a range of professionals and stakeholders, notably social scientists, will help to ensure not only that more examples of innovative water management approaches are shared, but that their long-term success is more likely to be sustained.

Significant improvements to the actual maintenance regimes of decentralised infrastructure can also be achieved. Modern approaches to asset management that focus on the application of sensors and streamlined data management have the capability to greatly improve operational efficiencies. For example, harnessing the power of these techniques can support faster and more precise leak detection and the cost-effective prioritisation of pipe repairs and replacements. They can also be applied to help operators minimise the extent of over-pressurised sections of pipe that are otherwise responsible for large water losses (Hering *et al.* 2013). It is also important to remember that with the right enabling environment in place, community-level schemes can succeed in galvanising a local sense of ownership that, in turn, leads to sustained system performance. In Australia for example, the number of homes with rainwater tanks increased by 37% between 1994 and 2007 (Moglia and Sharma 2013). This reflected policy initiatives and incentives to address the Millennial Drought but was also reliant on widespread public acceptance of the need to address water scarcity. Rates of domestic water consumption also fell significantly and, in contrast to many other global examples, these low rates were largely sustained when the drought eventually broke (Head 2014).

12.5 Circular Systems that Transform 'Wastes' to 'Resources'

The traditional, centralised approach to water management essentially considers all by-products of human interaction with water as wastes. As a result, society disregards and foregoes a whole host of valuable resources, for example nutrients, organic matter, embedded heat and kinetic energy. By 'mining' (rather than 'managing') the resources of the natural environment, societies have been able to ignore this profligacy, instead relying on rapidly depleting natural stores to meet our insatiable demand for resources. However, as our exploitation of these stores reaches and increasingly passes their respective sustainable limits, decision makers are recognising that the embedded resources of so-called 'wastewater' are too valuable to continue to ignore.

To better understand and exploit the resources present in wastewater, it is vital to first recognise that the wastes from one human activity can provide valuable inputs to many others. If this shift in mindset can be achieved, it quickly becomes apparent that the separately managed water and wastewater networks that typify most cities no longer meet the characteristics required of a sustainable urban water supply. The alternative is to consider our wastewater treatment plants as resource factories, harvesting and regenerating useful products that can provide the inputs to a wide variety of industrial, agricultural and domestic activities. Importantly, this recovery of resources doesn't just benefit the recipient industries. It also provides the responsible utility with the opportunity to diversify its revenue base, helping to make the establishment of a long-term financially viable business or operation far more likely.

New technologies and advancements in water treatment are increasingly providing the means to isolate specific components of wastewater so that they can be regenerated into valuable secondary products. Table 12.5 highlights a few of the increasing number of case studies of water resource recovery.

Table 12.5 Case studies of resource recovery.

Case study	Description
Treated wastewater reuse: various	In Europe in 2006, the percentage of treated wastewater reused in beneficial applications was just 2.4% (European Commission 2015). In Greece, Italy and Spain, water reuse constituted only 5–12% of treated urban effluents in 2006, while much higher rates are encountered in Cyprus (almost 100%) and Malta (about 60%). In Israel in 2011, 73% of treated wastewater was reused, principally to irrigate agriculture in peri-urban areas (Rygaard *et al.* 2011). Wong *et al.* (2014) estimate that the countries of the Middle East and North Africa reuse more than 50% of their treated wastewater for irrigation. In Orange County, southern California, the region's water supply and wastewater strategies have been integrated to support the development of a groundwater replenishment scheme which stores treated wastewater underground for use during times of scarcity.
Direct potable reuse: Windhoek, Namibia	Examples of direct potable reuse schemes for treated wastewater are rare. Windhoek, the capital of Namibia, is the most-often-cited example, and has been successfully practicing direct potable reuse of treated wastewater since 1968. The city's wastewater treatment plant serves a population of 220,000, reclaiming municipal wastewater to potable quality standards to meet around one-quarter of the city's drinking water needs (2030 WRG 2013). Unfortunately, other examples of direct potable reuse are rare, constrained by public distrust and a perceived lack of system control (Rygaard *et al.* 2011). In the last few years however, several small-scale direct reuse schemes have been implemented in California, fuelled by the persistent drought of the region. Large cities in the state are also seriously considering the option.
Anaerobic digestion: California, USA	Anaerobic digestion of sewage is less energy intensive than aerobic treatment and also results in the production of biogas that can be burnt to generate electricity. Through this process, a wastewater treatment plant in Oakland, California is the first in the USA to become a net producer of electricity. It does so by pooling domestic sewage and food waste from nearby farms, food processing facilities, restaurants and wineries (Fulcher 2014).
Phosphorus recovery: Slough, UK	A treatment module at Slough sewage works in the UK recovers struvite, a compound containing phosphorus and ammonia, from wastewater. The product, which would otherwise cause scaling of the treatment system, is recycled into a high-quality fertiliser for agricultural use. As well as providing 150 tonnes of fertiliser a year, the treatment module (the first of its kind in Europe) also reduces the amount of chemical dosing required at the facility, saving the utility up to £200,000 a year (Ostara 2013).
Heat recovery	Heat exchangers retrofitted to sewerage pipes can be used to extract heat that can then be applied to heat water or to supplement space heating. Pamminger *et al.* (2013) estimate there to be over 500 wastewater heat pumps in use around the world; several systems in northern Europe have been operating successfully for more than 30 years.

(Continued)

Table 12.5 (Continued)

Case study	Description
Kinetic energy recovery	Energy recovery hydro turbines can be used to generate power from excess water pressure in pipes. In an assessment of potential applications in Ireland, McNabola *et al.* (2014) found that at one site, installation of a turbine on a water supply pipeline could produce enough energy to power between 200 and 330 homes. Potential recovery rates at other sites were lower, although recovered energy could still provide an important benefit through the powering of remote telemetry systems. A turbine on a water tunnel supplying the town of Innsbruck, Austria generates up to 6 MW of electricity, enough to supply a few thousand homes (Choulot *et al.* 2012).
Water and resource reuse: Gippsland Water Factory, Victoria, Australia	Located in Victoria, Australia, the Gippsland Water Factory is an integrated municipal and industrial water reclamation plant producing high-quality recycled water for use by a local paper mill and treated secondary effluent for use in agriculture (A. Hodgkinson, pers. comm., 2014). The location of the facility ensures that influent wastewaters and recycled water travel the minimum distance necessary, thereby reducing operational pumping costs. The facility's anaerobic treatment process results in the cogeneration of heat and power from biogas, while a micro-hydropower station captures excess kinetic energy from water flowing through pipes. The electricity produced through these means meets up to 40% of the facility's demand (A. Hodgkinson, pers. comm., 2014).

Examples such as those presented in Table 12.5 are indicative of a gradual but growing trend away from the 'take-use-dispose' approach to resource exploitation, towards one that is far more regenerative and self-enhancing. The concept of a 'circular economy' embodies this ethos, aiming to facilitate the cyclical reuse and recycling of products in order to decouple economic growth from the unsustainable exploitation of natural resources. The benefits of a circular economy are numerous and wide-ranging, and include:

- cascading reuse of a given input through a variety of applications in order to derive maximum benefit for the minimum of waste; in the case of water, domestic grey water could be reused in the home to flush toilets and wash cars while spent cooling water from a power plant could be used to heat nearby homes and commercial premises. Through these regenerative cycles, only a very minor proportion of the original water supply is actually discarded; this means that the final water treatment facility has to work less hard to ensure that residues returned to the environment cause no harm;
- more widespread application and adoption of repair services, for clothes for example, as opposed to the 'throw-away' culture that characterises most modern consumption;
- recycling products into their component parts to be used in new products and applications (facilitated not just by more efficient recycling processes, but also through the design of the products themselves); and

- increased focus of business on providing services as opposed to just selling products. This could take the form of sharing and rental schemes for products that are typically used infrequently.

Through the adoption of these initiatives, resource consumption in a circular economy becomes defined by three primary characteristics:

1) durable products are repaired and reused through as many cycles as possible;
2) consumable products (those whose performance deteriorates more rapidly than durable products) are designed to be used in as many cascades as possible before being recycled into their component parts for reuse; and
3) all natural capital stocks are used only to the extent that they can be regenerated through natural processes in appropriate timeframes.

Importantly, these characteristics don't just sustain natural resources and the environment, they provide unique opportunities to enhance the efficiency of human activities and therefore represent important sources of competitive advantage in business. By adopting circular principles, businesses can shelter themselves from the increasing volatility of resource prices while also reducing operating expenditure, two of the primary factors governing financial performance. A circular approach also requires a business to develop an intimate understanding of its customers, enabling it to more effectively balance supply and demand, satisfy its customers and therefore grow its customer base. In the European Union, it has been estimated that every 1% increase in resource efficiency is worth as much as €23 billion for business and has the potential to create up to 200,000 jobs (Accenture 2014). Some commentators have argued that the only way to protect our environment for future generations is to shift from a capitalist to a steady-state economy that no longer prioritises economic growth (e.g. Alexander 2014). In a circular economy however, economic growth and the sustainable management of natural resources can both be achieved simultaneously.

In order to incentivise the development of a circular economy, improvements need to be made in the efficiency with which natural resources are supplied to their users. Because treatment occurs at a centralised facility in most water management systems, all water must be treated to the standard required by the most demanding user. This means that, although drinking water needs often make up only a small fraction of total water demand, all water is treated to this quality in most cases. A more efficient system might tailor the degree of water treatment to the needs of each group of users. In Los Angeles, USA a centralised treatment facility produces five different qualities of water that, in turn, are supplied to separate users that include cooling towers, industrial boilers, landscape irrigation, and recharge of groundwater (Hering *et al.* 2013).

Another alternative would be to provide only a basic level of water treatment at the centralised facility, equivalent to that required by the least demanding user. That quality of water would then be delivered to all customers where, at or close to the point of end use, small specialised treatment facilities would provide final treatment. Such an approach would minimise energy consumption and chemical inputs, but would also require a large number of specialised treatment units that, in turn, would incur more complex operation and maintenance regimes. Recent innovations in data and asset management (see Section 12.4.1) do however suggest that the cost-effective management of complex systems such as these is possible.

The more efficient supply of water would also be supported if different users with symbiotic resource needs chose to co-locate their facilities. For example, if a business is located in close proximity to another whose wastewater it can use, then no interim water treatment or, importantly, the significant energy expenditure associated with pumping water long distances, are required. This idea of industrial symbiosis was introduced in Chapter 4 and can be seen in several European case studies (e.g. Kalundborg in Denmark). The Gippsland Water Factory described in Table 12.5 is also a good case study of this approach.

12.6 Conclusions

The physical integration of infrastructure represents the final component of the three water management initiatives proposed in this book. It relies on society recognising the vital importance of water to almost all human activities, and on institutions to reflect this social stance in enabling policies, strategies and regulation. With these foundations in place, decision makers can make enhancements to physical infrastructure with the necessary confidence that they will receive widespread support.

This chapter has highlighted the increasing vulnerability and inefficient performance of traditional water networks. To address these limitations, it has advocated a new approach characterised by the targeted enhancement of legacy infrastructure with both green and grey alternatives. Collectively, these interventions combine to:

- slow down the passage of water across the landscape;
- capture excess water that would otherwise jeopardise downstream assets;
- improve the efficiency of water treatment in order to deliver water of a quality that is fit for purpose;
- harness the embedded resources in wastewater; and
- minimise the generation of waste.

By sustaining and enhancing nature's ability to provide ecosystem services, these initiatives enable significant improvements in the management of not just water, but also the resources of food, energy, land and biodiversity. They also represent a rare opportunity for businesses to secure competitive advantage, encourage innovation and, ultimately, to fuel economic growth.

Although delivering on this New Water Architecture represents a challenging ambition, the case studies presented in this and other chapters show that each component (the conceptual, institutional and physical) can be and have been achieved; albeit, these successes have so far typically occurred in discrete and isolated cases. To scale up and integrate the initiatives, we need to first secure a fundamental recognition by all stakeholders that water is absolutely the number one resource and catalyst for social wellbeing, economic growth and environmental health. It will then take the collective advancement of institutional and physical water management to ensure that future generations exploit their water resources in a manner that provides the necessary foundation for prosperous livelihoods while sustaining the consumption of natural resources within safe limits.

References

2030 Water Resources Group. 2013. Managing Water Use in Scarce Environments: A Catalogue of Case Studies. 2030 Water Resources Group, Washington DC, USA.

Accenture. 2014. Circular Advantage: Innovative business models and technologies to create value in a world without limits to growth. Accenture, Dublin.

Alexander, S. 2014. Degrowth and the Carbon Budget: Powerdown strategies for Climate Stability. Available at: http://simplicityinstitute.org/wp-content/uploads/2011/04/DegrowthandtheCarbonBudgetSamuelAlexander1.pdf (accessed February 2017).

Black & Veatch Corporation. 2014. Strategic Directions: U.S. Water Industry – a Black & Veatch Report. Black & Veatch, Kansas, USA.

Brown, R., Keath, N. and Wong, T. 2008. Transitioning to water sensitive cities: historical, current and future transition states. In Proceedings of 11th International Conference on Urban Drainage. Edinburgh, Scotland.

Choulot, A., Denis, V. and Punys, P. 2012. Integration of small hydro turbines into existing water infrastructures. In Samadi-Boroujeni, H. (ed.) Hydropower – Practice and Application. InTech, Croatia.

Danilneko, A., van den Berg, M. M. and Moffitt, J. 2014. The IBNET Water Supply and Sanitation Blue Book 2014. World Bank. Washington DC, USA.

Emerton, L. and Bos, E. 2004. Value: Counting Ecosystems as a Water Infrastructure. International Union for Conservation of Nature, Gland, Switzerland.

European Commission. 2015. Optimising Water Reuse in the EU – Final Report – Part I. Brussels, Belgium.

Foster, J., Lowe, A. and Winkleman. S. 2011. The Value of Green Infrastructure for Urban Climate Adaptation. The Centre for Clean Air Policy, Washington DC, USA.

Fulcher, J. 2014. California Wastewater Utility Makes History as Net Energy Producer. Available at: http://news.wef.org/california-wastewater-utility-makes-history-as-net-energy-producer/ (accessed February 2017).

Grant, S.B., Saphores, J.-D., Feldman, D.L., Hamilton, A.J., Fletcher, T.D., Cook, P.L.M., Stewardson, M., Sanders, B.F., Levin, L.A., Ambrose, R.F., Deletic, A., Brown, R., Jiang, S.C., Rosso, D., Cooper, W.J. and Marusic, I. 2012. Taking the "waste" out of "wastewater" for human water security and ecosystem sustainability. *Science*, 337, 681–686.

Harman, J. 2013. The Shark's Paintbrush: Biomimicry and How Nature is Inspiring Innovation. Nicholas Brealey Publishing, London, UK.

Head, B.W. 2014. Managing urban water crises: adaptive policy response to drought and flood in Southeast Queensland, Australia. *Ecology and Society*, 19, 2.

Hering, J.G., Waite, T.D., Luthy, R.G., Drewes, J.E. and Sedlak, D. 2013. A changing framework for urban water systems. *Environmental Science and Technology*, 47, 10721–10726.

Keller, A., Sakthivadivel, R. and Seckler, D. 2000. Water Scarcity and the Role of Storage in Development. International Water Management Institute Research Report 39, IWMI, Colombo, Sri Lanka.

Krchnak, K.M., Smith, D.M. and Deutz, A. 2011. Putting Nature in the Nexus: Investing in Natural Infrastructure to Advance Water-Energy-Food Security. International Union for Conservation of Nature and The Nature Conservancy, Gland, Switzerland.

Liu, J., Zang, C., Tian, S., Liu, J., Yang, H., Jia, S., You, L., Liu, B. and Zhang, M. 2013. Water conservancy projects in China: achievements, challenges and way forward. *Global Environmental Change*, 23, 633–643.

Marlow, D. and Tjandraatmadja, G. 2014. Challenges associated with the hybridisation of water systems. *Journal of the Australian Water Association*, 41(3), 60–66.

Matthews, J.H., Wickel, B.A.J. and Freeman, S. 2011. Converging currents in climate-relevant conservation: water, infrastructure and institutions. *PLoS Biology*, 9(9), 1–4.

McCartney, M. and Smakhtin, V. 2010. Water Storage in an Era of Climate Change: Addressing the Challenge of Increasing Rainfall Variability. International Water Management Institute, Colombo, Sri Lanka.

McCartney, M., Rebelo, L.-M., Xenarios, S. and Smakhtin, V. 2013. Agricultural Water Storage in an Era of Climate Change: Assessing Need and Effectiveness in Africa. International Water Management Institute, Colombo, Sri Lanka.

McNabola, A., Coughlan, P. and Williams, A.P. 2014. Energy recovery in the water industry: an assessment of the potential of micro-hydropower. *Water and Environment Journal*, 28, 294–304.

Moglia, M. and Sharma, A.K. 2013. The need for monitoring the social performance of decentralised water systems. CSIRO: Water for a Healthy Country National Research Flagship. Available at: https://publications.csiro.au/rpr/download?pid=csiro:EP104719&dsid=DS3 (accessed February 2017).

Muñoz, I., Milà-i-Canals, L. and Fernández-Alba, A.R. 2010. Life cycle assessment of water supply plans in Mediterranean Spain: the Ebro River transfer versus the AGUA programme. *Journal of Industrial Ecology*, 14(6), 902–918.

Nanninga, T.A., Bisschops, I., López, E., Martínez-Ruiz, J.L., Murillo, D., Essl, L. and Starkl, M. 2012. Discussion on sustainable water technologies for peri-urban areas of Mexico City: Balancing urbanization and environmental conservation. *Water*, 4, 739–758.

Norton, M. and Lane, A. 2012. 'New Water Architecture': an integrated water management model'. *Proceedings of the Institution of Civil Engineers: Management, Procurement and Law*, 165(3), 159–171.

Pamminger, F., Scott, D., Aye, L. and Jelbert, R. 2013. Heat energy recovery potential from sewers in Melbourne: a study to determine whether heat extraction from sewers is technically and financially viable. *Water Journal*, November 2013. Available at: http://www.ndynet.com/sites/default/files/Heat%20from%20sewers%20AWA%202013.pdf (accessed February 2017).

Pikaar, I., Sharma, K.R., Shihu, H., Gernjak, W., Keller, J. and Zhiguo, Y. 2014. Reducing sewer corrosion through integrated urban water management. *Science*, 345(6198), 812–814.

Rygaard, M., Binning, P.J. and Albrechtsen, H.-J. 2011. Increasing urban water self-sufficiency: new era, new challenges. *Journal of Environmental Management*, 92, 185–194.

Savić, D.A., Bicik, J., Morley, M.S., Duncan, A., Kapelan, Z., Djordjević, S. and Keedwell, E.C. 2013. Intelligent urban water infrastructure management. *Journal of the Indian Institute of Science*, 93, 319–336.

Shilling, J.D., Chomitz, K. and Flanagan, A.E. 2007. The Nexus Between Infrastructure and Environment. Evaluation Cooperation Group of the International Financial Institutions, World Bank, Washington DC, USA.

Ostara. 2013. Fertiliser from sewage saving the planet – and cash. Available at: http://ostara.com/project/fertiliser-from-sewage-saving-the-planet-and-cash/ (accessed February 2017).

United Nations Environment Programme. 2014. Green Infrastructure Guide for Water Management. UNEP, New York, USA.
United States Environmental Protection Agency. 2009. Drinking Water Infrastructure Needs Survey and Assessment: Fourth Report to Congress. US EPA, Washington DC, USA.
Wong, C., Hicks, L. and Clinton, L. 2014. Evaporating Asset: Water Scarcity and Innovations for the Future. SustainAbility, New York, USA.

13

A Way Forward

To address the water management challenges described in Parts I, II and III of this book, we have proposed here in Part IV a new systems-based framework of conceptual, institutional and physical integration that we term New Water Architecture.

We know that many other entities, writers and commentators are also making assessments of how to meet these challenges. While we know that our proposals are not a 'silver bullet' we, like many others, feel strongly that integrated water management must be better and more widely understood and more effectively implemented if water crises are to be averted. We do not want to be simply another shrill cry in the gathering frequency, pace and urgency of calls for water to be managed sustainably. Rather, we wish to propose a pragmatic framework within which progressive and real steps towards sustainable water management can be made.

To conclude this book, in this final chapter we describe what we consider to be a series of achievable (albeit challenging) steps towards New Water Architecture. We have tried to avoid the syndrome of a multiplicity of proposed actions (common to many international policies and strategies). Instead, we have identified nine steps , presented in Figure 13.1, that if successfully implemented would make a significant contribution to truly integrated water management and thereby lay the foundations for a New Water Architecture.

13.1 Conceptual Integration

At its most elemental, conceptual integration represents a societal understanding of the true role of water in how our planet functions: of water being rightly viewed as the 'bloodstream of the biosphere'. Currently, water is all too often managed and regulated in the silos of other sectors, and from a platform of undervaluation that fails to appreciate its wider roles. The potential power of IWRM and effective decision making is therefore compromised.

Conceptual integration refers to this mindset of the wider role of water. If we are to see a new appreciation by society and politicians of the fundamental role played by water, and if we are to see conceptual integration emerge, then water professionals have a pivotal role to play. Physical and institutional integration cannot evolve without first establishing this mindset. To address these barriers, water professionals must 'get out of the box' and enter the less comfortable world of politicians, policy makers, commercial

Water Resources: A New Water Architecture, First Edition.
Alexander Lane, Michael Norton and Sandra Ryan.

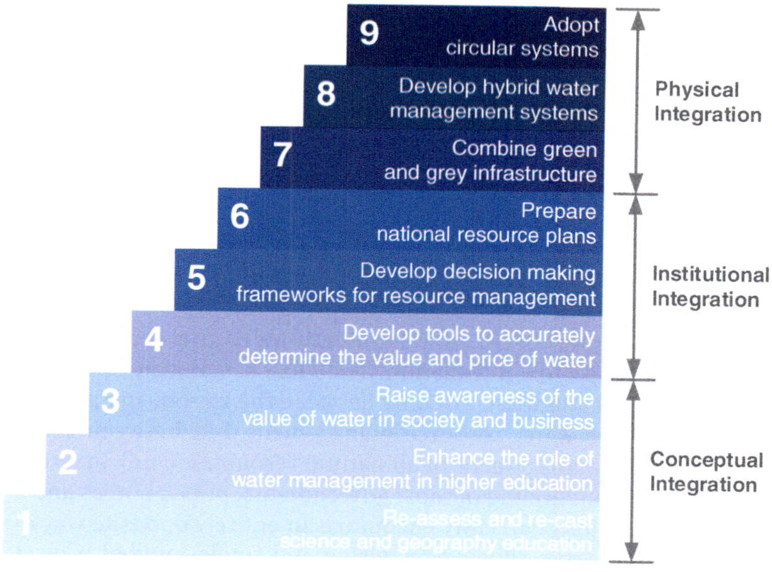

Figure 13.1 Nine Steps to a New Water Architecture.

business and influencers within which decisions over practical initiatives are made. We believe that all steps on the path to conceptual integration will require water professionals to make this commitment.

1) **Step 1 is to re-assess and re-cast the education syllabuses of science and geography** to reflect the role and value of water in the biosphere. This depends on the designers of syllabuses having sufficient knowledge of the role and importance of water and on them possessing the necessary capability to initiate change. Water professionals therefore need to work closely with the education sector and the relevant political powers to highlight and facilitate the changes.
2) **Step 2 is to enhance the role of water management within and across Higher Education courses** that currently largely consist of isolated syllabuses on hydrology, water quality, engineering and hydraulics. We can see examples of the required integration already happening (e.g. water-security-related masters courses at the Universities of Saskatchewan, Canada and East Anglia, UK), but we would like to see undergraduate courses in engineering, geography and environmental sciences take a much deeper look at water and its value to society, business and the biosphere. While the water professionals who work in the higher education sector should have sufficient insight to do this, we often find that their experience is too focused on specific technical disciplines, which in turn become the syllabus subjects.
3) **Step 3 is to raise awareness in society and business of the role played by water** and its true value in comparison to its current price. In some senses, this task is more difficult to achieve than Steps 1 and 2 because there is no captive audience for water professionals to engage. There are however many mediums through which knowledge transfer can take place, for example newspapers, blogs, radio, television and social media. The challenge is to harness communication pathways through social

scientists to ensure that society embraces a subject matter that, on the face of it, may offer little immediate interest to members of society or business managers. We believe that learned societies have an important role to play by engaging with the communications sector. We can learn much from the activities of those NGOs who have successfully engaged society and business such as the World Wildlife Fund and the Carbon Disclosure Project.

13.2 Institutional Integration

We have shown how the relationships between water and other natural resources are too complex to be managed successfully in isolation. The most sustainable outcomes can only be achieved through collaboration, not only between the various institutions and organisations responsible for water resource management, but between those organisations with a stake in the management of all natural resources. Unfortunately, we have found that the internal functioning of most governments, public bodies and a large number of private organisations currently encourage an entirely opposite way of working: one characterised by inertia, siloed strategising and single-minded project delivery.

To transform this approach, we believe that institutional structures must recognise the role and value of water in vertical directions (between different levels of governance) and horizontal directions (between different stakeholders). We have identified the five key enablers of such an approach

- *Transformational leaders*: that act out of a sense of responsibility, not just on the need to deliver a return on investment or to secure re-election.
- *Collective partnerships*: within which all stakeholders are driven by recognition that their own interests are best served through the achievement of collective goals.
- *Coordinated governance*: that harmonises the aims of different departments and organisations in different sectors (horizontal integration) and that balance bottom-up and top-down governance (vertical integration).
- *Genuine stakeholder engagement*: that moves away from 'public relations' to embrace continuous and transparent stakeholder engagement, ensuring that it becomes an everyday process, not just forced by regulation.
- *Technology transfer and decision support*: to ensure that integrated resource management becomes standardised as an everyday process and to ensure that success stories and solutions are shared openly and globally.

To put these enablers in place is a huge undertaking, but we propose three steps that will help make a start on this path. These steps are not sequential, though Steps 4 and 5 would help prepare for Step 6.

4) **Step 4 is to develop tools and methods that can accurately determine the value and price of water**, including the elusive ecosystem services component. We have cited several examples of progress already made towards this goal; however, to date, they have tended to represent niche initiatives, fringe papers or isolated reports. The aim must be to have this work perceived to be as important as the assessment of oil, sugar or copper prices.

5) **Step 5 is to develop decision support frameworks and tools** that allow rational decisions to be taken about resource management, appropriately informed by the true role and value of water in all sectors. The challenge is one largely of achieving consistency of approach vertically and horizontally within resource sectors, institutions and organisations, and this requires the effective exchange of knowledge and technology.
6) **Step 6 is to prepare Natural Resource Plans** at global, regional, national and subnational levels. This ambition sounds seductively simple but will be incredibly challenging, requiring at every level of governance a transformational leader willing to step up and bridge our current siloed planning frameworks. For example, developing a national plan will require ministries who deal with energy, minerals, agriculture, water, environment and finance to become much more aware of their interdependencies. Although we believe that Step 6 can be achieved now, the process will be smoothed and its outcomes improved if Steps 4 and 5 can be achieved first.

13.3 Physical Integration

Physical integration of infrastructure represents the third and final component of New Water Architecture. Its success will rely on recognition by society and business of the vital importance of water to almost all human activities, and on institutions to reflect this social stance in enabling policies, strategies and regulation. With these foundations in place (our Steps 1 to 6), decision makers will be able to make rational, effective and integrated enhancements to physical infrastructure which:

- slow down the passage of water across the landscape;
- capture excess water that would otherwise jeopardise downstream assets;
- improve the efficiency of water treatment in order to deliver water of a quality that is fit for purpose;
- harness the embedded resources in wastewater; and
- minimise the generation of waste.

Ultimately, physical integration will be the basis of an advanced form of water stewardship signalling the emergence of New Water Architecture. We propose three steps through which progress can be made on this journey:

7) **Step 7 is to invest in green and grey infrastructure that, together, deliver collective outcomes at the catchment scale.** These networks would both slow down the passage of water across land and cityscapes and store the high water flows that may otherwise damage downstream assets or be lost to the ocean. There are already many examples of successful green infrastructure, but their development at catchment scales remains hampered by difficulties associated with predicting their performance and by institutional barriers. Addressing these issues requires decision makers to possess a much more rounded appreciation of the benefits of working with nature and its associated ecosystem services.
8) **Step 8 is to speed up the hybridisation of traditional water infrastructure systems.** Examples such as retrofitting green infrastructure or including decentralised satellite networks within traditional systems are already emerging because no single

option or scheme, whether green or grey, is likely to simultaneously maximise water reuse, enhance system resilience and limit the generation of waste. To help us achieve this step, insight could be gained from the other centralised infrastructure networks that characterise our cities (e.g. electricity supply) and that, as is the case for water, are increasingly experiencing pressures that are forcing decision makers to look to alternative service models.

9) **Step 9 is to intensify research into the adoption of circular systems** that recognise the resource value of what were traditionally termed 'wastes'. While wastewater from households is currently treated to recover water for return to the environment, we envisage a future in which the value of wastewater is rightly perceived to lie in all its component parts: its water, energy, nutrients, and minerals.

In these three steps, we see significant potential to transform the management of not just water but also of the resources of food, energy, land and biodiversity. The steps also represent an opportunity for businesses to secure competitive advantage, encourage innovation and, ultimately, to fuel economic growth while establishing a vehicle that allows nature to provide society with ecosystem services at sustainable levels.

13.4 Summary

In this book we argue that the world faces water security challenges of a scale previously unseen and unsuspected by most of its population. In Earth's forty-fifth millionth century, a freshwater scarcity crisis is on our doorstep; this crisis is accelerating through our unbridled development, burgeoning demand for food and energy, and the effects of climate change. We are already withdrawing one-quarter of our accessible renewable water resource, much of which is already needed to sustain our ecosystems and biodiversity (themselves vital for our survival).

To confront these crises and to address their associated challenges, we argue that water professionals must emerge from their comfort zones and put themselves at the centre of water science, technology, politics, environment and economics. They must engage with politicians, decision makers and those with influencing power to articulate new models for truly integrated water management that appropriately address the complexity of society's collective water demands. While we know that we are not alone in arguing such a case, we believe that a new systems-based framework, a New Water Architecture, can provide the catalyst for real progress on the ground.

In our introduction we quoted Sir Martin Rees: 'This is a crucial century. The Earth has existed for 45 million centuries. But this is the first when one species, ours, can determine – for good or ill – the future of the entire biosphere'. We are inspired by the profundity of this statement. We believe that the future of the biosphere as a sustainable habitat for mankind in the twenty-second century will be framed by how effectively we manage our water. We believe that New Water Architecture can deliver that framework.

Index

Water Resources: A New Water Architecture, First Edition.
Alexander Lane, Michael Norton and Sandra Ryan.

Printed and bound by CPI Group (UK) Ltd, Croydon, CR0 4YY
17/09/2026
14964297-0005